COSMIC QUESTIONS

ANNALS OF THE NEW YORK ACADEMY OF SCIENCES

Volume 950

COSMIC QUESTIONS

Edited by James B. Miller

The New York Academy of Sciences
New York, New York
2001

∞ *The paper used in this publication meets the minimum requirements of the American National Standard for Information Sciences—Permanence of Paper for Printed Library Materials, ANSI Z39.48-1984.*

Library of Congress Cataloging-in-Publication Data has been applied for.

GYAT/B-M

Printed in the United States of America

ISBN 1-57331-346-7 (cloth)

ISBN 1-57331-347-5 (paper)

ISSN 0077-8923

ANNALS OF THE NEW YORK ACADEMY OF SCIENCES

Volume 950
December 2001

COSMIC QUESTIONS

Editor
JAMES B. MILLER

This volume is the result of a conference entitled **Cosmic Questions**, sponsored by the AAAS Program of Dialogue on Science, Ethics and Religion, held April 14–16, 1999 in Washington, D.C.

CONTENTS

Part II. Was the Universe Designed?

Cosmic Evolution and Design

Is the Universe Designed?

Religious Reflections on Design

Part III. Are We Alone?

Life in the Universe

Intelligent Life in the Universe: A Debate

Theological Reflection

Afterword

Financial assistance was received from:

- **THE JOHN TEMPLETON FOUNDATION**
- **OFFICE OF THE PROVOST OF THE SMITHSONIAN INSTITUTION**
- **SCIENCE AND SPIRIT RESOURCES, INC.**
- **NORTH AMERICAN MONTESSORI TEACHERS ASSOCIATION**
- **ROBERT AND JEAN FRI**
- **ARTHUR AND ANNE HALE JORDAN**

Preface

JAMES B. MILLER

Program of Dialogue on Science, Ethics, and Religion, American Association for the Advancement of Science, Washington, D.C. 20005, USA

Sometime in the dawn of human history our evolutionary ancestors began to look at the sky with wonder. The vastness of the daylight sky could be impressive, at times made curious with the dynamics of ever-changing clouds. But the clear night sky was even more impressive. Without the ambient light of civilization or the unintended filter of civilization's atmospheric pollutants, the night sky would have been almost literally silver with stars. This nightly spectacle eventually revealed both change and pattern.

The night sky was a stimulus for wonder in two senses: first, as a stimulus for awe; second, as stimulus for wondering, for curiosity. Our "cosmic questions" originate in this second sense of wonder. Has the world always been this way? Or, observing that there is change in the heavens, did the world have a beginning? Is the world simply here, or do the apparent patterns of the world point to a purposeful source? Are the celestial objects simply ornaments, or are they living entities themselves or the homes for other living beings? Answers to these questions in earlier times were both the products of and the context for the most profound understandings of the place of humankind in the scheme of the world.

The questions of cosmic origins, of cosmic purpose and of the status of life in the cosmos have been perennial questions across human cultures and across human history. In the past these questions were answered in the context of religion and natural philosophy. But today, for the first time in human history, the scientific revolution that began in Western culture in the 16th century has come to provide observational and analytic tools that may allow answers to these questions that are more definitive or more adequate than in any previous period. As in the past these answers will shape our culture's most profound existential questions about what it is to be a human being.

In the spring of 1999 a conference was held at the National Museum of Natural History of the Smithsonian Institution in Washington, D.C. It brought

Address for correspondence: Dr. James B. Miller, Senior Program Associate, Program of Dialogue on Science, Ethics, and Religion, AAAS, 1200 New York Avenue, NW, Washington, D.C. 20005. Voice: 202-326-7044; fax: 202-289-4950.

jmiller@aaas.org

together in a public setting scientists, philosophers, historians, and religious scholars to explore contemporary efforts to answer three questions: Did the universe have a beginning? Is the universe designed? Are we alone?

The auditorium in which the conference took place was the site of the classic debate in 1920 between Harlow Shapley and Heber D. Curtis on the scale of the universe. As an homage to that encounter, the conference was organized around three stylized "debates" related to the three cosmic questions. The following papers are one of the fruits of these encounters.

The papers begin with an introductory presentation by Joel Primack and Nancy Abrams that explores how human beings have sought in the past to depict the universe in which they lived. They show how contemporary developments in cosmology create the need for new cultural pictures of the universe and offer some suggestions as to what these might be.

To place the general discussion of cosmic questions in the context of Western cultural history, Jaroslav Pelikan discusses how the main strands of Western thought, one symbolized by Jerusalem and the other by Athens, have intertwined in classic efforts to understand our cosmic setting in terms of natural philosophy and religion. Owen Gingerich follows with a discussion of the interplay of scientific and religious interests in two of the great cosmological discoveries in Western history, Copernican heliocentrism and Big Bang cosmology.

The stage of the "debate" on cosmic origins is set scientifically with two presentations. First, Sandra Faber describes the evidence that has established the Big Bang as the standard cosmological model. Second, Rocky Kolb takes the reader through the process of the Big Bang and the early formation of the universe.

But Big Bang cosmology is about cosmic evolution not about the beginning. What happened before the Big Bang? Was there a cosmic beginning? Alan Guth of MIT and Neil Turok of Cambridge University in England offer alternative answers to this question. Guth, who is one of the originators of the inflationary model of Big Bang cosmology, proposes that even in an eternally inflating model, where there are many universes, each universe would have a beginning. Neil Turok, on the other hand, proposes a model of cosmic origins in which there is only one universe but it does not have a beginning point. It is a universe that is finite in the past but unbounded.

Two religious reflections are offered on this question of cosmic beginnings. Anindita Balslev discusses notions of "beginninglessness" in Indian and Hindu philosophical and religious thought as they relate to cosmology. In the article that follows Robert John Russell points out that the fertile opportunity for Christian reflection on God as the creator of the universe is less about the point, $t=0$, the beginning, than about the relationship of a creative God to a universe that is continuing to evolve.

Consideration of the second cosmic question, "Was the universe designed?" begins with John Leslie's discussion of the meaning of the term *design*. John Barrow then discusses the "anthropic principle,"which in some forms has been taken to imply that the universe has been "fine-tuned" for life and mind and so implies a transcendent designer. He suggests that the "anthropic coincidences" might also be accounted for in other ways than a design argument. Anna Case-Winters concludes these preliminary considerations with a discussion of the relative importance of the "argument from design" within Christian theology. She concludes that it is perhaps less important than it might seem.

The "debate" on the question of design occurs between Nobel Laureate in physics, Steven Weinberg, and John Polkinghorne, fellow physicist and Anglican priest. Weinberg argues that contemporary developments in cosmology and science generally do not lend comfort to those who would see the universe as the consequence of the actions of a benevolent deity. In this light he concludes that "one of the great achievements of science has been…to make it possible for [intelligent people] not to be religious." Polkinghorne proposes, on the contrary, that while contemporary cosmology does not prove the existence of God, an affirmation of God's existence can be "motivated belief" that does not ignore the picture of the cosmos drawn by the sciences. The papers by these scientists are followed by the transcript of their discussion and response to questions following their formal presentations in the conference.

There follow three religious reflections on the question of design in the universe. Philosopher David Ray Griffin offers a "yes and no" response drawing from a perspective informed by the philosophy of Alfred North Whitehead. Astronomer Trinh Xuan Thuan considers the issue of design from a Buddhist perspective. He points out that the Buddhist idea of interdependence rules out a "first cause" or *creatio ex nihilo*. The religious responses to the question of design conclude with Rabbi Lawrence Kushner's series of stories that illustrate the understanding of design in creation within the Jewish kabbalistic tradition.

The final "debate" concerns the question of whether it is likely that there is other intelligent life in the universe. Biologist Sara Via provides background to this question by discussing the process of evolution that has led to the emergence of intelligent life on Earth. Exobiologist, Kenneth Nealson discusses current efforts to seek extraterrestrial life. He points out that even on Earth life can be found in many very hostile environments. Astronomer David Latham reports on the efforts to identify other planets and planetary systems in the universe that might serve as "homes" for extraterrestrial life. Concluding these preliminary considerations, astronomer Jill Tarter describes the scientific "search for extraterrestrial intelligence" (SETI).

The "debate" on the likelihood of success in this search is between anthropologist Irven DeVore and SETI astronomer Seth Shostak. DeVore argues that while there may be many examples of life emerging throughout the universe, the process of evolution as we can observe it here on Earth is so fraught with contingency and happenstance that the likelihood that a species with sufficient intelligence to build communication devices that we could detect is vanishingly small. Shostak, on the other hand, argues that, even if the biological evolution of intelligence is rare, this would not mean that the likelihood of detecting extraterrestrial intelligence is small. He suggests that biological evolution can lead to the evolution of non-biological robotic intelligences that would be more durable and that could travel well beyond their planetary home of origin. He anticipates radio "contact" in the next several decades, but also suspects that the intelligence on the other end of the line will not be "squishy."

Theologian John Haught reflects on the religious significance of the discovery of extraterrestrial intelligence. He places this possibility within a larger context of cosmic evolution that seems to have a direction toward beauty. He draws his definition of beauty from Whitehead's view that it is the "harmony of contrasts" and the "ordering of novelty." If the evolution of the universe does have such a character, then the emergence of mind and culture on Earth or elsewhere are examples of this aim but not the end.

The volume ends with a brief reflection on the relation of these "cosmic questions" to an understanding of the relationship between science and religion. It is noted that these questions are of very different kinds and so bear on the relationship between science and religion in different ways.

Cosmic Questions

An Introduction

JOEL R. PRIMACK AND NANCY ELLEN ABRAMS

Department of Physics, University of California–Santa Cruz, Santa Cruz, California, USA

ABSTRACT: This introductory talk at the Cosmic Questions conference sponsored by the AAAS summarizes some earlier pictures of the universe and some pictures based on modern physics and cosmology. The uroboros (snake swallowing its tail) is an example of a traditional picture. The Biblical flat-earth picture was very different from the Greek spherical earth-centered picture, which was the standard view until the end of the Middle Ages. Many people incorrectly assume that the Newtonian picture of stars scattered through otherwise empty space is still the prevailing view. Seeing Earth from space shows the power of a new picture. The Hubble Space Telescope can see all the bright galaxies, all the way to the cosmic Dark Ages. We are at the center of cosmic spheres of time: looking outward is looking backward in time. All the matter and energy in the universe can be represented as a cosmic density pyramid. The laws of physics only allow the material objects in the universe to occupy a wedge-shaped region on a diagram of mass versus size. All sizes — from the smallest size scale, the Planck scale, to the entire visible universe — can be represented on the Cosmic Uroboros. There are interesting connections across this diagram, and the human scale lies in the middle.

KEYWORDS: cosmology; changing pictures of the universe; overthrowing vs. encompassing scientific revolutions, cosmic horizon

INTRODUCTION

Today cosmologists are telling each other at every conference that this is the golden age—or at least *a* golden age—of cosmology. It is a tremendously successful period because we are seeing the death of so many theories! Back in 1984 George Blumenthal, Sandra Faber, Martin Rees, and I published a paper[1] in which we created the theory of "cold dark matter" and worked out two versions of it in some detail. A couple of years later Jon Holtzman, a stu-

Address for correspondence: Joel Primack, Department of Physics, University of California–Santa Cruz, Santa Cruz, CA 95064. Voice 831-459-2580; fax: 831-459-3043.
joel@ucolick.org

dent with Sandra Faber and me, worked out 96 different variants of the cold dark matter scenario.[2] Now, I am proud to say, all but one of them are pretty convincingly ruled out. But the one that is left (which is closely related to one of the two in our original paper) may actually be right.[3] That is fantastic progress!

Now as we are trying to put together a picture of the whole universe, its origin, evolution, structure, and future—and that is what cosmology is all about—the question arises whether this has any broader implications. Does it matter to people as people, and not just as cosmologists (or other kinds of scientists)? I think it does, but that is for you to decide. I hope this *Cosmic Questions* volume of the *Annals* helps.

What I am going to do in the rest of this paper is to try to explain cosmology, not in a technical way, but rather through stories and pictures. That is the way most people throughout history have experienced cosmology: not as a series of scientific theories that are put forward as hopeful explanations to be tested against data, which is what we do as professional cosmologists, but rather as an understanding that one can grasp and visualize: a picture.

The joke about cosmology used to be that it was the only field of science in which the ratio of theory to data was infinite. But now the situation is reversed, and the ratio has gone to almost zero. Today data are flowing in so fast from new instruments that the question is whether a single one of the current theories can survive. If one theory survives the present onslaught of data, it will be revolutionary.

The advent of a new cosmology can radically change the culture of its time. In particular, the scientific revolution of the sixteenth and seventeenth centuries helped end the Middle Ages and bring about the European Enlightenment. But it also split scientific knowledge from human meaning, which was at the time largely determined by religion. How will a new picture of the universe at the turn of the twenty-first century affect global culture? That is one of the questions explored in this volume.

SHIFTING PICTURES OF THE UNIVERSE

Traditional

The first panel of FIGURE 1 is an example of one kind of the many kinds of traditional representations of the universe and of time: the snake swallowing its tail. It is a symbol that is found all over the world.[4]

Biblical

This is a representation of the cosmos of the ancient Near East, the *Genesis* cosmos. It is a three-part picture: the heavens, the flat earth, the underworld.

As it is described in *Genesis*, God separated the waters, providing a space. The firmament held up the upper waters, making space for dry land where animals, plants, and people could find a home. But the firmament had the possibility of breaking, and in the Noah story, the "chimneys" in the firmament and the "fountains" of the deep open up. The Flood was understood as not just a rainstorm but a cosmic catastrophe, a threat to recreate the primordial chaos.[5]

Medieval

The Medieval picture is based on the Ptolemaic and even earlier Platonic and Aristotelian conceptions of the crystalline spheres, with the spherical earth at the center and the whole pattern of spheres revolving around the earth every day. The spheres also revolve slowly against each other. The innermost sphere carries the moon, and then Mercury and Venus and the sun (with Mercury and Venus closely linked to the sun). Beyond the sun were Mars, Jupiter, and Saturn (the Seventh Heaven), the fixed stars, and then angels.[6] This basic picture was reflected in many ways in Medieval culture. For example, in a mystical Jewish kabbalistic representation, there are also the ten spheres, but the story is different. In the Lurianic Kabbalah, God creates the universe by withdrawing from the center, a process called *Tzimtzum* in Hebrew. The ten spheres—or *spherot,* the numbers—represent the emanations of God back into the universe. *Ein Sof*—the infinite God—surrounds all.[7]

Newtonian

Galileo's observations with the telescope provided the first convincing evidence that the Ptolemaic picture was wrong. This pulled the rug out from the entire Medieval conception of a hierarchical structure of the universe—including the human universe. Galileo's work, published in Italy in 1610, spread quickly throughout Europe. Already in 1611 in England John Donne writes:

> The new Philosophy calls all in doubt,
> The Element of fire is quite put out;[a]
> The Sun is lost, and th'earth, and no man's wit
> Can well direct him where to look for it …
> 'Tis all in pieces, all coherence gone;
> All just supply, and all Relation;
> Prince, Subject, Father, Son, are things forgot …[8]

Such was the impact of this change in cosmology.

[a]The sphere of fire, the highest of the spheres below the lunar sphere, does not exist.

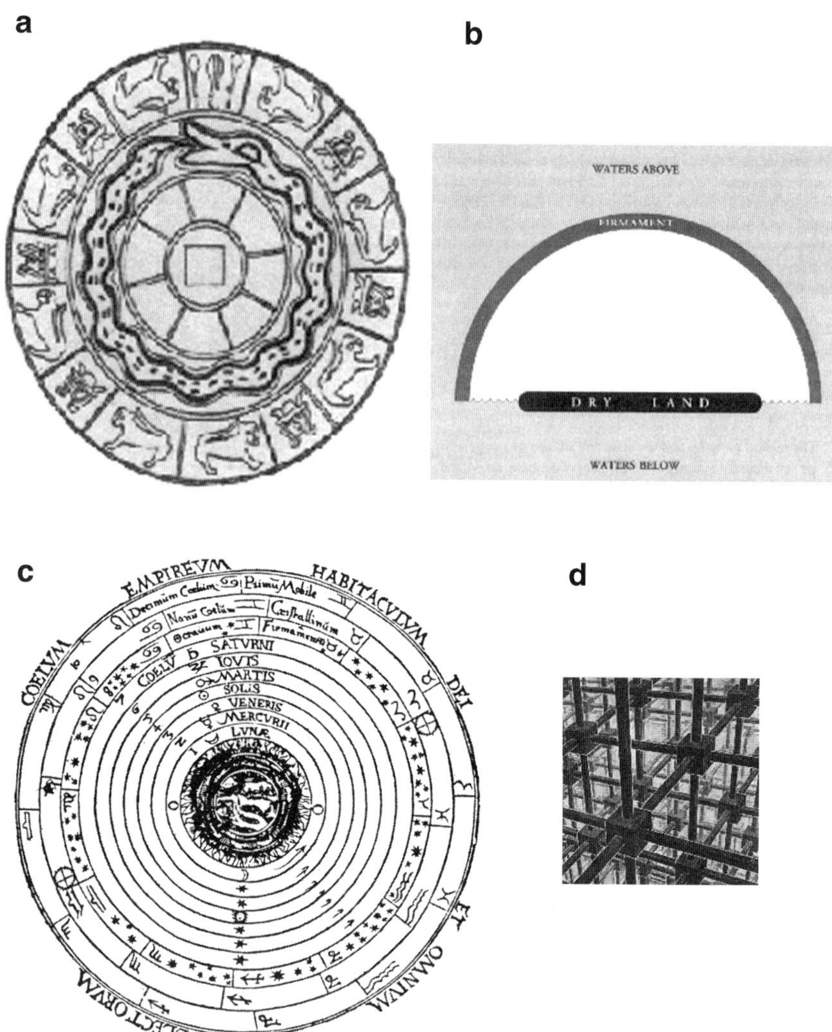

FIGURE 1. Views of the universe: traditional (**a**) the uroboros (snake swallowing its tail)**;** (**b**) biblical ; (**c**) medieval; (**d**) Newtonian.

What was the new picture? The Newtonian cosmos replaced the Medieval picture. There is simply empty space, the void, stretching on indefinitely in all directions. In the Middle Ages, when one went out at night and looked up, one saw majestic height, not infinite vastness. The statement in Pascal's *Pensées*, "the eternal silence of these infinite spaces alarms me,"[9] is a sort of statement that one simply never encounters in Medieval writings.[10] But it is

TABLE 1. Three cosmologies: Medieval, Newtonian, and Modern

Size R Age T Center	Composition	Unifying Ideas	God's Role?
	Medieval Cosmos		
Finite R Finite T Geocentric	Sublunary: Earth, water, air, fire Heavens: Ether	Circular motion Great Chain of being	Prime mover, hierarch, savior
	Newtonian Cosmos		
Infinite R? Infinite T? No center	Atoms, void ether?	Deterministic mechanics Universal gravitation;	Clockmaker
	Modern Cosmos		
$R = 10^{28}$ cm $T = 10^{10}$ yr Homogeneous & isotropic	Atoms, quarks, electrons Radiation Dark matter Vacuum	Gravity = space curvature Nondeterministic quantum theory Evolution	Before the Big Bang? Immanent?

a very common view once one is living in the Newtonian universe, represented in a work by Escher.[11]

Modern?

Our modern conception bears some elements of all these pictures, but it is very different from all of them. Let me start by summarizing the differences between the Medieval, Newtonian, and modern pictures.

The Medieval cosmos is of finite size: It began a finite length of time ago —which could be calculated by adding up the begats in *Genesis*—and it was geocentric. The physical part had to be finite in size because the whole thing goes around once every day. There is a distinction between the material contents of the sublunar world and of the perfect, unchanging heavens. The unifying ideas are constant circular motion and the Great Chain of Being: hierarchy, continuity, plenitude.[12] God—or gods—pervades the entire structure: pagan planetology coexisted with Christian cosmology.

Newton argued that if the cosmos were finite, then everything would fall to the center,[13] so it was probably infinite. But there were paradoxes associated with this: Kepler had already pointed out that the night sky would be bright as day in an everlasting infinite universe ("Olber's paradox"[14]). It also was not clear whether the Newtonian universe was created a finite length of time ago. The unifying ideas were deterministic local mechanics and universal gravitation: the laws of motion were the same on Earth as throughout the

FIGURE 2. Earthrise from Apollo 11 in lunar orbit.

universe. God's role was the creator of this clockwork universe at the beginning. For Newton at least, God also kept setting the clock right again every so often.

In the modern cosmos, we know how big the visible universe is—about 10^{28} centimeters (cm)—a distance called the "cosmic horizon." We know how long ago the universe started: about 14 billion years ago. We know that on large scales, the universe is homogeneous and isotropic (the same in all directions). It is made of atoms, dark matter, and radiation. Gravity is curvature of spacetime and can create horizons, and nondeterministic quantum mechanics and evolution are the key ideas. It is not clear whether there is a role for God.

SCIENTIFIC REVOLUTIONS

Now, what is the relationship among these three pictures? It is clear that the Copernican–Galilean–Keplerian–Newtonian revolution overthrew the Ptolemaic system, in the sense that Ptolemy is only taught as history and never as science. But I predict that Newton will always be taught, because Newton's picture is basically right—*on the scale of the solar system.* The scientific revolution that led to the modern cosmos, including the early twentieth century contributions of relativity and quantum mechanics, have encompassed Newtonian physics with physics that works at astronomical scales and velocities. But Modern cosmology reduces to the Newtonian treatment for

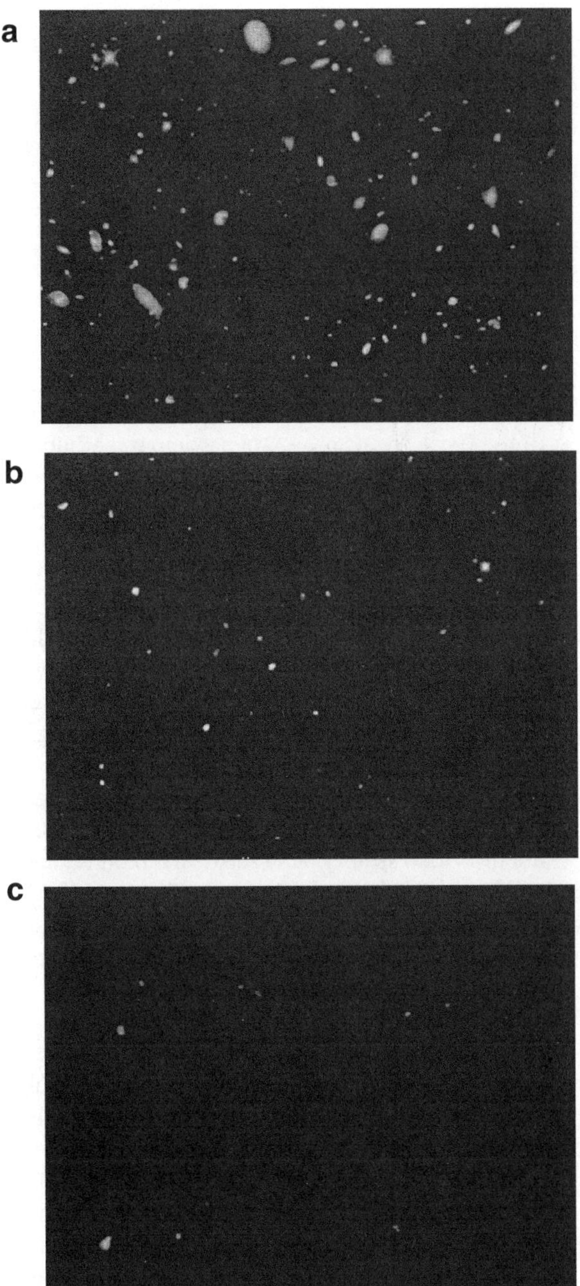

FIGURE 3. The Hubble Deep Field image. *Top:* a portion of the Hubble Deep Field. *Middle*: When the universe was about half its present age. *Bottom:* Back to the first two billion years.

the solar system, for normal-sized things on Earth, and generally for conditions where speeds are not too high and gravitational forces are not too great. So we can really only say we know that a theory is right when we know where it is wrong.[15] Inside the boundary where the theory begins to fail, the theory is correct (to some prearranged accuracy). Modern cosmology is undergoing an encompassing revolution as opposed to an overthrowing revolution.[16]

SOME MODERN PICTURES

Well, what is the modern picture? I do not think there is one— instead there are many, each of which captures part of the universe.

One of the modern icons, one that is engraved on every person's imagination, is the view of the Earth from space—in particular, perhaps, the view that the astronauts first had from orbiting around the Moon, with the dead gray lunar landscape in the foreground, and the gorgeous blue ever-changing Earth in the distance (FIG. 2).[17] Now we understand Earth viscerally as a small, fragile, very special planet, as most people did not until these pictures became available. I think this helps to show the power of a picture.

The Hubble Deep Field was the longest time exposure with the Hubble Space Telescope's camera of any region of the sky. For two weeks, the telescope looked at this same patch of sky continuously whenever the satellite was on the right side of the earth to see it. It is a very small region of the sky, about four arc-minutes across. That is the size of the intersection of two crossed sewing needles held at arms' length. A video of this image was made by Ken Lanzetta and his colleagues at the State University of New York at Stony Brook.[b] First the video simply pans across the picture, and every single bright spot you see is a galaxy. Many of these are giant galaxies like the Milky Way, big spiral galaxies that contain a hundred billion stars or so. The picture was taken looking out of the disk of the Milky Way so that there are very few stars in the way, but there are a few stars. The bright cross in the upper left corner of FIGURE 3a is what a nearby star looks like. Now the thing that one has to appreciate as one looks at an image like this is that one is seeing galaxies superimposed in front of other galaxies. We see all the galaxies that are reasonably bright, all the way out to the edge of the visible universe.

It would be wonderful to be able to see these galaxies not stacked on top of each other, but spread out in space and time, and that is what the video is about. It is possible to measure the red-shift of about five hundred of the galaxies and, based on those measurements, to estimate the red-shifts for all the rest from their colors. That is what the Lanzetta group did. And so what this

[b]This video is contained in the CD-ROM entitled *Cosmic Questions*, which recapitulates all the text in this volume with color illustrations among other features and which will be available in 2002 from the AAAS.

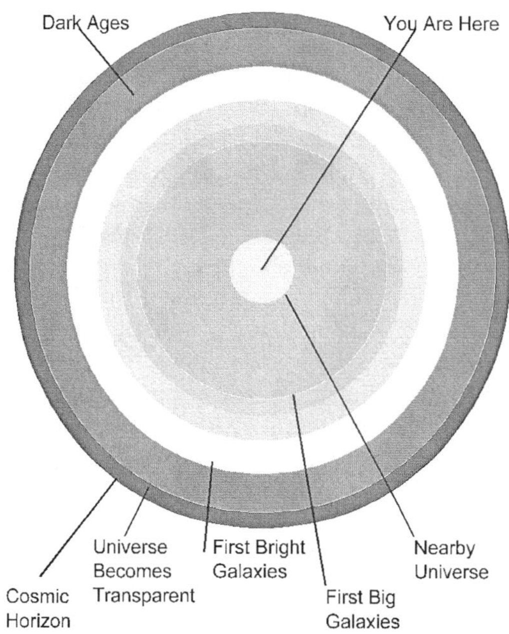

FIGURE 4. Cosmic spheres of time.

lets us do is to zoom into the picture. We see first the nearby galaxies go sliding away to the sides of the picture, and then, as we go farther and farther in, only the galaxies that are very far away from us are still visible.

The first galaxies to disappear are within the nearest billion light years or so. The colors are reasonably accurate, and the fact that the galaxies we can still see are now looking somewhat yellower reflects their red-shift: their light is shifted toward the red end of the spectrum because of the expansion of the universe since their light was emitted. Even about four billion light-years away, there are still plenty of big galaxies, but at about seven billion years back there are no more big galaxies visible in this field.

By the time we get back to about one billion years after the beginning, the sky is suddenly dark. If there were bright galaxies there, we would see them. We are now at the threshold of the real cosmic dark ages, before the cosmic night was pierced by the first beacons of bright starlight.

Now, how do we visualize the whole universe in our minds? We cannot paint a picture, because we cannot see it from outside—a picture is taken from outside the object, but we're inside the universe. We cannot see all times. As we look out in space, we look back in time. And most of the universe anyway is invisible stuff that we call dark matter. We do not know what it is. We know roughly *where* it is, but we cannot see it, so we cannot picture it in the sort of direct way that we are used to. An effective image should say something about the universe as a whole, but it does not need to say everything. Let me show you some examples:

Again, as we look out into space, we look back in time. The nearby universe surrounds us, and this is the part that we are busy exploring now, with the Sloan Digital Sky Survey and other very ambitious projects, but most of the volume of the universe remains only very partially explored. The way we do that exploration is by "drilling holes" through the distant universe in very narrow little images like the Hubble Deep Field. Then we study the pictures and try to take them apart to see the evolution of structure. Beyond a certain distance, we do not see any bright galaxies—there may be none!

The universe first became transparent about two hundred thousand years after the Big Bang. It is from this sphere—represented by the inside of the outermost band in FIGURE 4—that the Cosmic Background Radiation was emitted. This heat radiation from the Big Bang has been traveling to us through all of space ever since. And the very earliest stages of the Big Bang are concentric circles right at the edge of the figure that represent the eras of the great annihilations of the particles that initially populated the universe. There were initially almost equal amounts of matter and anti-matter, but now only the tiny remnant of matter survives. Even earlier were the eras of symmetry-breaking and cosmic inflation. We are surrounded by cosmic spheres of time—spheres somewhat different from the Medieval conception, however, and much larger.

There is another kind of picture (FIG. 5a), which represents not time but what the universe is made of.

FIGURE 5a. Cosmic density pyramid (top)

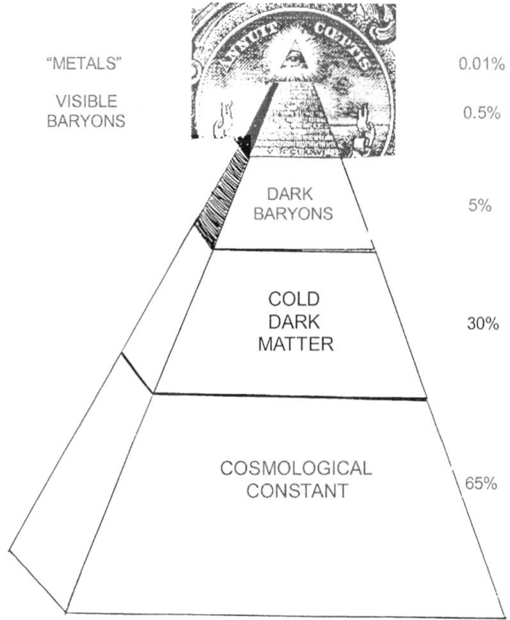

FIGURE 5b. The full cosmic density pyramid.

Now this is a picture that you probably all have in your pockets. It is the reverse of the Great Seal of the United States, which appears on the back of the dollar bill. It was an Egyptian symbol later adopted by the Masons (several of the Founding Fathers were Masons). The pyramid, I am told, symbolizes strength and solidity. Above is the all-seeing eye of God. Now, I am going to use this pyramid in a somewhat different way to represent all the visible matter in the universe. The part on top with the eye, which is about a quarter of the height of the full pyramid, represents the heavy elements, from lithium upward in atomic weight. Astrophysicists call all these elements—carbon, oxygen, nitrogen, silicon, iron, and so on—metals. They total about a hundredth of a percent of critical density. Critical density is the minimum density required for the expansion of the universe to be turned around by the gravitational attraction of all the matter in it. Hydrogen and helium, the two lightest elements, make up more than 99% of the mass of the elements in the universe. The volume of the rest of the pyramid represents all the hydrogen and helium we can see in the form of stars and gas, and this amounts to about half a percent of critical density. But there is much more mass than that. Invisible ordinary matter (atoms that are not lit up) amounts to about 5% of critical density, so ten times more than everything visible.

But there is still much more matter than that. Most of the mass of matter is dark matter, probably of the cold dark matter variety, possibly mostly made up of the lightest supersymmetric partner particles.[18] Although the key feature of dark matter is that it is *invisible*, not really "dark," the name dark matter has become standard for this mysterious stuff. It is gravitationally the most important matter in our own Milky Way galaxy and probably all other galaxies, and also in larger objects such as clusters of galaxies. It keeps the stars in their orbits around the galaxies, and keeps galaxies moving around inside the clusters. And the very same amount of dark matter is required to explain the bending of light around galaxies and clusters. The total mass of dark matter is probably about 30 percent of what we call critical density. Observations now convincingly indicate that the total density of all the matter, including dark matter, is significantly less than critical density.

The amount of matter is dwarfed by what seems to be the dominant stuff of the universe—whatever it is—which we call the "cosmological constant" or "dark energy." It makes up something like 65% of critical density.

This three-dimensional picture of the composition of the universe shows how very little there is of the metals we are made out of (and presumably all intelligent life could be made out of). The full Cosmic Density Pyramid shown in FIGURE 5b is also a picture of the universe— of an aspect of the universe, that is. Of course, we would love to know what the mysterious dark matter is, and also why there is a cosmological constant. These are two of the most important open questions in cosmology today.

Let us turn to one last pair of images: FIGURE 6 shows the possible densities of things. It is a plot of the mass of all things in the universe, from elementary particles up to the whole visible universe, versus their sizes. The ratio of mass to volume equals density. Interestingly, plants and animals, and stars, for that matter, all lie along one line. It is the water density line. As you can see, not all densities are allowed. The two great twentieth century laws of physics—general relativity and quantum mechanics—exclude two regions of the diagram.

The line that passes through points A and B (and continues in both directions beyond the figure) represents for every size on the X-axis the maximum mass that could possibly exist in it. If any more mass were crammed in, according to General Relativity it would collapse into a black hole.

The line through A and C (and beyond) represents the limit on sizes imposed by Quantum Mechanics (FIG. 6, top). The smallest physical size possible is the Planck size, and things close to that size are considered to be "on the Planck scale." It is a region 10^{-33} cm across. We cannot talk about, calculate, or conceptualize anything smaller in a way that has meaning in terms of our current concepts of physics

The largest size we know is that of the visible universe, but what exactly does this mean? Expansion of the universe means that space is expanding away from us faster and faster the further out we look. Since the velocity at

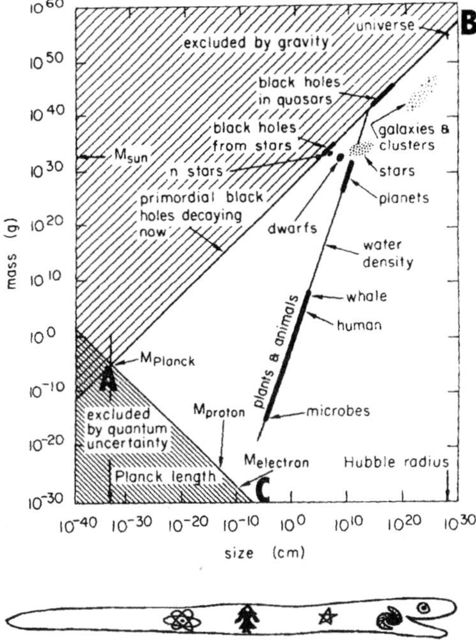

FIGURE 6. The wedge of material reality

which a point is moving away from us is always increasing with distance, far enough away space is expanding at the speed of light. That distance from us —which at this era in the history of the universe is about 10^{28} cm—is the radius of what we call the cosmic "horizon" and is the maximum distance from which we can receive information in principle. The horizon is a sphere, and we are at the center (FIG. 4). Although we have no reason not to believe space is just the same beyond our horizon, there is no way we can receive any direct information confirming it. Light cannot reach us from a region expanding away from us faster than the speed of light. On the figure there is an error bar for the density of the universe because we do not know more precisely than this the total amount of matter it contains and therefore its mass.

The wedge of material reality (FIG. 6) thus shows us that objects can only exist inside the wedge-shaped region of the plot. From the smallest size, the Planck size, to the largest, the horizon of the universe, is a difference of about 60 orders of magnitude. It is large, but not infinite.

Let us draw the possible size scales along the body of a snake instead of just the horizontal axis in the wedge of material reality (FIG. 7).[19] Sheldon Glashow was the first to draw an uroboros to represent the size scales in the universe.[20] The head swallowing the tail reflected his expectation that gravity controls on both the largest scales and also the smallest scales, and that there

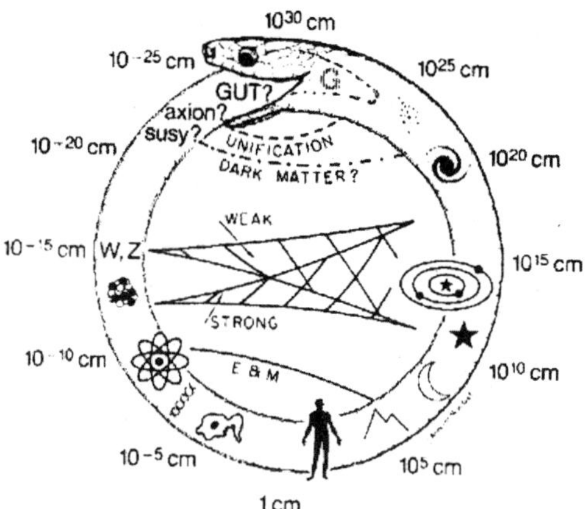

FIGURE 7. The cosmic uroboros.

will one day be a unification of all the laws of physics. A further thing that I find very interesting about this diagram is that there are connections across the Cosmic Uroboros. Electromagnetism controls on the scale from atoms up to mountains. Mountains are as high as they are on earth because of an interplay between the strength of materials—basically electromagnetic forces—and gravity. On a smaller planet, like Mars, the highest mountains are much higher, because they are made of essentially the same materials, but gravity is weaker and does not pull them down. There are also connections across the center of the diagram, from the very small to the very large. The weak and strong interactions, together with the electromagnetic interaction, control how stars burn, and thus also the compositions of planets. The processes at the center of the sun that ultimately generates sunlight involves conversion of two protons to two neutrons (that is a weak interaction) and their fusion (that is a strong interaction) to make a helium nucleus. On the still larger scales of galaxies and larger objects, dark matter is most important gravitationally, as we have seen. But dark matter is not associated with any of the forces that we know and understand on the scales we have probed so far, so we assume that it must be associated with laws of physics on still smaller scales—possibly supersymmetry ("SUSY") or other ideas such as "axions." We hope, as Glashow does, that maybe there is some unification of all the laws on the very smallest and the very largest scales.

Now, the laws that are important on different scales are different. The same physical laws apply on all scales, but they are not necessarily equally important. Electricity is much more important on small scales, gravity on large scales. Scale models can never work, because of the way the laws of physics work. Galileo pointed this out in the last of his great books, *The Discourses Concerning Two New Sciences*, where he showed that if the height of an animal were increased by a factor of three, it could no longer be the same shape. If its bones became three times longer in all directions, they would be 9 times thicker (because the cross-section is proportional to the area) but 27 times more massive (because the mass would scale as the volume, the *cube* of the height). To support so much weight, the bones would have to be much thicker.

We call the error of applying the laws and viewpoint appropriate to one size scale to phenomena on another scale, "scale incongruity." Imagining that the Big Bang can be understood using just commonsense physics is an example of scale incongruity. In the early universe, much of the material that we now see all the way out to the cosmic horizon was compressed into a much smaller volume, and such high densities and correspondingly high temperatures require relativistic quantum physics. Another example of scale incongruity is thinking of a molecule as ice or liquid. You need millions of molecules to make the tiniest piece of a snowflake. As one goes up in size scale, and thus complexity, such phase transitions show that one can get new "emergent" phenomena that are qualitatively different.

Many people believe the human size is insignificant compared to the cosmic scale. But perhaps you noticed on the Cosmic Uroboros that humans are essentially in the middle of the range of size scales in the universe. There is nothing arbitrary about that. That is where we are, and that is where creatures like us must be on this sort of diagram. Our brains must be as big as they are, to be as complex as they are. The universe is much bigger than we are, because it took billions of years for the universe to reach the level of complexity represented by human life, and it was expanding all that time. Just because we are so much smaller than the cosmic horizon does not mean that we are insignificant. After all, we may be the only creatures in the universe who are beginning to understand it.

NOTES AND REFERENCES

1. G. R. Blumenthal, S.M. Faber, J.R. Primack, and M.J. Rees, *Nature,* Vol. 311 (1984), pp. 517–525.
2. J. A. Holtzman, *Astrophysical Journal Suppl.*, Vol. 71 (1989), pp. 1–24.
3. J. R. Primack, *Reviews of Modern Physics*, in preparation.
4. E. Neumann, *Origins and History of Consciousness* (Princeton, NJ: Princeton University Press, 1995), Fig.6.
5. R. E. Friedman, *The Disappearance of God* (Boston: Little, Brown, 1995).
6. Cf. C. S. Lewis, *The Discarded Image* (London: Cambridge University Press, 1967).
7. Cf. D. S. Ariel, *The Mystic Quest* (New York: Schocken, 1988), especially pp. 169ff; D. C. Matt, *God & the Big Bang* (Woodstock, CT: Jewish Lights, 1996); J.R. Primack and N.E. Abrams, *Tikkun*, Vol. 10, No. 1 (Jan–Feb, 1995), pp. 66–73.

8. J. Donne, "An Anatomie of the World: First Anniversary," in *John Donne's Poetry*, A. L. Clements, ed. (New York: W.W. Norton & Co., 1992), p. 102. Cf. T. S. Kuhn, *The Copernican Revolution* (New York: Vintage Books, 1959), esp. pp. 193ff.
9. Blaise Pascal, *Pensées* (1670), Sec. III, no. 206.
10. Cf. C.S. Lewis, *The Discarded Image*, op. cit.
11. M.C. Escher, *The Graphic Work of M. C. Escher* (New York: Ballantine Books, 1971).
12. A.O. Lovejoy, *The Great Chain of Being: A Study of the History of an Idea* (Cambridge, MA: Harvard University Press, 1936).
13. I. Newton, letter to R. Bentley, in *Space, Time and Creation*, M. K. Munitz, ed. (New York: Free Press, 1957).
14. E.R. Harrison, *Darkness at Night: A Riddle of the Universe* (Cambridge, MA: Harvard University Press, 1987).
15. C.W. Misner, "Cosmology and Theology," in *Cosmology, History, and Theology*, W. Yourgrau and A. D. Breck, eds. (New York: Plenum Press, 1977), pp. 75–100.
16. N.E. Abrams and J.R. Primack, *Philosophy in Science*, Vol. 9 (2001), pp. 75–96.
17. Earthrise from Apollo 11 in orbit around the Moon, Fig. 45 in *The Home Planet*, Kevin W. Kelley, ed. (Reading, MA: Addison Wesley, 1988).
18. H. Pagels and J.R. Primack, *Phys. Rev. Letters*, Vol. 48 (1982), p. 223 ; J. Ellis, J.S. Hagelin, D.V. Nanopoulos, K. Olive & M. Srednicki, *Nuclear Physics*, Vol. B238 (1984), pp. 453–476.
19. J.R. Primack and G.R. Blumenthal, "What is the Dark Matter?—Implications for Galaxy Formation and Particle Physics," in *Formation and Evolution of Galaxies and Large Scale Structures in the Universe*, J. Audouze & J. Tran Thanh Van, eds. (Dordrecht, the Netherlands: Reidel, 1983), pp. 163–183.
20. S.L. Glashow, sketch reproduced in T. Ferris, *New York Times Magazine,* Sept. 26, 1982, p. 38. Cf. S. L. Glashow, *From Alchemy to Quarks: The Study of Physics as a Liberal Art* (Pacific Grove, CA: Brooks/Cole, 1994).

Athens and/or Jerusalem

Cosmology and/or Creation

JAROSLAV PELIKAN

Sterling Emeritus Professor, Yale University, New Haven, Connecticut, USA

ABSTRACT: For all three cosmic questions—"Did the universe have a beginning?" "Is the universe designed?" and "Are we alone?"—it was the conjunction as well as the divergence between Athens (Classical philosophy, especially Plato's *Timaeus* and Aristotle's *Physics*) and Jerusalem (the Bible, especially the Book of Genesis and the apostle Paul) that illumined the questions themselves, provided material for the answers, and set the terms for the subsequent discussion of them in later centuries.

KEYWORDS: angelology; chain of being; *creatio ex nihilo*; teleology

In this chapter it my task, my *daunting* task, to suggest some of the ways in which the three cosmic questions of the program—"Did the universe have a beginning?"/"Is the universe designed?"/"Are we alone?"—may have taken the form they have because of the historical interaction between Athens and Jerusalem (here I use the name "Jerusalem" to include also the New Testament, whose principal events did, after all, take place there). But first I need to point out again that it is to neither of these ancient cities but to Classical Rome that we must look for the most brilliant—and the most skeptical—formulation of these three cosmic questions in Classical Antiquity: I refer specifically to the *De rerum natura* of Lucretius,[1] for Lucretius stands with Goethe and Dante among those whom George Santayana early in the 20th century described as the "three philosophical poets" who embodied in a special way the combination of poet *and* philosopher *and* scientist[2]:

> Whence was a pattern for making things [*exemplum gignundis rebus*] first implanted in the gods, or even a conception of humanity [*notities hominum*], so as to know what they wished to make and to see it in the mind's eye?
>
> Or in what manner was the power of the first-beginnings [*uis principiorum*] ever known, and what they could do together by change of order, if nature herself did not provide a model for creation [*specimen creandi*]?[3]

In coping with the three cosmic questions asked by this symposium, however, and therefore also, directly or indirectly, with the devastating challenge

Address for correspondence: 156 Chestnut Lane, Hamden, CT 06518-1604.

of Lucretius to the theistic and teleological answers to them, both the Greek East and the Latin West, for a thousand years and more, pondered above all the heritage of Athens and Jerusalem. From Athens this meant Plato's most important treatise on cosmology, and the *only* dialogue of Plato that was known in the West during most of the Middle Ages, while the scientific works of Aristotle, which Byzantium did preserve along with the other works of Plato, were still unknown in the West: the *Timaeus*, which determined the vocabulary of science, philosophy, and theology, indeed, their way of framing questions. And from Jerusalem it meant the Book of Genesis: not in its original Hebrew, except when occasional contacts with Jewish biblical scholarship made it accessible,[4] but in the Septuagint, the Greek translation by Hellenistic Jews of Alexandria during the two centuries B.C.E.,[5] whose interpretation and, I believe, very vocabulary were significantly shaped by the *Timaeus*; and then in the Latin Vulgate, which still owed much to the Septuagint, even though Jerome had translated from the Hebrew. It should, moreover, be emphasized, especially in the light of subsequent history, that the effort to harmonize Athens and Jerusalem on all three of our cosmic questions had not originated with Christianity but with Judaism, more particularly with Philo of Alexandria.[6] (As a historical curiosity, I should add that some time ago I recently received from Hong Kong a Chinese translation of Philo's *On the Cosmogony of Moses*—speaking of "daunting tasks"!).

DID THE UNIVERSE HAVE A BEGINNING?

At first glance, the answer at least of Jerusalem to this question seems to be simple and direct: the very first word of the Hebrew Bible is *Brēshith*, and in Greek *en archēi*, "in the beginning," echoing the first sentence of Part II of *Timaeus*,[7] which contained the words: "*hē toude tou kosmou genesis... kat' archas*"[8]; and the phrase "heaven and earth" in any language, even in Lucretius's Latin,[9] means "universe." But already in the exegesis of the rabbis the prior question arose about the existence, or the metaphysical status, of this "beginning": the very first letter of the Hebrew Bible and the second letter of the Hebrew alphabet, the letter *Beth*, can mean not only "in" but also "by means of," so that the sentence could be rendered "By means of the primordial God made heaven and earth," prompting one to ask whether therefore this "primordial" had already (or always!) been there. If it had, what, if anything, could be said about it?

At this point Athens came to the aid of Jerusalem with the theory of the four elements, which Empedocles, calling them *rizōmata*, had enumerated as earth, air, fire, and water, and which Plato, especially in the *Theaetetus*[10] but also in the *Timaeus*,[11] called *stoicheia*,[12] the term that stuck. Lucretius's Latin word for them was either *elementa*, which he seems to have been the one to coin as a technical term in physics and chemistry that we still use in Western languages,[13] or *primordia*.[4] If, then, the opening sentence of the Book of Genesis were to be translated with the instrumental sense of its first letter that I suggested earlier, "By means of the primordial God made heaven and earth," the first of the three cosmic questions on this program— "Did the universe have a beginning?"—would be answered by the familiar philosophical (and theological and scientific) device of moving it back one notch: Did the elements of the universe have a beginning? The consideration of this device would lead Thomas Aquinas, drawing upon Aristotle, but not without remnants of Plato and of Timaeus still in his vocabulary (especially because of the influence of Augustine[15]), to argue already in Question Two of Part One of the *Summa Theologiae* against an "infinite regress" of motion or of causality or of possibility and necessity as basically inconceivable, and therefore for the Creator/*Primum Movens Immobile*/First Cause as a proposition that was not only an article of faith and a doctrine of revelation, but was also demonstrable by reason.[16]

It was out of such considerations that the definition of divine creation as specifically a *creatio ex nihilo* had come.[17] Lucretius was, again, the one who posed the issue, quoting and paraphrasing Epicurus: "The first principle of our study we will derive from this, that no thing is ever by divine agency produced out of nothing [*Principium cuius hinc nobis exordia sumet,/nullam rem e nilo gigni diuinitus umquam*]."[18] Thus the combination—or, as I have called it, the "counterpoint"—of Athens and Jerusalem led to a system in which the elements that were the building blocks of the universe had in turn been created. Out of what? (See how incurable, despite Aquinas's best efforts, the *regressus infinitus* is!) Once again, Athens came to the rescue with an answer: it was out of the Ideas, the Platonic Forms, which were already there! And when Jewish scholars like Rabbi Akiba began to speculate that there was a "Torah from heaven [*Torah min ha-shamayim*]," which had preexisted in God before it came to earth through Moses,[19] there were the "building blocks" for a combination of Athens and Jerusalem that would have it both ways: *creatio ex nihilo* and also preexistent reality—but in God. Orthodox Christianity gave that combination its special turn in the doctrine of the transcendent Trinity, according to which the "Torah in heaven," the preexistent Wisdom of God described by Solomon in the Book of Proverbs as the principle of creation,[20] which was also the Word of God, was the divine Logos, who already existed "in the beginning, *en archēi*"—the identical Greek phrase having been used both at the beginning of the Septuagint Genesis and at the beginning of the Gospel of John ("through whom all things were made," and who "became flesh and dwelt among us"[21]). Therefore Basil of Caesarea in the fourth century, drawing upon Athens to explain Jerusalem, was able to posit a prior creation of "this invisible world," an "order of things" above all of the Platonic Forms, and then the creation of "a new world" of the empirical realities, both of these creations having been taught by Moses in the Hexaemeron (rightly interpreted, of course).[22] For Basil's harmonization of Athens and Jerusalem in his interpretation of the *Hexaemeron*, then, "the form which God wished to give" to each creature had come first; it was "in harmony with" that preexistent form that God "created matter"; and "finally, God welded all the diverse parts of the universe by links of indissoluble attachment and established between them so perfect a fellowship and harmony that the most distant, in spite of their distance, appeared united in one universal sympathy."[23]

IS THE UNIVERSE DESIGNED?

So successfully has this harmonization of Jerusalem and Athens to produce a doctrine of cosmology and/or creation that came together in "one universal sympathy" penetrated our understanding, indeed our very language, that when the second of our three cosmic questions is raised, "Is the universe designed?" it is almost automatic to associate it primarily with the tradition of Jerusalem. That happens regardless of how one answers it: in the negative, and even if one dismisses it (although Lucretius did not so much dismiss it as rule it out of court as unanswerable); or in the affirmative—giving either answer on either theological or philosophical or scientific grounds (or, more usually, on some combination of these, whether one admits it or not). A search of the Hebrew Bible for its own answer to this cosmic question, however, comes up with significantly fewer proof texts than might initially have seemed obvious. The immediate contexts of two of the most familiar such texts, which have often been cited in discussions of this question of design—the words of the Psalmist, "When I consider thy heavens, the work of thy fingers, the moon and the stars, which thou hast ordained,"[24] and the divine version of the cosmic question addressed as a summons to Job out of the whirlwind, "Where wast thou when I laid the foundations of the earth?"[25]—show that both passages are intended primarily not to propound an argument from design, as in the traditional proofs for the existence of God, but to point to the epistemological and ontological chasm between the Infinite and the finite, between our "know[ing] in part"[26] and Ultimate Mystery.

Because the New Testament, by contrast with the Old, has Gentiles and nonbelievers in view as well as believers, it refers to design somewhat more often: in two appeals of Paul directly to the Greeks, arguing that God "left not himself without witness, in that he did good, and gave us rain from heaven, and fruitful seasons,"[27] and again, more familiarly, that "God that made the world and all things therein, seeing that he is Lord of heaven and earth, dwelleth not in temples made with hands..., though he be not far from every one of us: for in him we live and move and have our being: as certain also of your own poets have said, 'For we are also his offspring'"[28]; and, most familiarly of all, spoken not *to* the Greeks but *about* them, "The invisible things of him from the creation of the world are clearly seen, being understood by the things that are made, even his eternal power and Godhead"[29]—although the phrase "from the creation of the worlds [*apo tēs ktiseōs tou kosmou*]" here probably does not mean "*on the basis of* the creation of the world," as it has often been taken,[30] but "*ever since* the creation," as all the significant twentieth-century English versions have rendered it: the Revised Standard Version; the New English Bible; the New Jerusalem Bible; New Revised Standard Version; the Revised English Bible.

For a direct and explicit consideration of this second cosmic question, it is nevertheless necessary to turn first not to Jerusalem but to Athens, beginning, as one might expect, with the genius of the pre-Socratic natural philosophers. The brief but incisive chapter, "The Teleological Thinkers: Anaxagoras and Diogenes," in Werner Jaeger's Gifford Lectures connects their metaphysics to their physics:

> The idea of this preconceived world-plan is quite worthy of the rational physics of the fifth century [B.C.E.]; it is peculiarly fitting in a period that ascribes decided significance to *technē* in all realms of being and even finds it present in nature itself. The mechanism of the creative vortical motion is the ingenious device by which Anaxagoras, like other of his contemporaries, tried to explain the formation of the world. The fact that he made the divine Mind guide the vortex in a specific direction gave his physics its new teleological aspect. That is what caught Plato's attention.[31]

Despite such anticipations, however, Francis Cornford is correct in concluding, on the basis of the contrast both with pre-Socratic philosophy and with the cosmogonies of Homer and especially of Hesiod, that it was Plato in the *Timaeus* who "introduced, for the first time in Greek philosophy, the alternative scheme of creation by a divine artificer, according to which the world is like a work of art designed with a purpose."[32]

According to Plato in the *Timaeus*, "the supreme originating principle of Becoming and the Cosmos [*geneseōs kai kosmou... archēn kyriōtatēn*]," or, as he had called it a bit earlier, "the Cause wherefor He that constructed it constructed Becoming and the All," could be stated this way: "He was good, and in him that is good no envy [*phthonos*] ariseth ever concerning anything; and being devoid of envy He desired that all should be, as far as possible, like unto Himself."[33] But this answer to the second of our three cosmic questions—"Is the universe designed?"—was in turn based on the presupposition Plato had just formulated in the previous paragraph: "When the artificer [*ho dēmiourgos*] of any object, in forming its shape and quality, keeps his gaze fixed on that which is uniform, using a model of this kind, that object, executed in this way, must of necessity be beautiful [*kalon*]."[34] But Plato also insisted, on the basis of this relation between the *kalon* model and the *kalon* created object, that "if so be that this Cosmos is beautiful [*kalos*] and its Constructor good [*agathos*], it is plain that he fixed his gaze on the Eternal [*pros to aïdion*]."[35] That was also why, as he said much later, relating goodness, beauty, proportionality, and design, "All that is good is beautiful, and the beautiful is not void of due measure [*pān dē to agathon kalon, to de kalon ouk ametron*]."[36] Harking back to the pre-Socratic consideration of the relation between creation and *technē,* Plato argued in addition that although "the most of men" considered the empirical causes of empirical effects to be "primary causes," they were actually only "auxiliary Causes [*xynaitiai*]," by comparison with this "Form of the Most Good [*tēn tou aristou...idean*]."[37]

It is, nevertheless, more than slightly disingenuous to attach the credit (or, if you prefer, the blame) for our second cosmic question, "Is the universe designed?" so exclusively to Athens, and not to Jerusalem as well. For when God the Creator, in the Septuagint Genesis, completed one step of the creation after another, he saw that it was (in the key word of *Timaeus) kalon*[38]; for good measure, the Septuagint even added one such "God saw that it was *kalon*" to the account of the creation of the firmament, where there is no corresponding statement in the Hebrew text.[39] Thus to become the comprehensive cosmogony it was to be in the form that it eventually acquired, Plato's answer in the *Timaeus*, that this pattern for making things and model for creation, which he called *paradeigma*, was the design of which the cosmos was an image, which he called *eikōn*[40]—or even Aristotle's version of causality and teleology—required a further clarification of the monotheistic problem. And that clarification came not from Athens but from Jerusalem, from the primal creed of Israel as confessed in the sacred formula of the Shema, "Hear, O Israel: The Lord our God is one Lord,"[41] and then from the elaboration and defense of monotheism in the opening words of the only truly ecumenical creed of Christendom, the Niceno-Constantinopolitan Creed of 381, "We believe in one God the Father all-powerful, maker of heaven and of earth, and of all things both seen and unseen."[42] The generation of fourth-century Greek Christian thinkers who wrote that creed of 381 also systematized this transition "from *tychē* to *telos*."[43] In the Latin West it was above all Augustine who brought together Athens and Jerusalem, cosmology and/or creation, into his own distinctive version of the answer to the second cosmic question, "Is the universe designed?" by setting forth an interpretation in which the divine design resembled a poem or psalm that preexists as an entity in my mind but assumes linear temporality when I recite it.[44]

ARE WE ALONE?

The longest sustained *Denkexperiment* devoted to trying to answer this third cosmic question—how to conceive of other creatures in the universe who are not descended from our ancestors and therefore do not belong to the genus *Homo*, but who are rational and capable of movement, thought, and perhaps of free will—was the thousand-year Medieval and Byzantine (as well as Jewish and Muslim) investigation of angels.[45] However, I should immediately add the *caveat* that I am not considering the Jewish and Muslim developments here, as well as the stipulation that several of us who work on the history of Medieval and Byzantine thought have long posted a reward (or bounty) for anyone who could find a scholastic dissertation about how many angels could dance on the head of a pin! Both Athens and Jerusalem made significant contributions to this investigation. Athens provided "angelology" with ontological categories and a conceptual apparatus for locating angels within "the great chain of being," as Arthur O. Lovejoy put it in his celebrated William James Lectures of 1933 published under that title.

The accepted philosophical, as distinct from the dogmatic, argument for the existence of angels rested upon these assumptions of the necessary plenitude and continuity of the chain of beings; there are manifestly possibilities of finite existence above the grade represented by man, and there would consequently be links wanting in the chain if such beings did not actually exist. The reality of the heavenly hosts could thus be known *a priori* by natural reason, even if a supernatural revelation did not assure us of it.[46]

That assurance on the basis of what Lovejoy calls "supernatural revelation" came not from Athens but from Jerusalem. To fill the categories provided by Athens, it was possible to provide from Jerusalem a vast body of specific data: these included the Cherubim, who had been "placed at the east of the garden of Eden, and a flaming sword which turned every way, to keep the way of the tree of life" after the fall of Adam and Eve[47]; angels prominent in the life of Christ (Gabriel's Annunciation to Mary, the Christmas angels of Bethlehem, the angel who strengthened Christ during his prayer in the Garden of Gethsemane, and "the angel of the Lord [who] descended from heaven, and came and rolled back the stone from the door" of Christ's sepulcher at the resurrection[48]); and the starring role of the angels throughout the drama of the Apocalypse of John.

Perhaps nowhere does the joint contribution of Athens and Jerusalem to this experiment in answering the third cosmic question "Are we alone?" become more strikingly visible than in a pseudonymous sixth-century work carrying the name of Dionysius the Areopagite, the shadowy figure mentioned in the Acts of the Apostles as part of its account of Paul's visit to Athens,[49] *The Celestial Hierarchy,* which, together with all the other writings of Pseudo-Dionysius, has now finally become available in a careful and accurate translation into English, with several introductions and with notes.[50] At its very outset *The Celestial Hierarchy* quotes, from the New Testament, the Epistle of James and the Epistle to the Romans,[51] in order between these two proof texts to assert: "Inspired by the Father, each procession of the Light spreads itself generously toward us, and, in its power to unify, it stirs us by lifting us up. It returns us back to the oneness and deifying simplicity of the Father who gathers us in."[52] That Neoplatonic doctrine of procession by emanation from the Divine and of return to the Divine, buttressed by biblical quotations, provides the foundation for positing the existence of the angels and their "immaterial [nonmaterial] hierarchies" as the Neoplatonism of Athens had taught, as well as for explaining that the biblical writers of Jerusalem had "clothed these immaterial hierarchies in numerous material figures and forms so that, in a way appropriate to our nature, we might be uplifted from these most venerable images to interpretations and assimilations which are simple and inexpressible." This was necessary because, just as "the appearances of beauty are signs of an invisible loveliness," so "material means capable of guiding us as our nature requires" were needed to let human minds, "in any immaterial way, rise up to imitate and to contemplate the heavenly hierarchies."[53] Throughout the rest of this curious but highly influential work, the concrete, empirical information of the biblical narratives is called upon to supply such "figures and forms."

On that basis, Thomas Aquinas, who, it has been estimated, quoted the *Corpus Areopagiticum* nearly a thousand times in his writings,[54] devoted an entire block of fifteen questions in Part I of the *Summa Theologiae*, each question consisting in turn of several articles, to a metaphysical-cum-exegetical exposition of the doctrine of angels.[55] Although Part I was the section of *the Summa Theologiae* in which he examined what was knowable by reason without revelation, Athens and Jerusalem were in dialogue already in the first question of this treatise on angels:

> Plato says in the *Timaeus*: "O gods of gods, whose maker and father am I: You are indeed my works, dissoluble by nature, yet indissoluble because I so will it."[56] But gods such as these can only be understood to be the angels. Therefore the angels are corruptible by their nature.... [To the contrary, according to Thomas:] By the expression "gods" Plato understands [not the angels, but] the heavenly bodies, which he supposed to be made up of elements which are composite, and therefore dissoluble of their own nature; yet they are for ever preserved in being by the Divine will.[57]

Thus Thomas drew upon Athens, the words of the Demiurge in Plato's *Timaeus*, to explicate Jerusalem, but no less did he draw upon biblical teaching to explicate Greek philosophy. And when Aquinas subsequently took up "the local movement of angels," the quotations were almost all from Aristotle's *Physics*; but that should not be permitted to obscure the real source, in the usage of Jerusalem, of such terms and concepts in this question as "a beatified angel" or "the holy angels." The clinching argument of the first article on "the local movement of angels," moreover, did not come from Aristotle, nor from any other citizen of Athens, but from Jerusalem. First Thomas invoked the fundamental distinction of Aristotle's *Physics* between potentiality and actuality in order to set the stage for a biblical proof text. In an argument that fittingly illustrates the complex relationships in this "tale of two cities" that I have been trying to tell, he linked these two sentences:

> [The first, from Athens]: The motion of that which is in potency [potentiality] is on account of *its own need*, but the motion of that which is in act [actuality] is not for any need of its own, but for *another's need*. [The second, from Jerusalem]: In this way, because of our need, the angel is moved locally, according to Hebrews 1:14, "They are all ministering spirits, sent to minister for them who receive the inheritance of salvation."[58]

Therefore it followed from this combination of authorities that in order to carry out this ministry and mission to those "who receive the inheritance of salvation," as this had been documented throughout the Old and the New Testament, it was necessary that the angels be capable of local movement. It had, after all, been by means of a considerable amount of such "local movement" that "the angel of the Lord went out, and smote in the camp of the Assyrians [commanded by Sennacherib] an hundred fourscore and five thousand: and when they arose in the morning, behold, they were all dead corpses."[59]

The effort of Thomas Aquinas, but also of Moses Maimonides, to conceptualize the nature, the movement, and the knowledge of the angels was an ambitious consideration of our third cosmic question, "Are we alone?" In this, as in the first two cosmic questions, "Did the universe have a beginning?" and "Is the universe designed?" it was the conjunction as well as the divergence between Athens and Jerusalem that illumined the questions, provided material for the answers, and set the terms for subsequent discussions of all three cosmic questions—including (who knows?) perhaps even the discussions that will follow later in this volume.

NOTES AND REFERENCES

1. Jaroslav Pelikan, *What Has Athens to Do with Jerusalem? "Timaeus" and "Genesis" in Counterpoint* (Ann Arbor: University of Michigan Press, 1997), pp. 1–22.
2. George Santayana, *Three Philosophical Poets: Lucretius, Dante and Goethe* (Cambridge: Harvard University Press, 1910).
3. Lucretius, *De rerum natura,* 5. 181–186.
4. Beryl Smalley, *The Study of the Bible in the Middle Ages* (reprint edition; Notre Dame, IN: University of Notre Dame Press, 1964), pp. 149–156.
5. Elias Bickerman, "The Septuagint as Translation," *Studies in Jewish and Christian History* (Leiden: E.J. Brill, 1976), Vol. 1, pp. 167–200.
6. David T. Runia, *Philo of Alexandria and the Timaeus of Plato* (Leiden: E.J. Brill, 1986).
7. Plato, *Timaeus,* Part I, "The Works of Reason," beginning at 29D; Part II, "What Comes of Necessity," beginning at 47E; Part III, "The Co-Operation of Reason and Necessity," beginning at 69A. Throughout this paper, I shall fol-

low the translation by R.G. Bury (Cambridge, MA: Harvard University Press [Loeb Classical Library]).
8. Plato, *Timaeus,* 48A.
9. Lucretius, *De rerum natura*, 5.67–69.
10. Plato, *Theaetetus,* 201E.
11. Plato, *Timaeus,* 48B.
12. George Stuart Claghorn, *Aristotle's Criticism of Plato's "Timaeus"* (The Hague, 1954), pp. 20–38.
13. *Thesaurus Linguae Latinae,* 5-II: 341–343.
14. Lucretius, *De rerum natura,* 1: 753–754; 778–779; 753–754.
15. See M.-D. Chenu, *Toward Understanding St. Thomas*, translated by A.-M. Landry and D. Hughes (Chicago: Henry Regnery, 1964), pp. 111–113, 141–143.
16. Thomas Aquinas, *Summa Theologiae,* Ia, Q. 2, Art. 3.
17. Gerhard May, *Creatio ex nihilo: The Doctrine of "Creation out of Nothing" in Early Christian Thought*, translated by A. S.Worrall (Edinburgh, 1994).
18. Lucretius, *De rerum natura,* 1.149–150.
19. Abraham Joshua Heschel, *Torah min ha-Shamayim* (3 vols.; London and New York, 1962–90); English translation in preparation.
20. Proverbs 8: 22–31. Quotations from the Bible according to the Authorized ("King James") Version throughout.
21. John 1:1–14.
22. Basil of Caesarea, *Hexaemeron,* 1.5.
23. Basil of Caesarea, *Hexaemeron,* 2.2.
24. Psalm 8: 3.
25. Job 38: 4.
26. 1 Corinthians 13: 12.
27. Acts 14: 17.
28. Acts 17: 24, 27–28.
29. Romans 1: 20.
30. Jaroslav Pelikan, *Christianity and Classical Culture: The Metamorphosis of Natural Theology in the Christian Encounter with Hellenism* (New Haven: Yale University Press, 1993), pp. 65–66.
31. Werner Jaeger, *The Theology of the Early Greek Philosophers*: The Gifford Lectures for 1936 (Oxford: Oxford University Press, 1947), p. 163.
32. Francis MacDonald Cornford, *Plato's Cosmology: The "Timaeus" of Plato Translated with a Running Commentary* (reprint edition; New York: Liberal Arts Press, 1957), p. 31.
33. Plato *Timaeus* 29D-E; cf. Luc Brisson, *Le même et l'autre dans la structure ontologique du Timée du Platon; un commentaire systÿeamatique di Timée de Platon* (Paris, 1974), p. 155 n. 1.
34. Plato, *Timaeus,* 28A.
35. Plato, *Timaeus,* 30A.
36. Plato, *Timaeus,* 87C.
37. *Plato, Timaeus,* 46C-D.
38. Genesis 1: 4; 10; 12; 18; 21; 25 (LXX).
39. Genesis 1:8 (LXX).
40. Plato, *Timaeus,* 29B.
41. Deuteronomy 6:4.

42. Norman F. Tanner, *Decrees of the Ecumenical Councils* (2 vols.; London and Washington: Georgetown University Press, 1990), 1:24 [following his capitalization and punctuation].
43. Pelikan, *Christianity and Classical Culture*, chapter 10, "From Tyche to Telos," op. cit., pp. 152–165.
44. Jaroslav Pelikan, *The Mystery of Continuity: Time and History, Memory and Eternity in the Thought of Saint Augustine* (Charlottesville: The University of Virginia Press, 1986), pp. 52–69.
45. Among many others, James D. Collins, *The Thomistic Philosophy of the Angels* (Washington, D.C.: Catholic University of America, 1947), clearly evaluates the metaphysical significance of Thomas's doctrine of angels.
46. Arthur O. Lovejoy, *The Great Chain of Being: A Study of the History of an Idea* (Cambridge, MA: Harvard University Press, 1936), p. 80.
47. Genesis 3: 24.
48. Luke 1: 26-38; Luke 2: 8–14; Luke 22: 43; Matthew 28: 2.
49. Acts 17: 34.
50. Paul Rorem, ed., *Pseudo-Dionysius: The Complete Works*, translated by Colm Luibheid, introductions by Jaroslav Pelikan, Jean Leclercq, and Karlfried Froehlich (New York: Paulist Press, 1987).
51. James 1: 17; Romans 11: 36.
52. Pseudo-Dionysius, *The Celestial Hierarchy,* 1.1.
53. Pseudo-Dionysius, *The Celestial Hierarchy,* 1.3.
54. J. Durantel, *Saint Thomas et le Pseudo-Denys* (Paris, 1910).
55. Thomas Aquinas *Summa Theologica* Ia, QQ. 50–64.
56. Plato, *Timaeus,* 41A, in the Latin translation of Chalcidius.
57. Thomas Aquinas, *Summa Theologiae,* Ia, Q. 50, Art. 5, Obj. 1 and *ad* 1.
58. Thomas Aquinas, *Summa Theologiae,* Ia, Q. 53, Art. 1, *ad* 3; italics added.
59. 2 Kings 19: 35.

Scientific Cosmology Meets Western Theology

A Historical Perspective

OWEN GINGERICH

Professor of Astronomy and the History of Science, Harvard University, Cambridge, Massachusetts 02138, USA

ABSTRACT: Traditional sacred geography of Christendom met a challenge not so much from Copernicus' heliocentrism per se as from the greatly explanded vision of the cosmos that it ushered in. The twentieth-century view of the vastness of both space and time has brought revolutionary conceptual changes to the sacred landscape. From a theistic perspective, God is not simply the source of the Big Bang, but the Creator in the larger sense of designer and intender of the universe.

KEYWORDS: Copernican revolution; Johannes Kepler; Galileo; Edwin Hubble; Big Bang cosmology; Fred Hoyle; steady-state cosmology

As the Middle Ages began to wane, when the information explosion brought about by printing with movable type was about to change the intellectual face of Europe, one of the popular genres of books was entitled *cosmographia*—literally, the mapping of the cosmos. European Christendom had readily adopted a geocentric Aristotelian cosmology into its own sacred geography, with mankind at center stage and with God ruling from empyrean regions just beyond the shell of fixed stars. These printed, illustrated cosmographies reinforced the images found here and there on cathedral walls, but their pictures were about to meet a serious challenge.

The challenge was *not* to do away with a flat earth, because educated people already knew full well that the earth was round. The whole mythology that Columbus had to persuade Ferdinand and Isabella that the earth is round begins, as far as the English-speaking world is concerned, with Washington Irving's two-volume biography of Columbus published in 1828. One of his most graphic, but highly fictionalized scenes, is set at the Spanish court in

Address for correspondence: Harvard–Smithsonian Center for Astrophysics, 60 Garden Street, Cambridge, MA 02138. Voice: 617-495-7216; fax: 617-496-7564.
ginger@cfa.harvard.edu

Salamanca, where Columbus faced a panel of clerics, "an imposing array of professors, friars and dignitaries of the church," who "came prepossessed against him, as men in place and dignity are apt to be against poor applicants."[1] According to Irving, they ridiculed the idea of the roundness of the earth and quoted scripture to infer its flatness. Columbus, a profoundly religious man, found himself in danger of being convicted not only of error but also of heresy.

The historical reality is that the imposing array of professors and clerics told Columbus not that the world was flat, but that its circumference was much bigger than Columbus supposed, and that Japan was more than 2500 miles beyond the Azores. They were right, of course, and they would never have made the voyage.

No, the challenge to the traditional sacred geography came not from Columbus, but from the northernmost diocese of Poland, where a visionary canon conceived of an entirely different blueprint for the cosmos. Yet, the discoveries of Columbus helped set the stage for the coming Copernican revolution, for, at the end of the 15th century, the ancient Alexandrian, Claudius Ptolemy, was probably better known as a geographer than as an astronomer. The discovery of America not only shook traditional theology because the New World was populated with beings who had never heard of the Incarnation, the Crucifixion, and the Resurrection, but it also filled the map with lands never dreamt of by Ptolemy. If Ptolemy as geographer–cosmographer could be challenged, what about Ptolemy as astronomer–cosmologer?

It is necessary to be quite clear that the Copernican revolution was of an entirely different type from the Columbian discoveries. It was not a matter of Copernicus using his eyes and with fresh observations showing the universe was arranged in a way other than his predecessors had believed, for in his day there was no observation that could distinguish between a geocentric and a heliocentric universe. And, as Galileo later said, he could not admire enough those who accepted the heliocentric doctrine *despite* the observations of their senses. Why, in fact, did Copernicus choose to defend such an unorthodox arrangement?

There are two popular, but totally mythological, arguments given for why Copernicus adopted a sun-centered cosmology. In the first place, we know that the procedures for predicting planetary positions given in the *Almagest* were not very accurate, and Copernicus himself was aware that in 1503 Saturn was ahead of the tables by a degree and a half, whereas Mars lagged by two degrees. Could Copernicus have embarked on a new cosmology to improve this state of affairs? In fact, Copernicus never mentions that the calculations were inaccurate, and for good reason. Because of a close geometric equivalence between the two systems, the predictions come out essentially the same way, and merely converting to a heliocentric framework does not by itself buy greater accuracy.

An even more popular wrong explanation is the notion that during the Middle Ages astronomers added a increasing number of small circles to the Ptolemaic system in order to make it more accurate, and by the end of the 15th century, the whole scheme was ready to collapse under its own weight. This idea has become so firmly embedded in modern lore that hardly a year passes without an author in the *Astrophysical Journal* or the *Physical Review* remarking that "perhaps my theory has too many epicycles in it." Historically, this is complete nonsense.[2]

So why, then, did Copernicus come to espouse such a radical cosmology?

What Copernicus saw in mind's eye were two beautiful aesthetic results from the heliocentric layout. Since antiquity the most puzzling aspect of celestial motions had been the fact that, while planets normally drift eastward against the background of stars, periodically they come to a standstill, and then move westward for several weeks. Mars, for example, goes retrograde about every two years, becoming especially bright, as it was in April of 1999. The Copernican system offers a completely natural explanation. What was apparent even to Ptolemy, but as a completely unexplained detail, is that the retrogression always occurs when the planet is opposite the sun in the sky. For example, when Mars is retrograde, it rises at about the time the sun sets, in opposition to the sun. In the Copernican, sun-centered system, it becomes completely obvious why this is so. Mars will appear to be retrograde when the faster-moving earth is bypassing it, and this can happen only when the earth is closest to Mars and the sun is then necessarily opposite to Mars in the sky.

Now when Ptolemy's circles for the planets are all scaled and superimposed to create a compact, sun-centered system, it automatically happens that Mercury, the fastest planet, comes out with the smallest orbit, closest to the sun, and Saturn, the slowest planet, falls into place with the largest orbit, with the others falling rhythmically in between. As Copernicus explains in the central cosmological chapter in his book, "In no other way do we find such a wonderful commensurability and sure, harmonious connection between the movement and size of an orbit." It's too beautiful to be wrong, surely a theory "pleasing to the mind."

As the great Danish astronomer, Tycho Brahe, would say in the next generation, Copernicus' theory nowhere offends mathematics, yet it throws the earth, a lazy, sluggish body unfit for motion, into a motion as fast as the ethereal stars themselves. And when laying out the specifics, he repeatedly said that Copernicus' theory goes against physics and against Holy Scripture—always in that order.

The scriptural objections to a moving earth were quick to emerge. Georg Joachim Rheticus, the young Lutheran who came to visit the Catholic Copernicus at the cathedral in Frauenburg, who in 1540 published a "first report" on the new cosmology, and who persuaded Copernicus to let him take a copy of the manuscript back to Germany for publication, was reputed to have written a "second report" dealing with additional questions about the new system. His report was long believed to be lost, but two decades ago it was brilliantly rediscovered by the Dutch historian of science Reijer Hooykaas. In it Rheticus examines five scriptural passages that seem to speak against the mobility of the earth and several against the immobility of the sun. For example, Psalm 104 says that the Lord God laid the foundation of the earth that it not be moved forever. What the Psalmist means, Rheticus declares, is that the earth maintains its established course and attains its prescribed positions. "Therefore, he who assumes its mobility in order to bring about a reliable calculation of times and motions is not acting against Holy Scripture." The book of Joshua describes how, at the battle of Gibeon, Joshua commanded the sun, not the earth, to stand still. As for this battle in the valley of Ajalon, Rheticus asks his reader to distinguish between appearance and reality. "When right reason concludes that the sun is immobile, even though our eyes lead us to think it moves, we do not abandon the accepted way of speaking. We say that the sun rises and sets, even though we hold this to be true only in appearance."[3]

It is interesting that Johannes Kepler, without knowing of Rheticus' still-unpublished report, picked up on the same two scriptural objections. Concerning Joshua, he wrote, "If someone had suggested that the sun was not really moving toward the valley of Ajalon, would not Joshua have exclaimed that he was seeking to extend the daylight by any means whatsoever? He would have reacted the same way, therefore, if anyone had taken issue with him over the sun's permanent immobility and the earth's motion."[4] Concerning Psalm 104, Kepler wrote, "The Psalmist is very far from speculating on physical causes. He rejoices entirely in the greatness of God who has made all things. If you pay close attention, this is actually a commentary on the first six days of creation of Genesis." After an extensive analysis of the Psalm, Kepler urges his reader to "extol the bounty of God in the preservation of living creatures of all kinds by the strength and stability of the earth," and also to "acknowledge the wisdom of the creator in its motion, so abstruse, so admirable."[5]

The implications of Copernicus' novel cosmology were muted in part because an anonymous introduction had been added to his great book, *De revolutionibus*, when it was printed in 1543. This was done in collusion with the Nuremberg printer by the proofreader, Andreas Osiander, who was a learned Lutheran clergyman at the St. Lorenz Kirche. Osiander wrote (and I paraphrase it), "You may be troubled by the ideas in this book, fearing that all of liberal arts are about to be thrown into confusion. But don't worry. An astronomer should make careful observations, and then frame hypotheses so that planetary positions can be established for any time. This our author has done well. But such hypotheses need not be true nor even probable. Perhaps a philosopher will seek truth, but an astronomer will just take what is simplest, and neither will find anything certain unless it has been divinely revealed to him. So if you expect to find truth here, beware, lest you leave a greater fool than when you entered."

Some critics, beginning with the head of the cathedral where Copernicus worked, have decried this introduction as contrary to Copernicus' own views concerning the reality of his vision. But had Osiander's interpretation not been there so conspicuously, it seems clear that both the Lutherans and the Catholics would have independently invented it. The rearrangement of their traditional sacred geography would have been too severe to assimilate all at once. As it was, Osiander's introduction disarmed religious critics and made it possible for advanced students all over Europe to examine its ideas. Among them was Johannes Kepler who became the enthusiastic Copernican realist, and it was Kepler, together with Galileo who entertained similar opinions, who really set the Copernican pot aboil.

In 1609 Galileo turned his newly perfected spyglass to the heavens; in January of 1610 he discovered the satellites of Jupiter, and within a few months published his astonishing findings. In his *Sidereus nuncius* or *"Starry Messenger"* he allowed himself one Copernican remark. Some people had refused to accept the notion of the earth revolving about the sun because they could see no way to keep the moon attached in orbit about a moving earth, but everyone—geocentrist or heliocentrist—agreed that Jupiter was moving. This should give some pause, Galileo wrote, to those who refused to accept Copernicus' idea because a moving body couldn't carry along a moon.

Before the year was out Galileo had discovered the phases of Venus, a pattern that showed Venus traveled around the sun, contrary to the Ptolemaic arrangement. The observed phases of Venus proved Ptolemy wrong, but did prove Copernicus right. Like Copernicus before him, Galileo never found an observational proof of the earth's motion. Nevertheless, the fat was in the fire, so to say, and presently Galileo followed Kepler's footsteps in offering an explanation of certain scriptural passages such as the account of Joshua at the battle of Gibeon. He maintained that the Bible should not be taken literally as a scientific textbook, since its purpose was different, and he quoted Cardinal Baronius, librarian of the Vatican, that "the Bible teaches how to go to heaven, not how the heavens go."

As a consequence of Galileo's well-publicized opinions, conservative churchmen in Rome became worried that the standard interpretation, that Copernicus' system was just a calculating device, not to be mistaken for physical reality, was under attack. The issue was particularly sensitive because Western Christendom was no longer monolithic—north of the Alps the Protestants were interpreting other Scriptures their own way, and the Roman theologians, determined to maintain a unified front, were not prepared to put up with an amateur interpreter like Galileo. It is interesting to note that Pope John Paul II has recently declared that Galileo was a better theologian than the professionals with whom he was contending. In any event, just to make sure that everyone understood what was involved in the cosmological issue, the Congregation of the Index moved to place Copernicus' *De revolutionibus* on the *Index of Prohibited Books* until corrected, and in a rare action, the Congregation actually specified how a book was to be corrected. In this case ten changes were ordered, almost all to make the book appear more hypothetical. Toward the end of the glorious cosmological chapter Copernicus explained why the earth's swift motion around the sun does not produce any observable consequence in the positions of the stars themselves: they are simply too far away. "So vast, without any question, is the divine work of the Almighty Creator." Why would such a pious statement be excised? Because it contained a whiff of reality, as if this was the way the Creator had done it.

The drama that unfolded in 1632, when Galileo published his enthusiastically pro-Copernican *Dialogue on the Two Great World Systems*, and his trial by the Inquisition the following year, is well known. I don't have time here to rehearse the story with the careful nuances that it deserves. Suffice it to say that Galileo was vehemently suspected of heresy though not actually convicted of it, but, for disobeying orders in teaching the Copernican doctrine, he was placed under house arrest for the rest of his life .

Galileo remained a hero in Italy, and after a century a Venetian printer sought to include the *Dialogo* as the fourth volume in an edition of all of Galileo's works, though the book had been placed on the *Index* in 1633. The permission was actually granted, but with a requirement that an explanation be included about the difference between theses and hypotheses. Then, a few years later, Pope Benedict XIV ordered the removal from the new edition of the *Index* in 1758 of the indictment of books teaching the mobility of the earth. However, the point of the Pope's action was obscured by the fact that Copernicus' *De revolutionibus,* Kepler's *Epitome of Copernican Astronomy,* and Galileo's *Dialogo* nevertheless remained on the list. These books did not disappear from the *Index* until the edition of 1835.

The publication of Copernicus' *De revolutionibus* in 1543 was of course only the initial step in what Alexandre Koyré called the march from a closed world to an infinite universe. Displacing humankind from the unique center of the cosmos to an orbiting planet was not necessarily a bad thing. Kepler argued that the moving vantage point gave us a much better view of the heavens, and Bishop John Wilkins declared that the center was a pit of corruption—how much loftier to be out among the celestial bodies. In fact, it is the vastness of both space and time that proves to be a greater threat to the traditional Biblical view wherein we occupy a focal point in the purpose of Creation itself, as if the universe was made for mind, and we are the living embodiment of that fundamental idea.

If the purview of this volume were larger, I would examine more fully the human impulse to know our place in the physical universe, an exploration closely akin to our religious aspirations. In what way, if any, are we significant in the cosmic scheme of things? Such tacit questions moved the King of Denmark to provide a ton of gold to support Tycho Brahe's observatory on Hven, and today motivate funding for the Hubble Space Telescope and for the Search for Extraterrestrial Intelligence.

Constraints of time force me to skip about 300 years in the story of cosmology and theology, including Richard Bentley's 1693 sermons on *A Confutation of Atheism from the Origin and Frame of the World,* one of our best sources for the philosophical thinking of Isaac Newton. Neverthess, I wish to touch on some critical questions from the twentieth century that have particular relevance to our considerations.

It was here in Washington on New Year's Day of 1925, at a joint meeting of the AAAS and the American Astronomical Society, that Edwin Hubble (in absentia) announced that he had established the distance of the Andromeda Nebula as 930,000 light-years.[6] Within a few more years, he established something even more remarkable: that the fainter the spiral nebula, the greater its spectral Doppler shift. In other words, the farther the spiral nebula, the faster it appeared to be receding from us.[7]

Within a year of Hubble's announcement what was to become the Big Bang cosmology was born, largely in the hands of a young Belgium priest, Georges Lemaître. He envisioned the universe as exploding out of a primeval atom into an expanding universe, but he never really described that opening moment. Instead, he mused that "Standing on a well-chilled cinder, we see the slow fading of suns, and we try to recall the vanished brilliance of the origin of the worlds."[8]

Perhaps the California poet Robinson Jeffers, brother of the Lick Observatory astronomer Hamilton Jeffers, came closer to the indescribable with his lines from "The Great Explosion":

> There is no way to express that explosion…All that exists
> Roars into flame, the tortured fragments rush away from
> each other into all the sky, new universes
> Jewel the black breast of night; and far off the outer nebulae
> like charging spearmen again
> Invade emptiness.[9]

Whether Lemaître's Catholic background played any role in motivating his creation cosmology is hard to gauge. In 1933 he wrote, "Hundreds of professional and amateur scientists actually believe the Bible pretends to teach science. This is a good deal like assuming that there must be authentic religious dogma in the binomial theorem."[10] At a Solvay Conference 25 years later he declared, "As far as I can see, such a theory [of the pimeval atom] remains entirely outside any metaphysical or religious question."[11]

In the 1930s Hubble continued to explore the red shifts and distances of the galaxies and by 1936, in his book *The Realm of the Nebulae,* he was able to plot 34 points on his velocity–distance relation for these nebulae, going out to a distance of 6.5 million light-years. He determined what is today called the Hubble constant or Hubble parameter, whose reciprocal is related to the expansion age of the universe. Today it seems quite astonishing that he could get a value eight times too large, and a reciprocal value for the age of the expanding universe eight times too small. What went wrong?

Hubble himself held some suspicions that the number had problems. Writing in 1937, he stated that "the observations as they stand lead to the anomaly of a closed universe, curiously small and dense, and, it may be added, suspiciously young."[12] Part of the answer as to what went wrong, but only part, was an error in the calibration of the cepheid variable stars that had been used to establish the distance scale. As Walter Baade was to show in a brilliant piece of astronomical detective work in the early 1950s, this contributed just over a factor of two in the discrepancy.[13] Simply doubling the distances of the fainter galaxies was, of course, not enough to reduce the Hubble constant by a factor of eight. Using the 200-inch Hale reflector as well as red-sensitive plates, Allan Sandage recognized that many of the "stars" in the faint nebulae that Hubble had measured with the 100-inch telescope were in fact much more luminous H II regions. This discovery had an even greater impact on the Hubble constant than Baade's recalibration of the distance scale. By 1959 Sandage had established a Hubble constant reasonably close to the value widely accepted today.

In the period of confusion, the cosmological dark ages before these revisions, the expanding universe seemed too young to contain its ancient parts, and so various attempts were made to accommodate the short 1–2 billion-year age indicated by the reciprocal of the erroneously large Hubble constant. These included the possibility of an eternal, steady-state universe, which was proposed in 1948 by Fred Hoyle, Herman Bondi, and Thomas Gold. In this scheme the universe expands ever faster, but in an exponential way so that one cannot extrapolate back to an initial singularity when the universe sprang into being. It described a universe that had existed forever, without a beginning.

At first the steady-state universe received comparatively little notice, until 1949, when Fred Hoyle had the opportunity to give a series of popular science broadcasts on the BBC.[14] In the penultimate lecture he expounded on the steady-state theory in considerable detail, criticizing those interpretations of the expansion that included a point of origin. Taking a clearly partisan stance, he argued that, "On scientific grounds this big bang assumption is much the less palatable of the two. For it is an irrational process that cannot be explained in scientific terms." This was, by the way, the first use of the term "Big Bang," here used a somewhat pejorative sense. Finally, in a lecture entitled "A Personal View," Hoyle expanded his discussion to include ethical and religious opinions as well, making trenchant criticisms of several Christian beliefs.

For better or worse, Hoyle's outspoken statements linked the steady-state theory with atheism. Although the reviews of his book, *The Nature of the Universe,* rarely addressed a possible anti-religious connection with Hoyle's cosmology, it was not far beneath the surface.[15] Certainly by that time, 1953, the parallels between the Big Bang and the "Let there be light!" of Genesis I had become well known, for Pope Pius XII, a pontiff much interested in astronomy, had in 1951 presented a triumphalist address to the Pontifical Academy of Science, which concluded, "Thus, with the concreteness which is characteristic of physical proofs, it [the Big Bang theory] has confirmed the contingency of the universe and also the well-founded deduction as to the epoch when the world came forth from the hands of the Creator. Hence, creation took place. We say: therefore, there is a Creator. Therefore, God exists!"[16]

The Pope's address was not only featured in *Time* magazine, but was also mentioned in *The Physical Review,* when physicist George Gamow, in a characteristically whimsical mood, inserted it in a footnote as infallible proof of his own cosmology.[17] In contrast, Georges Lemaître, then the president of the Pontifical Academy, took a dim view of mixing metaphysics with an unproven scientific theory.[18]

In the 1960s several new phenomena were discovered, including the quasars and the background radiation left over from the primeval fireball. Such discoveries convincingly demonstrated that our observable universe has a history, and that it has not always presented the same face. These findings sounded the death-knell for the steady-state cosmology. This paved the way for widespread, although never universal, acceptance of the Big Bang cosmology.

In the late 1970s the astrophysicist Robert Jastrow recorded this state of affairs in a book entitled *God and the Astronomers.* There he argued that the convergence between Genesis and cosmology should cause agnostic scientists to sit up and take notice, and even be a little worried. In one memorable passage in his book Jastrow wrote:

> At this moment it seems as though science will never be able to raise the curtain on the mystery of creation. For the scientist who has lived by his faith in the power of reason, the story ends like a bad dream. He has scaled the mountains of ignorance; he is about to conquer the highest peak; as he pulls himself over the final rock, he is greeted by a band of theologians who have been sitting there for centuries.[19]

The band of theologians had an answer: God did it!

But for either a self-professed agnostic like Jastrow or for a believer, such an answer is unrevealing and even superficial, for it simply cloaks our ignorance beneath a name and tells us nothing further about God beyond the concept of an omnipotent Creator, something not especially helpful for theologians.

Should theists accept that moment of celestial fireworks as the whole work of the Creator, they are on rather thin ice. Scientific theories, especially cosmological views, are notoriously subject to change, and cosmologists have taken it as a special challenge to eliminate the singularity of point zero when space and time vanish as the universe becomes infinitely dense. For example, Stephen Hawking has proposed to treat time as one of the dimensions of the curved space–time in that opening sequence. He constructed a coordinate transformation that blended time into space and left a closed surface without any singular origin and thus, curiously, a cosmos with a pretty definite age but without a specific beginning! In his best selling book *A Brief History of Time*, Hawking has written:

> So long as the universe had a beginning, we could suppose it had a creator. But if the universe is really completely self-contained, having no boundary or edge, it would have neither beginning nor end: it would simply be. What place, then, for a creator?"[20]

It was this challenge, as much as anything in the literature, that caused the organizers of this symposium to include on its agenda the question, "Did the universe have a beginning?"

I suppose the question has two aspects. First, there is the curiosity we all share as inquiring members of our species. We would really just like to know the answer. Was Aristotle right in declaring that the whole cosmos had been here forever? Or did the correct answer come from Jerusalem? But from the theologians, there is quite another question: "Does it make any difference?"

From a theistic perspective, the answer to Hawking's question is that God is more than the omnipotence who, in some other space–time dimension, decides when to push the mighty **ON** switch.[21] God is the Creator in the much larger sense of designer and intender of the universe, the powerful creator with a plan and an intention for the existence of the entire cosmos.

Over the past 500 years the sacred landscape has seen revolutionary conceptual changes.

The stage on which mankind struts in his brief, flickering moment is today set in a vast, ancient cosmos without a center. Is it only sound and fury, or are there profounder purposes to gauged? These are timeless questions, as meaningful today as in classical Athens. Is the universe designed? Is it made for mind? Are we alone? Theist, atheistic, existentialist alike can ponder them. Perhaps the pondering of these questions *is* the purpose of the universe. So, let us take these days of discourse seriously.

NOTES AND REFERENCES

1. Washington Irving, *A History of the Life and Voyages of Christopher Columbus* (New York: 1828), p. 74.
2. See, for example, "Alfonso X as a Patron of Astronomy," pp. 115–128 in Owen Gingerich, *The Eye of Heaven: Ptolemy, Copernicus, Kepler* (New York: 1993), esp. p. 125.
3. Reijer Hooykaas, *G. J. Rheticus' Treatise on Holy Scripture and the Motion of the Earth* (Amsterdam: 1984), p. 98.

4. Johannes Kepler, “Introduction” to *Astronomia nova* (Prague, 1609), translated by Owen Gingerich in *The Great Ideas Today 1983*, ed. by M. J. Adler and J. Van Doren, (Chicago: Encyclopaedia Britannica, 1983), p. 319.
5. Ibid., pp. 320 and 322.
6. E. Hubble, *Publications of the American Astronomical Society*, Vol. 5 (1925), p. 261.
7. E. Hubble, *Proceedings of the National Academy of Sciences,* Vol. 15 (1929), p. 168 (reprinted in K. Lang and O. Gingerich, *A Source Book in Astronomy and Astrophysics* [Cambridge: 1979], p. 726).
8. Lemaître, *The Primeval Atom* (New York: 1950), p. 78.
9. Robinson Jeffers, “The Great Explosion,” from *The Beginning and the End and other poems* (New York: 1963).
10. Helge Kragh, *Cosmology and Controversy* (Princeton: 1996), p. 59.
11. Ibid., p. 60.
12. Edwin Hubble, *Monthly Notices of the Royal Astronomical Society, Vol.* 97 (1937), pp. 506-513.
13. Walter Baade, W., *Transactions of the International Astronomical Union,* Vol. 8 (1952), p. 397; *Publications of the Astronomical Society of the Pacific*, Vol. 68 (1956), p. 5
14. My account here follows Helge Kragh, *Cosmology and Controversy* (Princeton: 1996), pp. 191ff.
15. I remember attending an NSF summer school at the University of Michigan in 1953, and in the informal discussions, Gerard Kuiper, one of the lecturers, openly complained that Hoyle was opposed to the universe expanding from a singularity only because it was too much like the Genesis account of creation.
16. Kragh, p. 257
17. Gamow, *Physical Review,* Vol. 86 (1952), p. 251; *Time*, Vol. 58 (3 December 1951), pp. 75–77.
18. Kragh, p. 431, note 186, quoting Ernan McMullin.
19. Robert Jastrow, *God and the Astronomer* (New York: 1978), p. 116.
20. Stephen Hawking, *A Brief History of Time* (New York, 1988), pp. 140–141.
21. Several years ago I had an interesting discussion of this point with Freeman Dyson, one of the most thoughtful physicists of our day. “You worry too much about Hawking,” he said, “It’s rather silly to think of God’s role in creation as just sitting up there on a platform and pushing the switch.” Creation is a far broader concept that just the moment of the Big Bang. The very structures of the universe itself, the rules of its operation, its continued maintenance — these are the more important aspects of creation.

The Big Bang as Scientific Fact

S. M. FABER

Department of of Astronomy and Astrophysics, University of California–Santa Cruz, Santa Cruz, California 95064, USA

ABSTRACT: In the year 1900, humanity had barely a notion of our place on the cosmic stage, and no inkling at all of how we got here. The one hundred short years of the twentieth century sufficed to unravel 14 billion years of cosmic history and how those grand events, after 9 billions of years or so, set the stage for the birth of our own home, the Solar System. The key events in this history are not hard to comprehend; they can be sketched in a few brief pages. This precious knowledge is part of our shared heritage as human beings and is fundamental to the future prospects of our species. Without it, we are ignorant of the powerful forces that have shaped our past and that will shape our destiny in the future. Read here the cosmic history of humanity, beginning with the Big Bang.

KEYWORDS: Big Bang; cosmology; cosmic microwave background; inflation; galaxy formation; steady-state universe

Unlike some of the other contributors to this volume, I have the luxury of dealing with a well-defined topic for which a great deal of hard data are available. My charge is to argue why a reasonable person should accept the Big Bang as proven scientific fact. In his recent book, *Before the Beginning*,[1] the British cosmologist Sir Martin Rees wrote that he was "fairly sure" of the Big Bang—at the 90% level. Actually, I suspect that Martin was being a bit coy, for in fact it is impossible to do any constructive research in cosmology today without standing firmly on the Big Bang model. Most cosmologists, I suspect —and I among them—would side with the famous Russian cosmologist Jakob Zeldovich, who said in 1982, "I am as sure of the Big Bang as I am that the Earth goes around the Sun."[2]

My task is specifically to motivate and defend the Big Bang model, but what exactly do I mean by the "Big Bang"? In what follows I take a narrow definition of the Big Bang as a moment in the *finite* past at which our Universe had very high density and (as explained later) a very high temperature.

Address for correspondence: Department of Astronomy and Astrophysics, University of California–Santa Cruz, Santa Cruz, CA 95064. Voice: 831-459-2944.
faber@ucolick.org

This claim is concrete, and the evidence for it is overwhelming. In defining the Big Bang so narrowly, I am deliberately sidestepping such broader questions as: Did the Universe begin with the Big Bang? Did space and time begin with the Big Bang? What came before the Big Bang? These and other harder questions will be the topic of later chapters in this volume—I have the easy job of setting the stage.

THE BIRTH OF BIG BANG COSMOLOGY

The notion of a Big Bang implies that the Universe was once in a state of high density that it no longer occupies today—in short, that it has *changed*. The idea of a changing Universe (as opposed to a static, eternal one) got a big boost from several sciences during the nineteenth century. Geologists began to realize that the Earth's crust had evolved and that it had taken a very long time to create the surface of the Earth at present geological rates. Paleontologists saw countless species in the fossil record come and go and they concluded that the logbook of terrestrial species had profoundly changed. Charles Darwin capped it all by arguing persuasively that terrestrial life and its environment had both evolved massively over a very long period of time.

Although the idea of an evolving Earth was in the air, by the turn of the century astronomers had not yet measured the age of any celestial object, and few in fact had any inkling of a Universe out there beyond our Galaxy (most thought the Universe *was* the Galaxy). Two events early in the century shattered this landscape and ushered in the era of scientific cosmology.

The first was Einstein's discovery of general relativity (GR) in 1915–1917.[3] General relativity is central to cosmology because it is a theory of gravity, and gravity is unique among the four forces of nature for its *long-range* character—stars and galaxies pull on one another by gravity clear across the Universe.[a] Moreover, unlike Newton's theory of gravity, GR links the very geometry of space to the distribution of matter and energy within it. Matter and energy are each a source of gravity in GR, and the local curvature of spacetime is determined by their local densities. The large-scale topology of the Universe is thus determined by its contents, and the two have to be

[a]In contrast, the comparably elegant theory of electromagnetism developed by Maxwell a few decades earlier did not stimulate cosmological speculations because the effects of the electromagnetic force are local in character. This is because large regions of space must be electrically neutral—if they were not, charged-particle currents could stream through the vacuum of space to cancel out charge imbalances; the resulting neutral regions could not pull on one another electrically. The other two forces of nature, the weak and strong force, have an even smaller reach than electromagnetism.

solved for simultaneously. This imposes a degree of self-consistency not present in Newton, making for a much tighter cosmological model.

Gravity is inherently dynamic—it pulls on things and makes them move. A little reflection shows that the only form of motion that is consistent with large-scale uniformity of the Universe is either *global expansion or contraction*. Strangely, it took twelve years after the promulgation of GR for this idea to sink in. One reason for this was Einstein himself. In developing the basic equations of GR, he noticed a freedom to insert an *ad hoc* term that at that time corresponded to no known gravitational force. This famous term, called the cosmological constant Λ, corresponds to a large-scale *repulsive* form of gravity (repulsive if the right sign for Λ is chosen). I will return to Λ later, but for now I want to stress that the classical Λ-force of Einstein does not depend on the presence of matter or energy for its source—it just simply *is*, at the same strength throughout all space and time. Einstein was attracted to Λ because of his belief at the time—the year was 1917 and no contrary evidence was as yet in hand—that the Universe was static and unchanging. The repulsive "anti-gravity" of Λ could then be chosen precisely to balance the attractive force of normal gravity and "prop the Universe up."

Twelve years later, in 1929, Edwin Hubble announced that the Universe was in fact expanding. This was the second event that marked the beginning of scientific cosmology. Actually, Hubble himself spoke only of "the recession of the nebulae" and avoided the word "expansion" as an unwarranted extrapolation beyond the data. We owe the concept of expansion to the Belgian cleric Georges Lemaître. Lemaître had been playing around with solutions to Einstein's GR equations during the 1920's[4] and had discovered (with Eddington) that Einstein's static Λ-model was in fact unstable—it was balanced, but on a knife edge. The slightest mismatch in the value of Λ either way would trigger either a dizzying collapse or a wild expansion. In such evolving models, Lemaître noticed that the velocities of galaxies would vary in proportion to their distances, the same law of proportionality discovered two years later for the real universe by Hubble. Perhaps the famous "Hubble law" should really be called "Lemaître's law."

Primed by his earlier work, Lemaître seized on Hubble's discovery as evidence for a real physical expansion, and in 1931 he produced the first cosmological model based on actual data—Hubble's. His model[5] had a repulsive Λ and was just entering its rapid expansion phase at the present time. It also contained substantial ordinary matter, whose gravity, Lemaître showed, would generate a "singularity" at a *finite* time in the past. Lemaître termed this initial high-density state the "primeval atom." He even described the exit from this state as a "bang," although he did not embellish it with the word "big." (That came later, from the British steady-state cosmologist Fred Hoyle, who invented the term "Big Bang"[6] as a derisive dismissal of a, to him, aesthetically unappealing rival to his steady-state cosmology—more on this below.) Label aside, the concept of the Big Bang had entered cosmology.

A BARE-BONES BIG BANG MODEL

In motivating the Big Bang (BB), I start simply with a bare-bones bang emerging out of a moment of very high density at some finite time in the past. This picture has two inescapable implications. First, the Universe has a finite age, and therefore *all objects in it should be younger than it is*. The time since the BB is the expansion age of the Universe, which is simply how long the Universe has had to expand in order to get to its present size. Mathematically, a rough estimate of the expansion age is given by the inverse of Hubble's constant, $1/H_0 = d/v$, where d is the distance of a galaxy and v is its recessional velocity. Getting the age this way is precisely like estimating the time it would take you to drive from Washington, D.C., to New York City based on the distance to New York and your *present* speed. It turns out that galaxies are rather like cars—it's easy to measure their speeds using a speedometer that astronomers have at their telescopes called the "Doppler effect." But getting their distances is harder. In your case, you'd look up the distance to New York in some table of inter-city distances, but there is no such table for galaxies. The Hubble Space Telescope (HST) is making great contributions here by allowing us to measure distances to nearby galaxies to an accuracy of about ± 10%.[7,b]

The latest expansion-age measurements from the HST are converging on an expansion age of about 11–14 billion years. How does this compare to the ages of the oldest stars? For technical reasons, the ages of stars can be accurately measured only in groups—star clusters or galaxies. The oldest star clusters in our Galaxy are turning out to be almost 14 billion years old, near the upper edge of the allowed expansion-age window. Most astronomers consider this to be a promising agreement, others feel it is an age that is uncomfortably large for H_0. The point here, however, is that no *very* old stars have been found—the upper limit of roughly 14 billion years for stellar ages seems to be universal. This is in contradiction to an infinitely old steady-state universe (see below), which would contain stars over a wide range of ages including some exceedingly old ones. The statement that there are no very old stars in the Universe could have been criticized a short while ago, as accurate ages were limited to star clusters in just our own Milky Way, which might happen to be a relatively young galaxy in an old universe. However, the HST and a new wave of very large ground-based telescopes have succeeded in age-dating a wide array of stellar populations in many galaxies, and the 14-billion-year limit still holds.

[b]There is a further source of uncertainty in the age caused by the fact that the *present* speed of galaxies may not equal their *average* speed over all time; ordinary gravity retards the expansion and causes galaxies to slow down whereas a repulsive Λ makes them speed up. Hidden in every deduction of the age of the Universe from H_0 is always some assumption about the past speeds of galaxies. This assumption adds an extra uncertainty of about ± 20% to quoted ages, but is not usually mentioned in popular accounts.

The second implication of the bare-bones Big Bang model is that *the Universe should look different at large distances.* That is because looking out into space is also looking back in time, due to the finite travel speed of light. The most distant objects visible in our telescopes carry us back in time to within a couple of billion years of the Big Bang, when the Universe was much smaller and all the galaxies within it were much younger. It is reasonable to expect that galaxies and their contents have changed greatly over such a span of time.

The first clue that the Universe really does look different far away came in the 1950s and '60s, when it was found that radio galaxies and quasi-stellar objects (QSOs) seemed to be more frequent as one looked farther away. The prevailing picture now[8] is that QSOs and radio galaxies are both formed when gas rains down onto a massive central black hole at the center of a galaxy, surrounding it with a swirling disk of glowing white-hot gas on its way into the hole. Plausibly there was more loose gas early in the Universe, before it condensed into stars. This loose gas fell more copiously into black holes at that time, making them shine more brightly. The same holes still exist at the centers of galaxies today, but are dim now because their gas supply has dried up.

Another even clearer piece of evidence for a different Universe in the past again comes from the HST, whose 250-hour "Hubble Deep Field" exposure is the deepest picture of the Universe ever taken. As Joel Primack[9] discussed, Ken Lanzetta's beautiful video takes us on a fly-through down the Hubble Deep Field "core-drilling" through space–time. Large, mature galaxies are present in the foreground, giving way roughly 10 billion years ago to small, unformed lumpy proto-galaxies in the process of formation. Finally space becomes empty as we fly past the earliest galaxies and into the cosmic "Dark Ages," the pre-galactic era before any galaxies had formed.

My arguments so far do not yet capture the deep appeal that the Big Bang has for cosmologists. The real attraction of the bare-bones BB is that it can so easily be embellished with several other ideas which can explain a great many more phenomena beyond those already mentioned. The Big Bang is like a Christmas tree on which many theorists have hung ornaments. Over time the tree has grown steadily more beautiful until now it is well nigh irresistible. The next three sections explore these important additions.

A HOT BIG BANG

The first embellishment is to fill the Universe with a bath of *thermal photons* at some very early time just at or after the Big Bang ("thermal photons" is just a fancy way of saying "heat radiation"). Such a thermal bath cools down as the Universe expands (just as a jet of air spurting out of a tire feels cool), but the remarkable thing is that the bath stays perfectly thermal *at a single temperature*. In the jargon of physics, the spectral-energy curve of the radiation retains its classic single-temperature shape, but drifts downward in frequency toward longer wavelengths as the Universe expands.

The hot Big Bang model predicts that we should see these thermal photons today, but much reduced in temperature by the expansion. Of course, we do see this radiation—it was detected as microwaves in 1964 by Penzias and Wilson[10] and is known as the cosmic microwave background (CMB), or "primeval fireball radiation." Since this radiation is a gas of photons filling the whole Universe, it should not be found to emanate from individual stars or galaxies, but should instead cover the whole sky evenly, as it indeed does. If our eyes were sensitive to CMB photons, the sky would glow faintly but uniformly like the daytime sky. The CMB spectrum has a perfect thermal shape corresponding to a single temperature of exactly 2.728 degrees above absolute zero.

The thermal spectrum of the CMB is extremely telling. Many people have tried but failed to produce this radiation in sources other than the hot BB (for example, by stars, QSOs, or cool clouds heated by stars). The problem is that such objects inevitably have a broad range of temperatures, unlike the CMB radiation, which has a *single* temperature to better than one part in 10,000. The hot BB model is uniquely able to explain the CMB because of its high early density, which allowed heat to flow rapidly so as to even out even tiny temperature differences. No other satisfactory explanation for the CMB radiation has ever been found.

Let us now run the hot BB model backwards from today and observe how the Universe heats up. For every factor of 10 contraction in the size of the Universe, the temperature also increases by a factor of 10. Particles move fast, and photons become more energetic—the result is that at very early times the whole Universe turns into a colossal particle accelerator, the envy of every particle physicist on Earth.

Alan Guth[11] and Neil Turok[12] in subsequent chapters will discuss events very early in the Universe that are at or beyond the limits of present physical laws. Well after that—at a time of about 100 seconds when the temperature had fallen to 10^9 K and the density was that of water—a key series of particle reactions occurred. Protons and neutrons fused together to make deuterium (proton + neutron), He (2 protons + 2 neutrons), and other light elements. These are the same fusion reactions that drive stars and the hydrogen bomb, and they are exceedingly well understood.[13] The net result is that roughly 25% of the original protons and neutrons in the Universe were converted into helium plus small amounts of deuterium, lithium, etc. The predicted amounts closely match the observed cosmic abundances of these elements. Twenty-five percent turns out to be a lot of He, and the only other known source of He—nuclear burning in stars—falls short of producing enough of this element by a factor of 10 or so. This elegant argument was used by George Gamow, Ralph Alpher, and collaborators in the 1940s to *predict* the hot Big Bang and forecast the CMB radiation some 16 years prior to its discovery![14] In short, the CMB radiation and the high helium content of the Universe fit together like lock and key.

It is hard to overstate the impact of these two mutually reinforcing discoveries on the subsequent progress of cosmology. While the temperature and density of the Universe at nucleosynthesis are not extreme, the theory does entail the extrapolation of known physical laws over 10 billion years all the way back to only $t = 100$ seconds! The masses of the neutron and proton, the decay time of the neutron, the properties of small particles called "neutrinos," the nuclear reaction rates, and the value of the gravitational constant G must all be identical to what they are now. The fact that all these basic properties of the Universe were the same then as now inspired the notion that the Universe is inherently simple and understandable. The success of primordial nucleosynthesis emboldened cosmologists to take even more daring leaps to the earlier times and much higher temperatures that you will hear about at this symposium.

A HOT BIG BANG WITH DENSITY RIPPLES

The second key embellishment to the basic bare-bones model is to add small density ripples just after the original hot Big Bang. By "density ripples" I mean that the combined density of matter and energy varied slightly from place to place in the early Universe. We can think of ripples as small-scale peaks and valleys of density piled on larger-scale peaks and valleys, much as tiny ocean waves riffle the surface of long rollers. Since, according to Einstein, it is the combined matter–energy density that generates the gravitational field, even tiny ripples generate perturbations in the gravitational field. Consider, for example, the neighborhood around a matter–energy peak. Particles near the peak feel a net gravitational force pulling them toward its center and retarding their outward expansion. Ultimately they fall back into it, which increases its mass still more. The net result is a gravitational runaway in which growing peaks come to dominate and gobble up ever larger regions. By the same token, underdense regions, called "voids," expand faster than average, emptying out over time as their matter is stolen away by nearby peaks.

The physics of peak formation is quite well developed. Each region around a peak can be regarded as a "bound" mini-universe that expands for awhile, reaches a maximum radius, and then collapses back on itself to form a self-gravitating object. Such collapsed objects remain embedded in the Hubble flow and expand with the rest of the Universe. The first collapsed objects formed roughly a billion years after the Big Bang and had a mass of about a million solar masses, the size of a globular star cluster. They in turn were clumped irregularly (due to larger-scale density ripples), and clusters of them quickly began to merge. These clumps merged into yet larger clumps in a grand *clustering hierarchy.*

Gravitational collapse and clustering is the process that built all the structure in the Universe we see today. The first objects to form were star clusters and small galaxies. These soon merged to form larger galaxies, and finally groups and clusters of galaxies. The process is continuing today, building mammoth superclusters containing thousands of galaxies that are separated by empty voids up to 300 million light years across.

Key to this process is the origin and nature of the original density ripples; it is their sizes, locations, and peak heights that determine all subsequent gravitational clustering. I will say more about the origin of these ripples shortly, but for now I simply note that there exists a conceptually simplest theory for their strength versus mass, the so-called *Harrison–Zeldovich* spectrum. Their clustering behavior also depends on the nature of the matter in the Universe, whether it is mostly normal "baryonic" matter or some unknown form of "dark matter" particles, and, if dark, whether massive, "cold" particles or light, "hot" particles. Assuming that all of this is known, the whole collapse process can be predicted in great detail: when objects of a given mass will form, how large their radii will be, and how fast the particles inside them will orbit. The best-fitting theory says that the original ripples had a Harrison–Zeldovich spectrum and then grew in a Universe whose matter consists mostly of *cold* dark matter. Rather amazingly, this picture matches the masses, radii, and orbital speeds within galaxies and clusters over nine orders of magnitude in mass, from small galaxies up to huge superclusters.

A BIG BANG WITH INFLATION

The last and final embellishment of the Big Bang theory is the concept of "inflation." As developed by Alan Guth, Andrei Linde, and others,[15] inflation should be regarded as an enhancement of the Big Bang, not a replacement of it as some popular accounts have implied. To understand inflation, we need to start first with a "normally" expanding universe whose gravity is generated purely by matter. In such a universe, matter becomes more dilute with the expansion, and gravity weakens. The expansion is thus a measured, stately affair, and one can show that the size of the universe (i.e., the distance between any two freely expanding particles) varies only as some power of the time (t^n).

Inflation is produced when gravity is produced not by ordinary matter but by a repulsive Λ. This is equivalent to putting a *negative pressure* into the GR equations, and one that remains *constant* despite the expansion. Einstein's Λ was a mysterious term that arose, literally, out of nothing (i.e., the vacuum). Modern particle physics is teaching us, however, that a vacuum, despite being empty of matter, is far from inert. A revolutionary new concept is the idea that, under certain conditions, such as those reached in the early Universe, the vacuum can possess a finite energy density, even when empty of matter. Being disjoint from ordinary matter, this energy density does not become more dilute as the Universe expands, yet, being energy, it does generate a gravitational field. *Voilà*, modern physics has produced a physical *raison d'être* for a constant Λ! The resultant gravity turns out to be repulsive and propels the Universe in an eyeblink into uncontrolled, stupendous expansion—hence the term "inflation." The expansion rate is exponential, that is, the size of the Universe varies as 10^{Kt}, where K is in proportion to the energy density of the vacuum. The presence of t in the exponent produces an expansion that is far more powerful than the power-law expansion of a normal matter-dominated universe.

This early epoch of inflation was very brief. The early vacuum of high-energy particle physics proves to be unstable, and the inflation phase lasted from only 10^{-35} seconds to 10^{-32} seconds (in the classic picture). Nevertheless, in this briefest of instants our entire visible Universe (now 30 billion light years across) inflated by a factor of 10^{60} or so, from an indescribably small speck to about the size of a grapefruit. Thereafter it expanded "normally." All of the space we see today was created out of virtually nothing during this brief flash of inflation.

Though totally at odds with common sense, the inflationary theory is taken seriously because it solves, in one fell swoop, several conceptually difficult puzzles that had been dogging the standard hot Big Bang model. I will mention just three of these:

(1) Why do all parts of the visible Universe look the same and obey the same laws of physics? Though clumpy on small scales, the Universe on large scales looks statistically the same everywhere. The ultimate test of this is the cosmic microwave background radiation, whose temperature of 2.7 degrees is the same wherever we look. A classic question in cosmology is how the temperature of the CMB got to be so uniform over the whole sky. CMB photons coming to us today from opposite sides of the sky were emitted long ago by patches of matter that, in the classic Big Bang picture, were so far apart that they never could have exchanged energy with one another. How then could they have reached a common temperature? Historically this equality of temperature everywhere had to be built in by an *ad hoc* assumption, a scientific *deus ex machina*. Inflation solves this problem by creating the entire visible universe (and much more) out of a tiny speck that was so small that there was ample time before inflation for temperature fluctuations to even out. Born entirely within this uniform speck, the Universe we see today naturally looks uniform.

This explanation for the CMB can be expanded to answer the deeper question of why the properties of matter and values of the physical constants are also the same in all parts of the visible Universe. This is a question that has plagued natural philosophers for centuries. A plausible answer is that the same equilibrating processes acting to smooth out temperature irregularities before inflation could have ironed out any irregularities in basic physics, too. Einstein's famous quote says "the most incomprehensible thing about the Universe is that it is comprehensible." Inflation goes a long way toward answering the implied question by providing a means whereby the Universe manages to present a consistent face at all times and locations—in short, why physics works.

(2) Where did the matter and energy in the Universe come from? The vacuum of the early Universe that gave rise to inflation had the property that it was unstable and decayed, hence its brief lifetime from 10^{-35} seconds to 10^{-32} seconds. The decay of the vacuum, or exit from inflation, involved some complicated and still poorly understood physics. The transition is thought to be a "phase change" analogous to the freezing of water. Like freezing water, the Universe actually cooled below the nominal phase-change temperature, but in this case by a much larger amount. The energy that was stored up in the vacuum tumbled out as the vacuum gave way, like water bursting through a dam. This freely available flood of energy spawned all the matter and energy that we see in today's Universe. By creating space filled with matter and photons, inflation was the original "nourishing mother" of our Universe, or, as Alan Guth has described it, the ultimate "free lunch."

(3) What spawned the density ripples? Students of galaxies, like myself, can also credit inflation with mothering the density ripples needed to kick off galaxy formation. An inflating universe literally expands faster than light in the sense that particles very close to you are swept off and accelerated to speeds faster than light. This is rather like particles falling into a black hole except that here they fall outwards to the rest of the universe rather than into the black hole. Analogous to the hole, every observer in an inflating universe is surrounded by an "event horizon," which marks the boundary where accelerating particles reach the speed of light and disappear from view. Event horizons emit "Hawking radiation," a bath of thermal (black-body) radiation, which in the case of black holes causes them eventually to evaporate. The Hawking radiation energy of an inflating universe is not quite uniform but varies slightly from place to place—this is the origin of the tiny energy-density ripples.

An alternative way of looking at the ripples is also instructive. High-energy physics tells us that vacua, including that of inflation, are busy places. According to the *Heisenberg uncertainty principle*, "virtual" pairs of photons and matter–anti-matter particles are spontaneously created out of nothing and are annihilating one another even in so-called "empty" space. This microscopic seething of the vacuum normally goes unnoticed because the fluctuations are short-lived: pairs appear and disappear quickly and on average have no measurable effect. Circumstances are essentially altered in inflation, however; now pairs are torn apart by the rapid expansion of space and, once separated, cannot find one another to annihilate. They therefore must become "real," frozen into the fabric of the Universe for all time. Their associated energy fluctuations are the tiny ripples—seized by inflation and blown up to what will ultimately become, in our day, galaxies and superclusters.

This whole scenario for the birth of structure can be summed up breathtakingly as follows: the Milky Way and all other galaxies, clusters, and superclusters, were born as weak, microscopic quantum fluctuations some 10^{-35} to 10^{-32} seconds after the Big Bang. Pumped up quickly by inflation, they continued to expand with the Universe until gravity corralled their expansion. As galaxies collapsed, stars began to be born within them, and galactic beacons visible across billions of light years switched on. Gravity continued to shepherd the flight of galaxies, weaving a lacy filigree of voids and superclusters that is now spangled over hundreds of millions of light years across the Universe. Simply put, galaxies and large-scale structure are the quantum noise of the Big Bang writ large—this picture surely represents one of the most remarkable insights of human science.

THE STEADY-STATE UNIVERSE

In conclusion I would like to return to a point glossed over earlier. I invited you to run the cosmic movie backward and mentally envision the resulting "first" moment of supremely high density as the Big Bang. This was convenient but dishonest: expansion by itself does not imply a Big Bang. To see this, consider an *exponentially* expanding universe shrinking backwards into the past. Such a universe shrinks by the *same* factor, let us say 2, for every tick of the clock. It gets continually smaller and denser but never reaches zero radius—like Xeno's runner, it never gets to the finish line in a finite time. The Big Bang of this Universe has been pushed infinitely far into the past.

Such an exponential expansion is the basis of the non-evolving, mathematically elegant steady-state model (SSM) of the Universe by Bondi, Gold, Hoyle, Burbidge, and Narlikar.[16] To appreciate why exponential expansion is central to a steady-state model, observe that the basic equation of General Relativity has two sides. The left side describes the topology, or curvature, of space-time. An important term here is the Hubble constant, given by R'/R. This term does not change for exponential expansion. The other terms on this side can also be kept constant by choosing a flat universe with zero spacetime curvature. With these choices, the whole left side can be made unchanging, that is, in steady state.

We still need to deal with the right side, which expresses the GR source terms for the gravitational field. In the classic steady-state model, gravity comes only from ordinary matter, which becomes more dilute with the expansion. If this were uncompensated, gravity would weaken, negating the notion of a steady state. Matter must therefore be *spontaneously created* from empty space to maintain gravity constant. The proponents of the model suggested that this occurred via the appearance of protons, at a rate just adequate to keep the average density of matter constant in time. In practice, matter must be created at a rate of about 1 atom of hydrogen per century in a volume equal to that os the Empire State Building. It is imagined then to collect into galaxies, where it forms stars in the usual way.

Statistically speaking, a steady-state universe presents the same face to observers over all time. There is no Big Bang; the Universe existed infinitely far into the past and will exist infinitely far into the future. Galaxies expand out of a given volume, to be replaced by young galaxies newly formed from matter spontaneously generated within the same volume. Galaxies in any one place have a mixture of ages, the typical value being near the expansion age $1/H_0$ (a few billion years), with a long tail to much older ages. In the classic steady-state model, the only radiation is emitted by galaxies: principally by stars, interstellar dust heated by stars, and the active nuclei of Seyfert galaxies, quasars, and radio galaxies.

SSM was being unhorsed when I was in graduate school, and my fellow graduate students and I enjoyed kicking it while it was down. Many data can be brought to bear against the conventional steady-state model, among which I will mention three favorites:

(1) SSM does not easily account for the high helium abundance of the Universe, some 25% by mass (see above). The problem is that the abundance of elements heavier than He is much smaller, only 1% by mass. If He and heavier elements were co-produced together in stars, we should either have ten times less He or ten times more heavy elements. This problem is solved in the Big Bang picture by making He separately via primordial nucleosynthesis when the Universe was about 100 seconds old; the steady-state model has no such easy mechanism.

(2) Classic SSM says that each volume of the Universe is statistically unchanging over time. It therefore predicts that all volumes should look the same, including those seen far away and back in time. In the 1960s, there was already strong evidence that quasars and radio galaxies were more frequent at large distances and earlier times (see above), and this was sufficient to sink the SSM in most astronomers' opinion. Such evidence has now become incontrovertible from the HST—the Hubble Deep-Field fly-through shows clearly the changes in (and ultimate disappearance of) galaxies at sufficiently

early eras. The appearance and numbers of distant galaxies from the HST are generally quite consistent with predictions from the BB hierarchical-clustering theory for galaxy formation, but are highly inconsistent with the classic steady-state model.

(3) Finally, classic SSM cannot easily account for the precisely thermal spectrum of the cosmic microwave background (CMB). As noted, the preponderant sources of radiation in the SSM are a mixture of stars, hot galactic dust, and active galactic nuclei such as QSOs. These do not combine naturally to make a purely thermal spectrum at a *single* temperature. A recent drastic alteration of the SSM model by Burbidge, Hoyle, and Narlikar[17] attempts to solve this problem by imagining a "quasi"-steady state universe punctuated by successive collapses and re-expansions. The authors claim that intergalactic dust can intercept and re-thermalize the diffuse stellar radiation field during each collapse phase, producing uniform CMB radiation during the subsequent re-expansion, although no detailed calculations for this are presented. Regrettably the model also *evolves* over many collapse/expansion cycles and thus cannot be said to be truly in steady state. In my view, little has been gained, and at the loss of the clean elegance that was the hallmark of the original model.

CONCLUSION

Let me close by suggesting that the steady-state model may have lost the battle but has in fact won the war, if winning means leaving the more enduring mark on intellectual history. The debate between the Big Bang and steady state has been one of the most productive in the history of science. First, in looking for an alternative source of He production, proponents of SSM were led to examine nucleosynthesis carefully in stars, discovering the basic rules that explain the origin of the entire periodic table beyond helium, a giant achievement. Second, the most novel, if off-putting, feature of the steady-state theory is its spontaneous creation of matter out of nothing. With appropriate re-interpretation, this can now be seen as a forerunner of the false vacuum of inflation: both mechanisms maintain constant gravity despite expansion—both create matter and energy out of apparently nothing. Indeed, a remarkably prescient paper in 1951 by McCrea[18] equated the matter-creation field of steady state to the Λ of GR by invoking a negative pressure, precisely the modern interpretation! The only change in the ensuing decades, it seems, is that particle physics has come up with a plausible physical mechanism for Λ, that is, the false vacuum. Without such a prop, the proponents of steady state looked ridiculous for their time, but it now might be argued that they were merely too far *ahead of their time*.

In fact, it is steady-state universes, not matter-dominated universes, that are the most intensely studied today. In addition to the inflating Universe, which was in a steady state for a brief but crucial period, evidence suggests that we may be entering another steady-state era today. This comes from measurements of the brightness of distant supernovae, which, when coupled with the size scale of irregularities in the CMB radiation, seem to imply that Λ is nonzero and repulsive.[19,20] If true, we are entering another exponential expansion phase, with the generation once more of event horizons and Hawking radiation. What wholly new physical phenomena lie behind this door are as yet unknown—potentially another false vacuum, phase change, and a new hierarchy of (low-energy) physical phenomena? Who can say? At any rate, the commonly voiced predictions of the inevitable "heat-death" of the Universe may be premature if we are indeed rescued by another steady state.

Finally, I note that Linde's theory of "eternal" inflation,[21] to be described more fully by Alan Guth in this volume, is really also just a steady-state model in a different guise. In eternal inflation, an exponentially expanding vacuum field spawns baby universes spontaneously and sporadically in much the same fashion statistically that the classic steady-state model spawned hydrogen atoms. The result is an endlessly unfolding hierarchy of universes that some have called the Meta-universe. An ensemble of such universes, if each had different physics, would provide the necessary ensemble of varied universes required by the anthropic principle, also described by John Barrow later in this volume.

In the history of modern cosmology, the only credible alternative to the Big Bang theory has been the steady-state model. I began this paper by siding firmly with the Big Bang. While my faith in our own Big Bang is undiminished, I muse on the possibility that steady-state and quasi-steady state universes unlock more and deeper secrets of fundamental physics.

NOTES AND REFERENCES

1. Rees, M., *Before the Beginning: Our Universe and Others (*Reading, MA: Perseus Books, 1997), p. 58.
2. Zeldovich, Ya.B., Lecture to the International Astronomical Union, August 1982, Patras, Greece.
3. Einstein, A. "Kosmologische Betrachtungen zur allgemeinen Relativitätstheorie," *Sitzungsberichte de Königle Preussische Akademie der Wissenschaften zu Berlin* (1917), p. 142 .
4. Lemaître, G., *Annales de Societe Scientifique de Bruxelles*, Vol. 47 (1927), p. 49.
5. Lemaître, G., *Monthly Notices of the Royal Astronomical Society*, Vol. 91 (1931), p. 490.
6. From Hoyle's five-part radio program on cosmology for the BBC in 1949; the scripts were published as a popular monograph, *The Nature of the Universe*, (Oxford: Blackwell, 1951).

7. Freedman, W., "Measuring the cosmological parameters," *Proceedings of the National Academy of Sciences*, Vol. 95 (1998), p. 2.
8. For a readable account of black holes and quasars, see Kip S. Thorne, *Black Holes and Time Warps* (New York: W. W. Norton, 1994).
9. See the first chapter in this volume, "Cosmic Questions: An Introduction," by James Miller.
10. Penzias, A.A., and R.W. *Astrophysical Journal*, Vol. 142 (1965), p. 419.
11. See Guth, A.H., "Eternal Inflation," in this volume.
12. See Turok, N., "Inflation and the Beginning of the Universe," in this volume.
13. A famous and readable popular account of primordial nucleosynthesis is Steven Weinberg's *The First Three Minutes (*New York: Bantam Books, 1984).
14. Alpher, R., H. Bethe, and G. Gamow, *Physical Review*, Vol. 73 (1948), p. 803.
15. For a highly readable account of inflation, see Alan H. Guth's *The Inflationary Universe (*Reading, MA: Addison Wesley, 1997).
16. For a comprehensive account of the steady-state model, see Helge Kragh's *Cosmology and Controversy* (Princeton, NJ: Princeton University Press, 1996).
17. Burbidge, G., F. Hoyle, and J.V. Narlikar, *Physics Today*, Vol. 52 (1999), p. 38.
18. McCrea, W.H., *Proceedings of the Royal Society*, Series A, Vol. 206 (1951), p. 562.
19. S. Perlmutter *et al.*, *Astrophys. J.,* Vol. 517 (1999), p.565.
20. S. Perlmutter, M.S. Turner, and M. White, *Physical Review Letters*, Vol. 83 (1999), p. 670.
21. Linde, A. D., *Physics Letters*, Vol. 129B (1983), p. 177.
22. See Barrow, J.D., "Cosmology, Life and the Anthropic Principle," in this volume.

A Recipe for Primordial Soup

EDWARD (ROCKY) KOLB

NASA/Fermilab Astrophysics Group, Fermi National Accelerator Laboratory, Batavia, Illinois, USA

ABSTRACT: If the universe emerged from a state of high temperature and high density, we can study the origin of the universe by recreating those conditions in the laboratory. While we can not reproduce the Big Bang, we can recreate its conditions and study the physical processes responsible for producing the rich and varied universe we observe today.

KEYWORDS: cosmology; Big Bang; origin of the universe

In this presentation I will describe events that occurred in the first second of the life of the universe. There have been approximately four-hundred-thousand-million-million seconds since the beginning of the universe, so to concentrate on only one of them might seem the ultimate degree of overspecialization. But the very first second was really something special.

In some sense cosmology is history; it is the science of history. If *history of science* is a well-established field, I see no reason why there can't be something known as *science of history.* Indeed, my approach to cosmology is that of a historian, and certainly not that of an antiquarian. An antiquarian is interested in old things simply because they are old. To an antiquarian, a laundry list from 1215 is just as significant as the Magna Carta since it is equally old. A historian, on the other hand, is interested in the past because past events shape the present. Of course, not every past event is equally significant. It is the job of an historian to sort through the past to find the most important events, those that shape the future. Studying the past helps us understand the present.

I am interested in the first second of the universe because events during that time are responsible for shaping the structure of the present universe. Just as in history, in cosmology the past shapes the present. We will never have a complete understanding of the present universe without at least a rudimentary understanding of the origin of the universe.

Address for correspondence: Dr. Edward W. Kolb, Fermi National Accelerator Laboratory, 209 (WH 3W), P.O. Box 500, Batavia, IL 60510-0500. Voice: 630-840-4695; fax: 630-840-8231. rocky@fnal.gov

This discussion is intended to be nontechnical. Although some of the ideas may seem difficult, and nomenclature (such as "quarks," "photons," or "gravitons") may be unfamiliar, two important themes can be appreciated without a mastery of the exotica of modern physics.

The first theme is that today, about 12 billion years after the Bang (AB), we have a pretty good idea what the universe was like earlier than 1 second AB,[1] because we can understand the very early history of the universe by studying the present universe. All around us we see the evidence of what the universe was like at earlier times because the early universe left behind something analogous to a fossil record for us to examine. The fact that the universe is comprehensible at all is astounding. That we can speak with confidence about the universe 1 second AB is one of the proudest intellectual accomplishments of our species.

The second theme that runs through this discussion regards the unity of science. Although the study of science at schools and universities is organized into different departments for physics, chemistry, biology, and so on, nature is not so easily categorized. A single part of nature cannot be studied in isolation. This was eloquently stated by the great American naturalist and conservationist John Muir: "When one tugs at a single thing in nature, he finds it hitched to the rest of the universe."

We will see that one cannot understand the largest things in the universe without understanding the smallest things, and to truly understand nature on the smallest scales, one has to understand it on the largest scales.

We call this link between the large and the small the "inner space/outer space" connection. In a very real sense, the largest telescope is also a microscope, and the most powerful microscope is also a telescope.

Before describing the basics of our view of the universe circa 1999, it is interesting to reflect on how much our view of the universe has grown in the twentieth century. One-hundred years ago an astronomer would have probably described the universe as a flattened disk, 30,000 light-years in diameter and 6,500 light-years thick, with our solar system located "at or very near" the center. This model universe, named the *Kapteyn Universe* after the Dutch astronomer Jacobus Cornelius Kapteyn (1851–1922), does not extend beyond our own MilkyWay galaxy. In 1918, the American astronomer Harlow Shapley demonstrated that we do not live in the center of the Milky Way, and in 1924 another American astronomer, Edwin Hubble, proved that as our sun is just one of billions of stars in our home galaxy, the Milky Way is just one of billions of galaxies. After this discovery of Hubble, we realized that the universe is of far greater proportions than imagined in 1899 (FIG. 1).

Not only did Hubble extend our cosmic horizon to encompass billions and billions of galaxies beyond the Milky Way, but his 1929 discovery of the expansion of the universe was the first of two great cosmological discoveries of the twentieth century. The second great discovery was in 1964, when Arno Penzias and Robert Wilson discovered the existence of a cosmic background

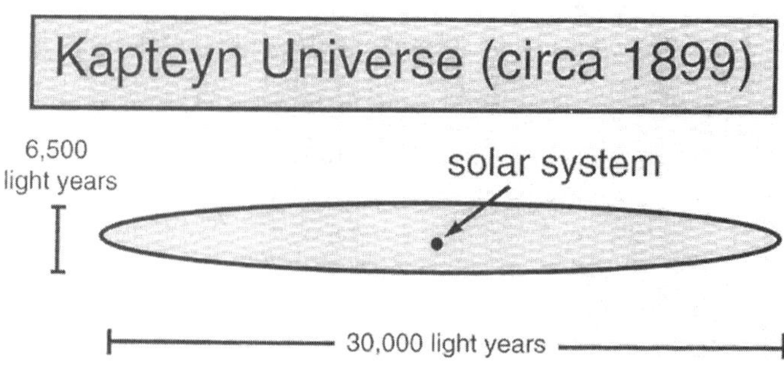

FIGURE 1. The Kapteyn Universe

of radiation. We now understand that the cosmic background radiation, currently at a temperature of about three degrees Kelvin (3°K) above absolute zero (equivalent to –270°C or –454°F), is the remnant of the hot radiation from the early universe. The discovery of the expansion of the universe and the observation of the cosmic background radiation are the cornerstones of our modern model of the universe, the Big Bang model.

The body of evidence supporting the Big Bang model is so overwhelming that it is impossible to escape the conclusion that the universe was once denser and hotter than it is today. We no longer question that the Big Bang is a good description of the universe for most of its history. The real question is how far back we can extrapolate the Big Bang. When (if ever) will the Big Bang model fail? Can we trust the Big Bang model under the extreme conditions near the time of the Big Bang? Let's journey through the complete history of the universe, running the film of the Big Bang backwards, pausing occasionally to examine the conditions we encounter.

Today the universe is about 12 billion years old, with a cosmic temperature of 3°K. When our sun formed, some 5 billion years ago, the universe was 7 billion years old, with a cosmic temperature of 4°K. We can see what the universe was like 7 billion years AB, because we can look back in time by looking out into space.

Since light has a finite velocity, when we observe distant objects we see them as they existed in the past, when the light we now observe was emitted from the object. For instance, the sun is 8 light-minutes away from Earth, so when we stand on Earth and look up, we observe the sun as it existed eight minutes before we "see" it. The nearest large galaxy is Andromeda, some 2 million light-years distant. When we observe Andromeda, we see it as it existed 2 million years ago, when the light reaching us now started its journey. The lesson is simple: looking out in space is looking back in time. The more distant the object, the further back in time we are able to probe.

Looking out in space, looking back to the time in the history of the universe when our sun formed, we find that the spatial density of galaxies was about three times the present density of galaxies in space (although the internal densities of galaxies have not changed). The universe was indeed denser in the past. We can look much deeper into space than the distance corresponding to 7 billion years AB. The most distant astronomical objects we observe are quasars. Quasars are very bright objects that we believe signal the birth of galaxies. When we look at very distant quasars we can sample the universe when it was only 1 billion years old and the cosmic temperature was 16°K.

Although 16°K is hot compared to the present cosmic temperature, it is still incredibly cold. We have to journey back much earlier in the history of the universe to find warmer conditions. About 14 million years AB the temperature of the universe was about the freezing point of water. Two million years earlier than that, about 12 million years AB, the primordial fireball was at a temperature very pleasant for humans (although humans were not around to enjoy it). But the comfort zone only lasted a few million years. Earlier than nine-million years AB the temperature was above the boiling point of water.

Three-million-three-hundred-thousand years AB the temperature of the universe was truly hellish, assuming the temperature of hell is around the boiling point of brimstone, T = 445°C = 833°F. Cosmology suggests that hell was in our past, not in our future (FIG. 2).

The universe was really cooking 300,000 years after the Bang, when the temperature of the universe was about 3000°K. This epoch in the history of the universe was the true birth of the atomic age. Earlier than 300,000 years AB, when it was hotter than 3000°K, atoms could not exist because there was enough hot thermal radiation to ionize any atoms (strip the negatively charged electrons from the positively charged nuclei). For the first 300,000 thousand years the universe was a hot plasma of positively charged nuclei and negatively charged electrons. Electromagnetic radiation (e.g., light) cannot travel unimpeded through a dense plasma, so the universe was opaque for its first 300,000 years. The universe became transparent to electromagnetic radiation only after atoms formed. The cosmic background radiation last interacted with matter during the formation of atoms, so the microwave photons are a direct relic of the universe at this time and we can "see" the universe 300,000 years AB by studying the microwave background radiation. Since the universe was opaque for the first 300,000 years of its life, we cannot use light of any wavelength to look out in space to a time earlier than 300,000 thousand years AB (FIG. 3).

Although we cannot observe directly the universe for its first 300,000 thousand years, we can simulate the early universe by recreating the conditions of the early universe in the laboratory. If we can discover the fundamental forces and particles relevant for the extreme conditions of density and pressure present in the early universe, we can understand its evolution.

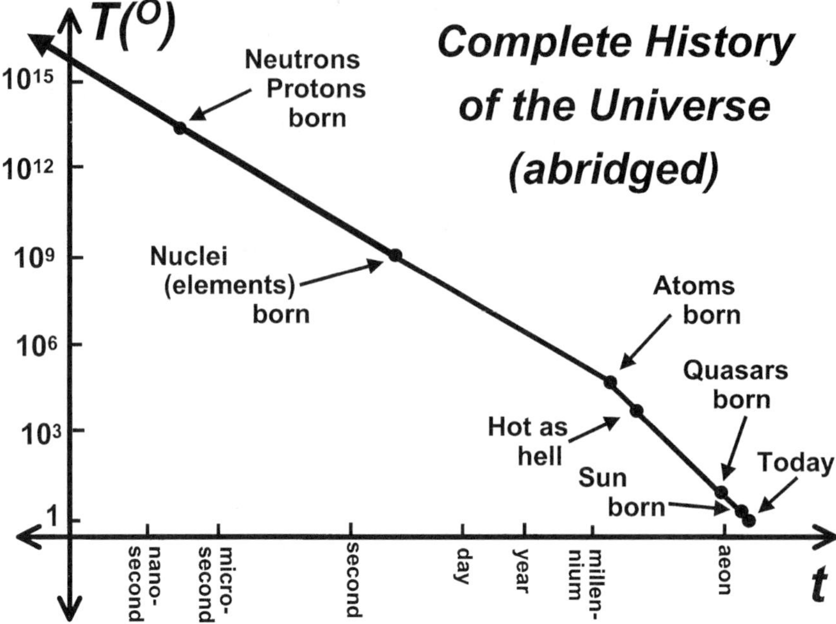

FIGURE 2. An abridged history of the universe.

The simulation of the early universe uses the familiar equation of Einstein's theory of gravity:

$$R_{\mu\nu} - \tfrac{1}{2} g_{mn} R = \mathbf{8} n G T_{\mu\nu}$$

One does not have to be a physicist to grasp the pertinent feature of this equation: it has a left-hand side and a right-hand side. The way I have cleverly written the equation (also the way Einstein wrote it), the left-hand side describes space and time—the curvature of space and the expansion of space. The right-hand side describes the fundamental particles and forces, and how matter and energy are distributed throughout the universe. The left-hand and the right-hand sides talk to each other. The curvature of space and the expansion of space tell matter how to move, and the fundamental forces and the arrangement of the fundamental particles tell space how to curve and how to expand. So if we can understand the right-hand side of the equation, the fundamental particles and forces, we can use the Einstein equation to study the evolution of the early universe. Here we see the inner-space/outer-space connection at work. Microphysics determines the evolution of the universe.

Since we use Einstein's equation to study the origin of the universe, the ten non-linear partial differential equations contained in the Einstein equation can be regarded as the ten commandments of modern genesis.[2]

Following the ten commandments of modern genesis, we can study the early universe so long as we can reproduce the conditions of temperature, density, and pressure expected for the epoch in question. If we extrapolate the temperature–time relation from Einstein's equation to 3 minutes AB, we expect that the temperature was about 1,000,000,000°K. This is an enormous temperature, but because we can reproduce the same conditions in the laboratory, we have a good understanding of the behavior of matter in this extreme environment.

Earlier than 3 minutes AB the universe was so hot that nuclei could not exist. Just as any atom in the universe would have melted if the temperature was hotter than 3000°K, nuclei melt at temperatures hotter than 1,000,000,000°K. Three minutes AB was the birth of the nuclear age; before then the universe was made of the constituents of nuclei: neutrons and protons.

Our extrapolation back in time is rewarded, because we can predict what sort of nuclei were produced 3 minutes AB when neutrons and protons formed nuclei. Using what we learn from studying nuclear reactions in the laboratory, and using the modern laws of genesis to predict how the universe cooled as it expanded, we predict that 3 minutes AB the primordial neutrons and protons should form 75% hydrogen and 25% helium, with just a trace of lithium.[3] Furthermore, the Big Bang model predicts the abundance of different isotopes of hydrogen (normal hydrogen and heavy hydrogen) and helium (helium-4 and helium-3). The agreement between predictions and observations is an astounding success. The fact that the predictions agree so well with observations of the primordial elemental abundances gives us a lot of confidence that we understand the universe 3 minutes AB.

There is no reason to stop the extrapolation back in time at 3 minutes AB. We have a very good knowledge of the types of interactions expected in a universe of electrons, neutrons, protons, and photons, and so long as we can understand the relevant constituents of matter and its behavior at high temperature, we can reasonably speak of the evolution of the universe.

We know that neutrons and protons are not elementary particles, but rather are made of quarks and gluons. At sufficiently high temperature and density the neutrons and protons themselves should melt. Earlier than a microsecond (one-millionth of a second) after the Bang, the universe should have been a primordial soup of quarks and gluons.

Since our goal is to simulate the extreme conditions of the early universe, we must find a way to produce the corresponding temperatures in the laboratory. The most powerful tool for generating high temperatures is the particle accelerator. The accelerator is used to examine nature on the smallest scales and to discover the fundamental forces and particles. Accelerators are the world's most powerful microscopes. But the most powerful microscope is also a telescope, because by colliding particles at high energy we can recreate the environment of the early universe and understand how it evolved into the present universe.

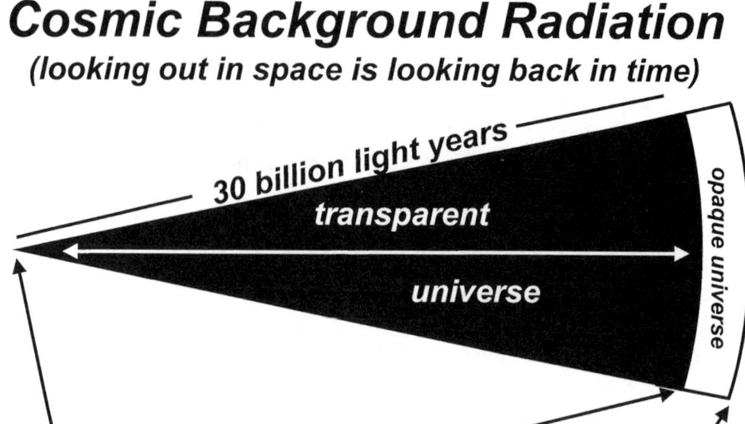

FIGURE 3. Looking out in space is looking back in time.

The world's most powerful accelerator is at Fermi National Accelerator Laboratory (Fermilab) in Batavia, Illinois. Collisions of protons and antiprotons at Fermilab can produce temperatures of 3,000,000,000,000,000°K, which existed in the universe 4 picoseconds (four-millionth-millionth of a second) after the Bang. We can study the early universe by recreating in terrestrial laboratories a little piece of the primordial soup. The ingredients of the primordial soup are the elementary particles produced and studied at accelerators (TABLE 1).

The most abundant ingredients in the primordial soup were quarks and antiquarks. They are so numerous because there are six types of quarks and each

TABLE 1. The known ingredients in the Primordial Soup (circa four picoseconds AB)

Percent	Ingredient	Comment
56%	Quarks	Up, down, strange, charm, top, bottom
16%	Gluons	Strong nuclear force carriers
9%	W and Z particles	Weak nuclear force carriers
9%	Electron-like particles	Electron, muon, tau
5%	Neutrinos	Electron-type, muon-type, tau-type
2%	Photons	Electromagnetic force carriers
2%	Gravitons	Gravitational force carriers
1%	Higgs bosons	Still undiscovered

type of quark comes in three "colors." The next most abundant particle species in the primordial soup was gluons, the carriers of the strong force. Quarks, antiquarks, and gluons make up 72% of the primordial soup. The electrons, muons, photons, and all the other elementary particles were the flavorful croutons in the primordial soup.

The recipe for the universe calls for a hot primordial soup of elementary particles 4 picoseconds after the Bang, cooling in the expansion of the universe to form neutrons and protons from the quarks and gluons a microsecond after the Bang, which then form nuclei from the neutrons and protons 3 minutes after the Bang, which finally make atoms, 300,000 thousand years AB.[4]

So in a very real sense, everything we see in the universe came from the primordial soup, and we know about the early universe because we can cook a little bit of it in the laboratory and sample it. With this taste of success, we can try to understand the universe before the era of primordial soup.

One of the wonderful things about science is that every answer seems to lead to a deeper and more fundamental question. If everything comes from the primordial soup, it is irresistible to ask about the origin of the ingredients.

In the last few years cosmologists have developed a compelling theory for the origin of the primordial soup: it came from nothing. Since nothing is not a very scientific sounding word, I usually refer to nothing as the vacuum. This idea, first proposed by Alan Guth and known as the inflationary universe, assumes that before primordial soup the universe consisted of nothing but empty vacuum.

To the casual observer the vacuum is boring, but to the true cognoscenti, there is a beauty, richness, and texture to nothing. Perhaps understanding nothing will allow us to understand everything in the universe. To grasp the basics of the inflationary origin of the primordial soup it isn't necessary to understand everything about nothing, just three little facts:

1. Nothing is *something*.
2. Nothing has *energy*.
3. Nothing can *change*.

1. Nothing is something. In fact, nothing is a lot. In a region of space one can remove all matter and radiation, but it is impossible to remove quantum uncertainty. Because of Heisenberg's uncertainty principle, at any point in space it is possible for a particle and antiparticle to emerge from the vacuum, seemingly violating conservation of energy momentarily before annihilating and returning to the vacuum. On submicroscopic scales the vacuum is a seething, writhing foam of particle-antiparticle pairs popping in and out of a virtual existence.

2. Nothing has energy. There is another important aspect to nothing: the Higgs field associated with unification of forces. All natural phenomena can be described by the action of four fundamental forces: the gravitational force, the electromagnetic force, the strong nuclear force, and the weak nuclear force. What a simple, beautiful picture. It is not necessary to invent different

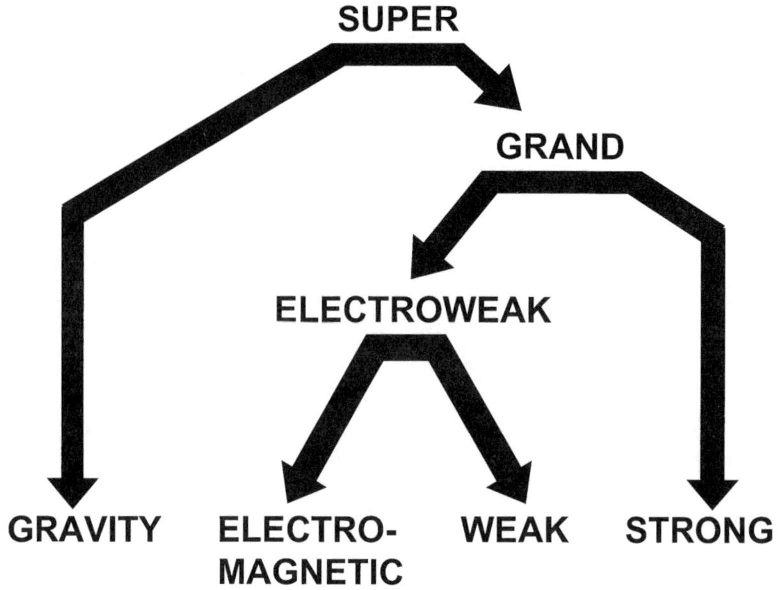

FIGURE 4. The four fundamental forces of nature.

forces for different phenomena. We know that the force of the wind, the force of the tides, all of the forces at work on Earth, in the sun, and throughout the universe involve one or more of just four fundamental forces (FIG. 4).

It was once imagined that there were many more forces at work in nature. But nature seems to be simpler than that. Phenomena once thought to be caused by distinct forces are actually different manifestations of the same force. There are two famous examples of this "unification" of forces. In the seventeenth century Isaac Newton realized that the force causing an apple to fall on his head was the same force responsible for keeping the moon in orbit about Earth, and Earth and the other planets in orbit about the sun. This unification started the inner space/outer space connection, because it implied that the force responsible for motions of celestial bodies is amenable to study in the laboratory.

The second example of unification of forces is the synthesis of a single electromagnetic force out of the electric force and the magnetic force. Before the work of Michael Faraday and James Clerk Maxwell in the nineteenth century, electric forces and magnetic forces were thought to be distinct. The shocking realization that electricity and magnetism were different manifestations of a single phenomenon known as electromagnetism led to a great simplification of physics and a deeper understanding of nature, as well as illuminating the world with electricity.

We have witnessed another unification in our century. The work of many physicists in the 1960s and 1970s, most notably Sheldon Glashow, Steven Weinberg, and Abdus Salam, led to the realization that the electromagnetic force and the weak nuclear force are actually different manifestations of a single *electroweak* force.

There is strong evidence that there is a further unification of forces, and that perhaps all of the four forces are different aspects of a single *superforce.* If this idea of unification of all the forces is correct, at temperatures far exceeding what we are capable of producing in present-day accelerators, the four forces would behave in a similar way. Superforce temperatures may have been reached near the beginning of the early universe, and perhaps true unification was the initial state. If so, the temperatures required for superforce unification have not existed anywhere in the universe since the instant of the Bang.

If the laws of nature have a deep and profound unifying symmetry, the symmetry is not manifest in the universe today since the observed character of the four forces is very different. One of the most beautiful ideas in modern physics is that the laws of nature can have symmetries that are hidden from us because the vacuum need not respect all the symmetries. The process by which the vacuum hides symmetry is known as the Higgs mechanism, named for the Scottish physicist Peter Higgs. In the Higgs mechanism there is a Higgs potential energy in the vacuum. For instance, electroweak unification involves a Higgs potential energy of 246 billion volts in the vacuum. Even this tremendous energy may be small compared to the energy that was once in the vacuum.

3. Nothing can change. True superunification was fleeting. As the universe expanded and cooled, the forces took on different characters as the nature of the vacuum changed. About 10^{-43} seconds after the Bang the gravitational force became distinct from the other forces. Then, around 10^{-35} seconds later, the strong nuclear force split from the electroweak force. Finally, 10^{-20} seconds AB, the electromagnetic force became distinguishable from the weak nuclear force.

If the strong nuclear force is unified with the electroweak force (grand unification), then the vacuum once possessed a tremendous potential energy, ap-

TABLE 2. The Cosmic Symphony (a modern Harmonice Mundi)

Tempo	Movement	Epoch
pizzicato (seconds)	String dominated	$t \approx 10^{-43}$ s
prestissimo (seconds)	"Vacuum" dominated (inflation)	$t \approx 10^{-35}$ s
presto (years)	Radiation dominated	$t < 10{,}000$ yr
andante (years)	Matter dominated	$t > 10{,}000$ yr
largo	"Vacuum" dominated (inflation)	$t \approx$ yesterday

proximately 10^{-43} billion volts higher than today. When the Higgs potential energy changed as the electroweak force split from the strong nuclear force approximately 10^{-35} seconds AB, the tremendous potential energy in the vacuum would have been released to produce the hot primordial soup.

We are not sure of the exact time the primordial soup emerged from the vacuum pressure cooker, but we have a pretty good idea. In the spirit of Johannes Kepler's use of musical harmonic ratios to describe the velocities of planets orbiting the sun,[5] one can describe the different epochs in the history of the universe as movements of a cosmic symphony. The symphony started at a frenzied pace, but the tempo of expansion decreased as the universe cooled (TABLE 2).

The first movement of the Cosmic Symphony may be dominated by the string section if on the smallest scales there is a fundamental stringiness to elementary particles. If this is true, then the first movement in the cosmic symphony would have been a *pizzicato* movement of vibrating strings about 10^{-43} seconds AB. The inflationary movement probably followed the string movement, lasting approximately 10^{-35} seconds. The primordial soup that came out of the inflationary vacuum was so hot that its radiation dominated the rest mass of all the particles in the universe for the first 10,000 years. Our fossil record of this era is the relative abundances of the different elements produced in the Big Bang, 3 minutes AB. About 10,000 years after the Bang the radiation finally cooled enough for the rest mass of the particles in the universe to dominate the thermal energy of the radiation, and the universe entered a matter-dominated era. Finally, if recent observations of the acceleration of the expansion of the universe prove correct, then the cosmic score may be marked *da capo,* and we may return to another inflationary movement.

When the origin of the ingredients of the primordial soup from nothing was first proposed in the early 1980s, it seemed like an idea destined to remain forever outside of the experimental or observational arena. But it was soon realized that the inflationary epoch would leave behind an fossil imprint in the form of small variations in the temperature of the cosmic background radiation temperature. These variations, first detected in 1992 by the Cosmic Background Explorer Satellite, may be a fossil record of the conditions in the universe 10^{-34} seconds AB. If this is confirmed by detailed study of the background radiation, then we will then have proof that everything comes from nothing.

I believe that in the first decade of the next millennium, detailed measurements of the temperature fluctuations of the background radiation, new surveys of the arrangement of matter in the universe, and possibly the discovery of the nature of dark matter will reveal unmistakable evidence that the ingredients of the hot primordial soup came from the frozen vacuum of the early universe.

So if we can understand the nature of nothing, as well as the character of the fundamental forces and particles, we will be able to use the machinery of the Big Bang model to account for everything we see in the universe.

If this comes to pass, then we would have witnessed yet another example of the true unity of science: the largest things in the universe, the outer space of cosmology, cannot be understood without knowledge of the smallest things in the universe, the inner space of fundamental particles and forces.

But this bridge between the large and the small is not the end of the journey.

If everything came from the hot primordial soup, and in turn, if the hot primordial soup came from the frozen vacuum of the expanding universe, we must now learn the origin of the vacuum, and the origin of space and time itself.

As always in science, as we finish one chapter, we begin writing the opening sentences of the next. Perhaps one day we will write the final chapter, and our voyage of discovery of understanding the origin of the universe will be over. But until then, the journey itself could not be more wondrous.

NOTES AND REFERENCES

1. I will not address the issue of whether the universe had a beginning or not, which in the modern cosmological context concerns the issue of whether inflation is eternal. For the purpose of this discussion, time zero of the Bang can be taken as some time before the end of inflation in the region of the universe we observe.
2. Technically, only six of the ten equations have to be satisfied, so it might be said that four of the ten commandments of modern genesis are optional. Since most people now regard at least four of the biblical Ten Commandments as optional, the analogy is still valid.
3. Apparently the universe needed a little lithium to make it through the first day, but it wasn't very much, only about one part in 10 billion.
4. It seems that the primordial soup had a yet-to-be-discovered secret ingredient: the dark matter that makes up most of the mass of the present universe. The identity of the mysterious and ubiquitous dark matter is a matter of intense theoretical speculation, as well as the subject of ongoing experimental searches.
5. Johannes Kepler, *Harmonice Mundi* (1619).

Eternal Inflation

ALAN H. GUTH

Department of Physics, Massachusetts Institute of Technology, Cambridge, Massachusetts, USA

ABSTRACT: The basic workings of inflationary models are summarized, along with the arguments that strongly suggest that our universe is the product of inflation. It is argued that essentially all inflationary models lead to (future-)eternal inflation, which implies that an infinite number of pocket universes are produced. Although the other pocket universes are unobservable, their existence nonetheless has consequences for the way that we evaluate theories and extract consequences from them. The question of whether the universe had a beginning is discussed but not definitively answered. It appears likely, however, that eternally inflating universes do require a beginning.

KEYWORDS: cosmology; inflationary cosmology; eternal inflation; universe; origin of universe

INTRODUCTION

The question of whether or not the universe had a beginning is, of course, by no means an easy question. When you ask a scientist a question that is not easy, he never gives just one answer, but instead gives a succession of answers. In this case, I would like to offer two levels of answers.

At the first level, I would argue that the answer to the question is *yes*, the universe had a beginning in the event that is usually referred to as the Big Bang.

I think that at least 99.9% of the people working in scientific cosmology today believe that the universe evolved from a hot dense state, exactly as Sandra Faber describes in another chapter in this volume. This theory is strongly supported by the direct observation of the expansion of the universe via the redshift of the light from distant galaxies, by the measurement of the abundances of the light chemical elements, and by the now very precise measurements of both the spectrum and the very small nonuniformities of the cosmic microwave background radiation. Thus, most scientists (including me) be-

Address for correspondence: Dr. Alan Guth, Department of Physics, Massachusetts Institute of Technology, 77 Massachusetts Avenue, Cambridge, MA 02139. Voice: 617-253-6265. guth@ctp.mit.edu

lieve that the universe as we know it began in a "Big Bang" some 11 to 16 billion years ago.[1]

However, as Sandra Faber has already emphasized, there is another level to the question of beginning. When cosmologists say that they are persuaded that the Big Bang theory is valid, they are using a rather precisely defined and restricted interpretation of the term "Big Bang." As it is used by scientists, the term refers only to the expansion of the universe from an initially hot, dense state. But it says nothing about whether the universe really began there, or whether there was something else that preceded what we call the Big Bang.

So, beyond the standard Big Bang, there is now a very significant body of research concerning the possibility of cosmic inflation.[2–5] Here I want to talk about inflation, and in particular I want to talk about a very likely ramification called eternal inflation. As you will see, the theory of inflation does not give a clear answer to the question of whether the universe had a beginning, but it does provide at least a context for discussing this question.

To begin, I would like to highlight the distinction between the questions that the standard Big Bang theory answers and the questions that inflation is intended to answer.

The standard Big Bang theory is, of course, a very significant scientific theory. It describes how the early universe expanded and cooled from an initially very hot, dense state. It describes how the light chemical elements that we observe today were synthesized during the first 200 seconds or so of this expansion period. And finally, although work in this area is still in progress, it seems to describe very well how the matter in the universe eventually congealed to form the stars, galaxies, and clusters that we observe in the universe today.

There is, however, a key issue that the standard Big Bang theory does not discuss at all: It does not tell us what banged, why it banged, or what happened before it banged. Despite its name, the Big Bang theory does not describe the bang at all. It is really only the theory of the *aftermath* of a bang.

So, in particular, the standard Big Bang theory does not address the question of what caused the expansion; rather, the expansion of the universe is incorporated into the equations of the theory as an assumption about the initial state—the state of the universe when the theory begins its description.

Similarly, the standard Big Bang theory says nothing about where the matter in the universe came from. In the standard Big Bang theory all the matter that we see here, now, was already there, then. The matter was just very compressed, and in a form that is somewhat different from its present state. The theory describes how the matter evolved from one form to another as the universe evolved, but the theory does not address the question of how the matter originated.

While inflation does not go so far as to actually describe the ultimate origin of the universe, it does attempt to provide a theory of the bang: a theory of what it was that set the universe into expansion, and at the same time supplied essentially all of the matter that we observe in the universe today.

HOW DOES INFLATION WORK?

I will begin by giving a quick rundown of how inflation works. Some of these issues are discussed by Sandra Faber, who gives an excellent description. For completeness, however, I will start my explanation at the beginning, but I will try to go more quickly when discussing points that Dr. Faber has already explained.

The key idea—the underlying physics—that makes inflation possible is the fact that most modern particle theories predict that there should exist a state of matter that turns gravity on its head, creating a gravitational repulsion. This state can only be reached at energies well beyond those that we can probe experimentally, but the theoretical arguments for the existence of the state are rather persuasive. It is not merely the prediction of some specific theory, but it is the generic prediction for a wide class of plausible theories. Thus, gravity does not always have to be attractive.[6]

The gravitational repulsion caused by this peculiar kind of material is the secret behind inflation. Inflation is the proposition that the early universe contained at least a small patch that was filled with this peculiar repulsive-gravity material.[7] There are a variety of theories about how this might have happened, based on ideas ranging from chaotic initial conditions to the creation of the universe as a quantum tunneling event. Despite the ambiguity of this aspect of the theory, there are two things to keep in mind. First, the probability of finding a region filled with this repulsive-gravity material need not be large. I will come back to this point later and argue that it is only necessary that the probability is nonzero. Second, the resulting predictions do not depend on how the initial patch was formed. Once the patch exists, inflation takes over and produces a universe that ends up inevitably looking very much like the one that we live in.

The initial patch can be incredibly small. It need be only about one-billionth the size of a single proton. Once the patch exists it starts to rapidly expand because of its internal gravitational repulsion. The expansion is exponential, which means it is characterized by a doubling time, which for a typical inflationary theory might be in the neighborhood of 10^{-37} seconds. So every 10^{-37} seconds the diameter of the patch doubles, and then it doubles again and again during each 10^{-37}-second interval. The success of the description requires about a hundred of these doublings, but there could have been many more. In the course of this expansion, the patch went from being a tiny speck to a size at least as large as a marble.

So the patch of repulsive-gravity material expanded by a huge factor. Whenever a normal material expands, its density goes down, but this material behaves completely differently. As it expands, the density remains constant. That means that the total amount of mass contained in the region increased during inflation by a colossal factor.

The increase in mass probably seems strange at first, because it sounds like a gross violation of the principle of energy conservation. Mass and energy are equivalent, so we are claiming that the energy of the matter within the patch increased by a colossal factor. The reason this is possible is that the conservation of energy has sort of a loophole, which physicists have known about at least since the 1930s,[8] but haven't talked about very much. Energy is always conserved; there are no loopholes to that basic statement. However, we normally think of energies as always being positive. If that were true, then the large amount of energy that we see in the universe could not possibly have gotten here unless the universe started with a lot of energy. However, this is the loophole: energies are not always positive. In particular, the energy of a gravitational field is negative. This statement, that the energy of a gravitational field is negative, is true both in the context of the Newtonian theory of gravity and also in the more sophisticated context of general relativity.

So, during inflation, total energy is conserved. As more and more **positive** energy (or mass) appears as the patch expands at constant density, more and more **negative** energy is simultaneously appearing in the gravitational field that fills the region. The total energy is constant, and it remains incredibly small because the negative contribution of gravity cancels the enormous positive energy of the matter. The total energy, in fact, could very plausibly be zero. It is quite possible that there is a perfect cancellation between the negative energy of gravity and the positive energy of everything else.

For the theory to be successful, there has to be a mechanism to end the period of inflation—the period of accelerated expansion—because the universe is not undergoing inflation today.[9] Inflation ends because the repulsive-gravity material is fundamentally unstable. So it doesn't survive forever, but instead decays like a radioactive substance. Like traditional forms of radioactive decay, it decays exponentially, which means that the decay is characterized by a half-life. During any period of one half-life, on average half of the repulsive-gravity material will "decay" into normal attractive-gravity material.

In the process of decaying, the repulsive-gravity material releases the energy that has been locked up within itself. That energy evolves to become a hot soup of ordinary particles. Initially the decay produces a relatively small number of high-energy particles, but these particles start to scatter off of each other. Eventually the energy becomes what we call *thermalized*, which means that it produces an equilibrium gas of hot particles—a hot primordial soup—which is exactly the initial condition that had always been assumed in the context of the standard Big Bang theory.

Thus, inflation is an add-on to the standard Big Bang theory. Inflation supplies the beginning to which the standard Big Bang theory then becomes the continuation.

EVIDENCE FOR INFLATION

So far I have tried to describe how inflation works, but now I would like to explain the reasons why many scientists—including certainly myself—believe that inflation really is the way that our observed universe began. There are six reasons that I will discuss, starting with some very general ideas and then moving to more specific ones.

The first reason is the obvious statement that the universe contains a tremendous amount of mass. It contains about 10^{90} particles within the visible region of the universe. I believe that most non-scientists are somewhat puzzled to hear anyone make a fuss over this fact since they think, "Of course the universe is big—it's the whole universe!" However, to a theoretical cosmologist who is hoping to build a theory to explain the origin of the universe, this number seems as though it could be an important clue. Any successful theory of the origin of the universe must somehow lead to the result that it contains at least 10^{90} particles. The fundamental theory on which the calculation is based, however, presumably does not contain any numbers nearly so large. Calculations can, of course, lead to factors of 2 or π, but it would take very many factors of 2 or π to reach 10^{90}. Inflation, however, leads to exponential expansion, and that seems to be the easiest way to start with only small numbers and finish with a very large one. With inflation the problem of explaining why there are 10^{90} or more particles is reduced to explaining why there were 100 or more doubling times of inflation.[10] The number 100 is modest enough so that it can presumably arise from parameters of the underlying particle physics and/or geometric factors, so inflation seems like just the right kind of theory to explain a very large universe.

The second reason is the Hubble expansion itself—the fact that the universe is observed to be in a state of uniform expansion. An ordinary explosion, like TNT or an atomic bomb, does not lead to expansion that is nearly uniform enough to match the expansion pattern of the universe. But the gravitational repulsion of inflationary models produces exactly the uniform expansion that was first observed by Edwin Hubble in the 1920s and '30s.

Third, inflation is the only theory that we know of that can explain the homogeneity and isotropy of the universe—that is, the uniformity of the universe. This uniformity is observed most clearly by looking at the cosmic microwave background radiation, which we view as the afterglow of the heat of the Big Bang. The intensity of this radiation is described by an effective temperature, and it is observed to have the same temperature in every direction to an accuracy of about one part in a hundred thousand, after we correct for our own motion through the cosmic background radiation. In other words, this radiation is incredibly smooth. As an analogy we can imagine a marble that has been ground so smoothly that its radius is uniform to one part in a hundred thousand. The marble would then be round to an accuracy of about a quarter of the wavelength of visible light, about as precise as the best optical lenses that can be manufactured with present-day technology.

In the standard Big Bang theory, there is no explanation whatever for this uniformity. In fact, one can even show that within the context of the standard Big Bang theory, no explanation for this uniformity is possible. To see this, we need to understand a little about how this cosmic background radiation originated. During the first approximately 300,000 years of the history of the universe, the universe was hot enough so that the matter was in the form of a plasma—that is, the electrons were separated from the atoms. Such a plasma is very opaque to photons, which are constantly scattered by their interactions with the free electrons. Although the photons move, of course, at the speed of light, they change directions so rapidly that they essentially go nowhere. During the first 300,000 years of the history of the universe, the photons were essentially pinned to the matter.

But after 300,000 years, according to calculations, the universe cooled enough so that the plasma neutralized. The free electrons combined with the atomic nuclei to form a neutral gas of hydrogen and some helium, which is very, very transparent to photons. From then on these photons have traveled in straight lines. So, just as I see an image of you when I observe the photons coming from your face, when we look at the cosmic background radiation today we are seeing an image of the universe at 300,000 years after the Big Bang. Thus, the uniformity of this radiation implies that the temperature must have been uniform throughout this whole region by 300,000 years after the Big Bang.

To think about whether the temperature of the observed universe could have equilibrated by this early time, we could try to imagine fancifully that the universe was populated by little purple creatures, whose sole purpose in life was to make the temperature as uniform as possible. We could imagine that each purple creature was equipped with a little furnace, a little refrigerator, and a cell phone so that they could communicate with each other. The communication, however, turns out to be an insurmountable problem. A simple calculation shows that in order for them to have achieved a uniform temperature by 300,000 years, they would have needed to be able to communicate at about a hundred times the speed of light. But nothing known to physics allows communication faster than light, so even with dedicated purple creatures we could not explain the uniformity of the cosmic background radiation.

So in the standard version of the Big Bang theory, before inflation is introduced, one simply has to hypothesize that the universe started out uniform.

The initial uniformity would then be preserved, since the laws of physics are by assumption the same everywhere. This approach allows one to accommodate the uniformity of the universe, but it is not an explanation.

Inflation gets around this problem in a very simple way. In inflationary theories the universe evolves from a very tiny initial patch. While this patch was very small, there was plenty of time for it to become uniform by the same mechanism by which a slice of pizza sitting on the table cools to room temperature: things tend to come to a uniform temperature. Once this uniformity is established on the scale of the very tiny patch, inflation can take over and magnify the patch to become large enough to encompass everything that we observe. Thus, inflation provides a very natural explanation for the uniformity of the universe.

Reason number four is known as the flatness problem. It is concerned with the closeness of the mass density of our universe to what cosmologists call the *critical density*. The critical density is best defined as that density which would cause the universe to be spatially flat. To understand what this means, one must understand that, according to general relativity, the geometry of space is determined by the matter that it contains. If the mass density of the universe is very high, the space will curve back on itself to form a closed universe, the three-dimensional analogue of the two-dimensional surface of a sphere. The sum of the angles in a triangle would exceed 180°, and parallel lines would meet if they are extended. If the mass density is very low, the space would curve in the opposite way, forming an open universe. The sum of the angles in a triangle would then be less than 180°, and parallel lines would diverge if they were extended. But with just the right mass density (for a given expansion rate) the spatial geometry will be exactly Euclidean, just like what we all learned in high school—180° in every triangle, and parallel lines remain parallel no matter how far they are extended. This borderline case is the critical density.[11]

Cosmologists use the Greek letter Ω (Omega) to denote the ratio of the average mass density of the universe to the critical density:

$$\Omega \equiv \frac{\text{actual mass density}}{\text{critical mass density}}.$$

Today there is growing evidence that Ω is equal to 1 to within about 10%,[12] but the issue is not completely settled. For purposes of this discussion, therefore, I will begin with the uncontroversial statement that Ω lies somewhere in the interval between 0.1 and 2.

You would probably not expect that much could be concluded from such an noncommittal starting point, but in fact it tells us a lot. When one looks at the equations describing the evolution of the universe, it turns out that $\Omega = 1$ is an unstable equilibrium point, like a pencil balanced on its tip. If the pencil is started exactly vertical and stationary, the laws of Newtonian mechanics imply that it will remain vertical forever. But if it is not absolutely vertical, it will rapidly start to fall in whichever direction it is leaning. Similarly, if Ω began exactly equal to 1, it would remain 1 forever. But any deviation from 1 will grow rapidly as the universe evolves. Thus, for Ω to be anywhere in the ballpark of 1 today, it must have started extraordinarily close to 1. For example, if we extrapolate backwards to 1 second after the Big Bang, Ω must have been 1 to an accuracy of 15 decimal places.[13] While 1 second may sound like an extraordinarily early time at which to be discussing an 11- to 16-billion-year-old universe, cosmologists have pretty much confidence in the extrapolation. The nucleosynthesis processes, which are successfully tested by present measurements of the abundances of the light chemical elements, were already beginning at 1 second after the Big Bang.

In fact, particle theorists like myself are attempting to push the history of the universe back to what we call the Planck time, about 10^{-43} seconds, which is the era when quantum gravity effects are believed to have been important. Since our understanding of quantum gravity is still very primitive, we generally make no attempt to discuss the history of the universe at earlier times. But if we attempt to extrapolate the history of the universe back to the Planck time without invoking inflation, we find that Ω at the Planck time must have been equal to 1 to an accuracy of 58 decimal places.

Without inflation, there is no explanation for the initial value of Ω. The Big Bang theory is equally self-consistent for any initial value of Ω, so one has no *a priori* reason to prefer one value over another. But if the theory is to agree with observation, one must posit an initial value of Ω that is extraordinarily close to 1.

With inflation, on the other hand, during the brief inflationary era the evolution of Ω behaves completely differently. Instead of being driven away from 1, during inflation Ω is driven very strongly towards 1. So with inflation you could assume that Ω started out as 1, 2, 10, 10^3, or 10^{-6}. It doesn't matter. As long as there was enough inflation, then Ω would have been driven to 1 to the extraordinary accuracy that is needed.

The fifth reason for believing the inflationary description is the absence of magnetic monopoles. Grand unified particle theories, which unify all the known particle interactions with the exception of gravity, predict that there should be stable particles that have a net magnetic charge. That is, these particles would have a net north pole or a net south pole, which is very different from an ordinary bar magnet, which always has both a north pole and a south pole. These magnetic monopoles are, according to our theories, extraordinarily heavy particles, weighing about 10^{16} times as much as a proton. In the traditional Big Bang theory, without inflation, they would have been copiously produced in the early universe. If one assumes a conventional cosmology with typical grand unified theories, one concludes that the mass density of magnetic monopoles would dominate all other contributions by an absurdly large factor of about 10^{12} (Ref. 14). Observationally, however, we don't see any sign of these monopoles.

Cosmic inflation provides a simple explanation for what happened to the monopoles: in inflationary models, they can easily be diluted to a negligible density. As long as inflation happens during or after the era of monopole production, the density of monopoles is reduced effectively to zero by the enormous expansion associated with inflation.

The sixth and final reason that I would like to discuss for believing that the universe underwent inflation is the prediction that the theory makes for the detailed structure of the cosmic background radiation. That is, inflation not only makes very definite predictions for the uniformity that we see around us, but it also predicts that there should be small deviations from that uniformity due to quantum uncertainties. The magnitude of these deviations depends on the details of the underlying particle theory, so inflation will not be able to predict the magnitude until we really understand the particle physics of very high energies. However, the shape of the spectrum of these nonuniformities—that is, the way that the intensity varies with wavelength—depends only

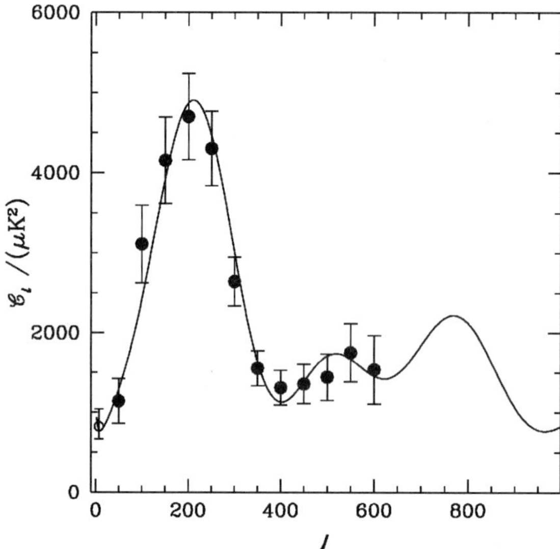

FIGURE 1. Spectrum of the cosmic background radiation anisotropies, as measured by the BOOMERANG experiment. The intensity of fluctuations is shown as a function of the angular size parameter ℓ, where the angular size of a fluctuation is roughly $180°/\ell$. The black line is a theoretical curve corresponding to a standard inflationary model with $\Omega = 1$. The mass density in the model is composed of 5% baryons, 25% cold dark matter, and 70% cosmological constant. The data and theoretical curve were taken from Lange *et al.*[15]

slightly on the details of the particle physics. For typical particle theories, inflationary models predict something very close to what is called the Harrison-Zel'dovich, or scale-invariant spectrum. These nonuniformities are viewed as the seeds for the formation of structure in the universe, but they can also be seen directly in the nonuniformities of the cosmic background radiation, at the level of about one part in 100,000. FIGURE 1 shows a graph of the recent data from the Boomerang experiment, plotted against a theoretical curve derived for an inflationary model.[15]

ETERNAL INFLATION: MECHANISMS

Having discussed the mechanisms and the motivation for inflation itself, I now wish to move on the main issue that I want to stress in this article: eternal inflation—the questions that it can answer and the questions that it raises.

Before going on, I should clarify that the different topics that I am discussing have various levels of certainty. The standard Big Bang theory, as far as cosmologists are concerned, appears to be essentially certain to all but a few of us. Inflation seems to be by far the most plausible way that the Big Bang could have started, but it is not so well established as the Big Bang itself. I should also admit that inflation is vague. It is not really a theory, but a class of theories, so there is a significant amount of flexibility in describing its predictions. Eternal inflation, which I am about to describe, seems to me to be an almost unavoidable consequence of inflation. This point, however, is somewhat controversial. In particular, I believe that Neil Turok, in another paper in this volume, argues that either eternal inflation does not happen, or that it is in any case not relevant to understanding the properties of the observable universe. I, however, will argue that eternal inflation *does* happen, and *is* relevant.

By eternal inflation, I mean simply that once inflation starts, it never ends.[16–18] The term "future-eternal" would be more precise, because I am not claiming that it is eternal into the past; I will discuss that issue later in this paper.

The mechanism that leads to eternal inflation is rather straightforward to understand. Recall that we expect inflation to end because the repulsive-gravity material is unstable, so it decays like a radioactive substance. As with familiar radioactive materials, the decay of the repulsive-gravity material is generally exponential: during any period of one half-life, on average half of it will decay. This case is nonetheless very different from familiar radioactive decays, however, because the repulsive-gravity material is also expanding exponentially. That's what inflation is all about. Furthermore, it turns out that in essentially all models, the expansion is much faster than the decay. The doubling-time for the inflation is much shorter than the half-life

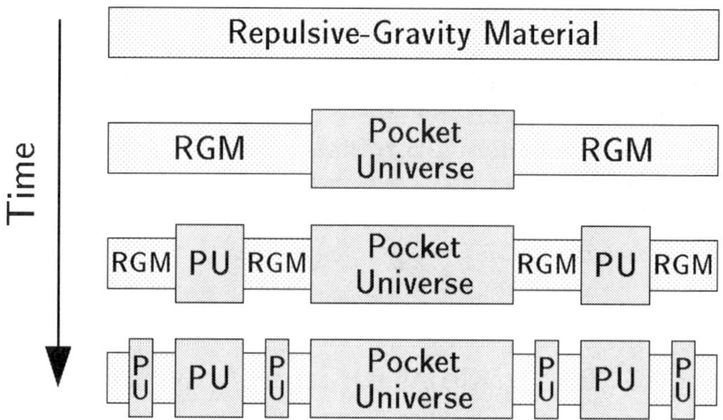

FIGURE 2. A schematic diagram illustrates the fractal structure of the universe created by eternal inflation. The four horizontal bars represent a patch of the universe at four evenly spaced, successive times. The expansion of the universe is not shown, but each horizontal bar is actually a factor of three larger than the preceding bar, so each region of repulsive-gravity material is actually the same size as the others. During the time interval between bars, one-third of each region of repulsive-gravity material decays to form a pocket universe. The process repeats *ad infinitum*, producing an infinite number of pocket universes.

of the decay. Thus, if one waits for one half-life of the decay, half of the material would on average convert to ordinary matter. But meanwhile the part that remains would have undergone many doublings, so it would be much larger than the region was at the start. Even though the material is decaying, the volume of the repulsive-gravity material would actually grow with time, rather than decrease. The volume of the repulsive-gravity material would continue to grow, without limit and without end. Meanwhile, pieces of the repulsive-gravity material decay, producing a never-ending succession of what I call *pocket universes.*

In FIGURE 2, I show a schematic illustration of how this works. The top row shows a region of repulsive-gravity material, shown very schematically as a horizontal bar. After a certain length of time, a little less than a half-life, the situation looks like the second bar, in which about a third of the region has decayed. The energy released by that decay produces a pocket universe. The pocket universe will inflate to become huge, so to its residents the pocket universe would look like a complete universe. But I will call it a pocket universe because there is not just one, but an infinite number of them.

On the second bar, in addition to the pocket universe, we have two regions of repulsive-gravity material. On the diagram I have not tried to show the expansion, because if I did I would quickly run out of room on the page. So you are expected to remember on your own that each bar is actually bigger than the previous bar, but is drawn on a different scale so that it looks like it is the same size. To discuss a definite example, let us assume that each bar represents three times the volume of the previous bar. In that case, each region of repulsive-gravity material on the second bar is just as big as the entire bar on the top line.

The process can then repeat. If we wait the same length of time again, the situation will be as illustrated on the third bar of the diagram, which represents a region that is three times larger than the second bar and nine times larger than the top bar. For each region of repulsive-gravity material on the second bar, about a third of the region decays and becomes a pocket universe, leaving regions of repulsive-gravity material in between. Those regions of repulsive-gravity material are again just as big as the one we started with on the top bar. The process goes on literally forever, producing pocket universes and regions of repulsive-gravity material between them, *ad infinitum*. The universe on the very large scale acquires a fractal structure.

The illustration in FIGURE 2 is of course oversimplified in a number of ways: It is one-dimensional instead of three-dimensional, and the decays are shown as if they were very systematic, while in fact they are random. But the qualitative nature of the evolution is nonetheless accurate: Eternal inflation really leads to a fractal structure of the universe, and once inflation begins, an infinite number of pocket universes are produced.

ETERNAL INFLATION: IMPLICATIONS

The pocket universes other than our own are believed to be completely unobservable, so one can question whether it makes any scientific sense to talk about them. I would argue that it is valid science, because we are pursuing the consequences of a theory for which we already have other evidence. Of course, the theory of inflation has to rest on the evidence that we can observe, but once we are persuaded by these observations, then I think that we should also believe the other implications, even if they involve statements that cannot be directly confirmed.

If one accepts the existence of the other pocket universes, then one can still question whether they have any relevance to the pursuit of science. I will argue that, even though these other universes are unobservable, their existence nonetheless has consequences for the way that we evaluate theories and extract consequences from them.

One question for which eternal inflation has relevance is the question of the ultimate beginning of the universe—what can be learned about it, and how can we learn it? In the following paper, Neil Turok will describe his work with Stephen Hawking[19] and others on the origin of the universe as a quantum event. He will argue that hypotheses about the form of the initial wave function lead to statistical consequences for our universe that can in principle be directly tested. However, if eternal inflation is a valid description of the universe (as I think it is), then I would expect that all such hypotheses about the ultimate beginning of the universe would become totally divorced from any observable consequences. Since our own pocket universe would be equally likely to lie anywhere on the infinite tree of universes produced by eternal inflation, we would expect to find ourselves arbitrarily far from the beginning. The infinite inflating network would presumably approach some kind of a steady state, losing all memory of how it started, so the statistical predictions for our universe would be determined by the properties of this steady-state configuration, independent of hypotheses about the ultimate beginning. In my opinion, theories of the ultimate origin would remain intellectually interesting, and, with an improved understanding of the fundamental laws of physics, such theories might even eventually become compelling. But I expect that any detailed consequences of such a theory would be completely washed out by the eternal evolution of the universe. Thus, there would be no way of relating the properties of the ultimate origin to anything that we might observe in today's universe.

Although I believe that the inflating network would approach a steady state, I should admit that attempts to pursue this idea quantitatively have run into several technical problems. First, the evolution of eternally inflating universes leads to physics that we do not understand. In particular, quantum fluctuations tend to drive the repulsive-gravity material to higher and higher energy densities, where the poorly understood effects of quantum gravity become more and more important.[20,21] Second, even if we impose enough assumptions so that the evolution of the eternally inflating universe can be described, we still do not know how to define probabilities on the infinite set of pocket universes that is produced.[22] The problem is akin to asking what fraction of the integers are odd.[23] Most people would presumably say that the answer is one-half, since the integers alternate between odd and even. However, the ambiguity of the answer can be seen if one imagines other orderings for the integers. One could, if one wished, order the integers as 1,3, 2, 5,7, 4, 9,11, 6, … , always writing two odd integers followed by one even integer. This list includes each integer exactly once, but from this list one would conclude that two-thirds of the integers are odd. Thus, the answer seems to depend on the ordering. For eternally inflating universes, however, there is no natural ordering to the regions of spacetime that comprise the entire universe. There are well-founded proposals for defining probabilities,[24] but at least in my opinion there is no definitive and compelling argument for any one proposal.

A second implication of eternal inflation is that the probability for inflation to start—the question of how likely it is for an initial speck of repulsive-gravity material to form—becomes essentially irrelevant. Inflation only needs to begin once, in all of eternity. As long as the probability is nonzero, it does not seem relevant, and perhaps it is not even meaningful, to ask whether the probability is large or small. If it is possible, then it will eventually happen, and when it does, it produces literally an infinite number of universes. Unless one has in mind some competing process, which could also produce an infinite number of universes (or at least an infinite space-time volume), then the probability for inflation to start has no significance.

The third and final implication of eternal inflation that I would like to discuss pertains to the comparison of theories. I would argue that once one accepts eternal inflation as a logical possibility, then there is no contest in comparing an eternally inflating version of inflation with any theory that is not eternal.

Consider the analogy of going into the woods and finding some rare species of rabbit that has never before been seen. You could either assume that the rabbit was created by a unique cosmic event involving the improbable collision of a huge number of molecules, or you could assume that the rabbit was the result of the normal process of rabbit reproduction, even though there are no visible candidates for the rabbit's parents. I think we would all consider the latter possibility to be far more plausible. Once we become convinced that universes can eternally reproduce, then the situation becomes very similar, and the same logic should apply. It seems far more plausible that our universe was the result of universe reproduction than that it was created by a unique cosmic event.

DID THE UNIVERSE HAVE A BEGINNING?

Finally, I would like to discuss the central topic of this session, the question of whether or not the universe had a beginning.

The name *eternal inflation*, as I pointed out earlier, could be phrased more accurately as *future-eternal inflation*. Everything that has been said so far implies only that inflation, once started, continues indefinitely into the future. It is more difficult to determine what can be said about the distant past.

For the explicit constructions of eternally inflating models, the answer is clear. Such models start with a state in which there are no pocket universes at all, just pure repulsive-gravity material filling space. So there is definitely a beginning to the models that we know how to construct.

In 1993 Borde and Vilenkin[25] proved a theorem that showed under fairly plausible assumptions that every eternally inflating model would have to start with an *initial singularity*, and hence must have a beginning. In 1997, however, they[26] noted that one of their assumed conditions, although valid at the classical level, was violated by quantum fluctuations that could be significant in eternally inflating models. They concluded that their earlier proof would not apply to such cases, so the door was open for the construction of models without a beginning. They noted, however, that no such models had been found.

At the present time, I think it is fair to say that it is an open question whether or not eternally inflating universes can avoid having a beginning. In my own opinion, it looks like eternally inflating models necessarily have a beginning. I believe this for two reasons. The first is the fact that, as hard as physicists have worked to try to construct an alternative, so far all the models that we construct have a beginning; they are eternal into the future, but not into the past. The second reason is that the technical assumption questioned in the 1997 Borde-Vilenkin paper does not seem important enough to me to change the conclusion, even though it does undercut the proof. Specifically, we could imagine approximating the laws of physics in a way that would make them consistent with the assumptions of the earlier Borde-Vilenkin paper, and eternally inflating models would still exist. Although those modifications would be unrealistic, they would not drastically change the behavior of eternally inflating models, so it seems unlikely that they would change the answer to the question of whether these models require a beginning.

So, as is often the case when one attempts to discuss a deep question scientifically, the answer is inconclusive. It looks to me that **probably** the universe had a beginning, but I would not want to place a large bet on the issue.

ACKNOWLEDGMENTS

This work is supported in part by funds provided by the United States Department of Energy (DOE) under cooperative research agreement No. DF-FC02-94ER40818.

NOTES AND REFERENCES

1. An excellent semi-popular-level book on the standard Big Bang theory is Steven Weinberg, *The First Three Minutes*, 2nd updated edition (New York: Basic Books, 1993). For a technically more sophisticated approach, see Andrew Liddle, *An Introduction to Modern Cosmology* (New York: John Wiley & Sons, 1999); or Michael Rowan-Robinson, *Cosmology*, 3rd edition (Oxford: Clarendon Press, 1996). For a treatment at the level of a graduate course, see John A. Peacock, *Cosmological Physics* (Cambridge: Cambridge University Press, 1999); P.J.E. Peebles, *Principles of Physical Cosmology* (Princeton, NJ: Princeton University Press, 1993); or Edward W. Kolb and Michael S. Turner, *The Early Universe* (Redwood City, CA: Addison-Wesley, 1990).
2. A.H. Guth, *Phys. Rev.* D, Vol. 23 (1981), pp. 347–356.

3. A D. Linde, *Phys. Lett.,*Vol. 108B (1982), 389–393.
4. A. Albrecht and P.J. Steinhardt, *Phys. Rev. Lett.*, Vol. 48 (1982), 1220–1223.
5. For a semi-popular-level description of inflation I recommend Alan H. Guth, *The Inflationary Universe* (New York: Perseus Books, 1997). For a technical treatment, see Andrei Linde**,** *Particle Physics and Inflationary Cosmology* (Chur, Switzerland and New York: Harwood Academic Publishers, 1990). Inflation is also discussed in the sources listed in Ref. 1.
6. The possibility of repulsive gravity arises because, according to Einstein's theory of general relativity, gravitational fields are produced not just by energy or mass densities, but also by pressures. The direction of the field caused by pressure is what you would probably guess: a positive pressure—the kind that we normally see—produces an attractive gravitational field. But the peculiar state of matter that I'm talking about produces a negative pressure, which you might also call a suction. It is in fact a very large negative pressure, resulting in a repulsive gravitational field which is stronger than the attractive field produced by the mass density of the matter. The result is a net gravitational repulsion, which is the driving force behind inflation.
7. The name for this peculiar gravitationally repulsive state is not well established. Sandra Faber referred to it as a vacuum with a finite energy density, and sometimes as a false vacuum. In my own technical articles I call it a false vacuum, although I have to explain that for most inflationary models this usage stretches the meaning for which the phrase had previously been used in particle physics. In this article I will refer to it as a repulsive-gravity material. It may seem strange to see the words "vacuum" and "material" used to describe the same thing, but keep in mind that this stuff is **strange**. The word "vacuum" is used to emphasize that it is different from ordinary matter, while I am calling it a material to emphasize that it is different from an ordinary vacuum!
8. R.C. Tolman, *Phys. Rev.*, Vol. 39 (1932), p. 320.
9. Actually there is strong evidence that the expansion of the universe is accelerating in the present era, and the mechanism for this acceleration is believed to be very similar to that of inflation. This acceleration, however, is much slower than the acceleration that inflationary models propose for the early universe, so in any case the rapid acceleration of the early universe must have come to an end.
10. Since the volume is proportional to the cube of the diameter, during 100 doublings the volume increases by a factor of $(2^{100})^3 = 2^{300}$ approximately equal to 2×10^{90}.
11. To clear up a possible source of confusion, I mention that the critical density has often been described in the semi-popular literature as the borderline between eternal expansion and eventual collapse. If Einstein's cosmological constant is zero, as most of us thought a few years ago, then this definition is equivalent to the one given above. Recent evidence, however, suggests that the cosmological constant may be nonzero, in which case the two definitions are not equivalent. In that case the one given in the text agrees with the definition used in the technical literature and is also the definition that is relevant to the current discussion.
12. See, for example, P. de Bernardis, *Nature,* Vol. 404 (2000), 955–959.

13. R.H. Dicke and P.J.E. Peebles, in *General Relativity: An Einstein Centenary Survey*, eds. S.W. Hawking and W. Israel (Cambridge: Cambridge University Press, 1979).
14. J. P. Preskill, *Phys. Rev. Lett.*, Vol. 43 (1979), pp. 1365–1368.
15. A.E. Lange *et al.*, [Boomerang Collaboration], *Phys. Rev.* D., Vol. 63 (2001) 042001 (8 pages), astro-ph/0005004.
16. A. Vilenkin, *Phys. Rev.* D, Vol. 27 (1983), pp. 2848–2855.
17. P.J. Steinhardt, in *The Very Early Universe*, Proceedings of the Nuffield Workshop, Cambridge, 21 June–9 July, 1982, eds. G. W. Gibbons, S. Hawking, and S.T.C. Siklos (Cambridge: Cambridge University Press, 1983), pp. 251–266.
18. A.D. Linde, *Mod. Phys. Lett.*, Vol. A1 (1986), p. 81; A.D. Linde, *Phys. Lett.*, Vol. 175B(1986), pp. 395–400; A.S. Goncharov, Linde, A.D., and Mukhanov, V. F. *Int. J. Mod. Phys,*. Vol.A2 (1987), pp. 561–591.
19. S.W. Hawking and Turok, N.G., *Phys. Lett.*, Vol. B425 (1998), 25–32, hep-th/9802030.
20. A. Linde, Linde, D. and Mezhlumian, A. *Phys. Rev.* D, Vol. 49 (1994), 1783–1826, gr-qc/9306035.
21. J. Garcia-Bellido and Linde, A. *Phys. Rev.* D, Vol. 51(1995), pp. 429–443, hep-th/9408023.
22. A. Linde, Linde, D., and Mezhlumian, A. *Phys. Lett.*, Vol. B345 (1995), 203–210, hep-th/9411111.
23. For an elaboration of this analogy, see A.H. Guth, in "Models and Connections to Particle Physics," *astro-ph*/0002188, to be published in the proceedings of the Pritzker Symposium on the Status of Inflationary Cosmology, Chicago, Illinois, 29–31 January 1999, eds. N. Pritzker and M.S. Turner (Chicago: University of Chicago Press, 2001).
24. A. Vilenkin, *Phys. Rev. Lett.*, Vol. 81 (1998), pp. 5501–5504 hep-th/9806185.
25. A. Borde and A. Vilenkin, *Phys. Rev. Lett.* 72 (1994), pp. 3305–3309, gr-qc/9312022.
26. A. Borde and A. Vilenkin, *Phys. Rev.* D, Vol. 56, pp. 717–723 (1997), gr-qc/9702019.

Inflation and the Beginning of the Universe

NEIL TUROK

Department of Applied Mathematics and Theoretical Physics, Cambridge University, Cambridge, England, U.K.

ABSTRACT: The question of whether there was a Beginning of the Universe is a truly challenging puzzle for physics and for philosophy. Classical general relativity implies that our observable universe originated in a singularity fifteen billion years ago, but this may be merely a reflection of the incompleteness of the theory. Inflationary theory and quantum cosmology provide our best current attempt to describe these early moments. It has been claimed that the inflationary mechanism renders moot the question of exactly how the universe began. I argue that, on the contrary, if one asks the question of what was in the past, we find ourselves staring directly back at a putative Beginning after all.

KEYWORDS: quantum cosmology; beginning of time; eternal inflation

INTRODUCTION

I have been asked to debate Alan Guth on the question "Did time begin?" We have both chosen to do so within the context of inflation, a theory of the very early universe, developed by Guth and others in the 1980s.[1,2] Inflation is the best theory we currently have of the origin of the cosmic expansion, of the apparent flatness of the Universe, and of the primordial density inhomogeneities which gave rise to galaxies, stars, and planets. So it must be taken seriously. It is also a good theoretical playground in which to attempt to discuss even deeper questions such as the beginning of time. But we should make clear at the start that inflation is still a scenario, not yet on the same footing as those parts of theoretical physics we now accept as "proven": electrodynamics, quantum mechanics, or general relativity. For one thing, inflation lacks a final theoretical formulation. There are many different models and no clear criterion to choose between them. And the observational tests, whilst rapidly improving, are still rather limited.

Address for correspondence: Dr. Neil Turok, Department of Applied Mathematics and Theoretical Physics, Cambridge University, Silver Street, Cambridge CB3 9EW, U.K. Voice: 44-1223-337872; fax: 44-1223-337918.

N.G.Turok@damtp.cam.ac.uk

Alan Guth thought to change the title of his talk from "Eternal Inflation" to "Semi-Eternal Inflation" in the run-up to this meeting. He concedes that inflation could probably not be infinite in the negative as well as positive time directions: it had to begin at some time. That means he has already conceded the debate: so I should begin by claiming victory!

But it would be a hollow victory if his main claim were true. For what he is saying is that inflation, once begun, is eternal to the future.[3] And second, he is implicitly arguing that we are most likely to exist at an epoch infinitely far to the future of the beginning of inflation. Since the universe by now would be infinitely large, the information regarding the beginning would be infinitely dispersed, and any theory of the beginning would be untestable in any conceivable, even "Gedanken," experiment. This is a convenient argument for inflationary theorists such as Alan, who would like inflation to be a complete theory which does not rest on a more fundamental theory of what came before inflation. But the argument has always seemed deeply unsatisfactory to me in the sense that whilst admitting a theory of initial conditions is needed, one is denying that it would have any consequences. In fact, I think the argument is wrong and I shall endeavor to axplain why.

Before doing so, it would be appropriate to explain to the reader that both Alan Guth and I regard these questions as ones which we hope to settle in the manner traditional in theoretical physics—namely by careful definitions, by mathematically precise formulations and calculations, by demonstrating that counter-arguments are nonsensical or violate some physical principle established through other means. There are, of course, philosophical assumptions underlying this type of work, which may be regarded as simplistic or reductionist by outsiders, although they are to a very large extent shared by practitioners. We have no profound objections to trying to construct a mathematical theory of how time and the Universe began, and believe the litmus tests should be mathematical consistency, and observational realism. These are extremely powerful criteria, so powerful that none of our current theories fully satisfy them.

EVIDENCE FOR INFLATION

The standard Big Bang theory provides a successful account of the broad features of the Universe we observe. Built on the equations of general relativity, and nuclear and particle physics, which have been independently checked in many ways, it has three main successes The first is a relation between the density and expansion rate of the Universe and the large-scale spatial geometry. Based on the assumption of a hot, smooth, expanding early Universe, it successfully predicted the existence and thermal spectrum of the relic cosmic microwave background. The most compelling success is the successful fit to the relative abundances of the five lightest elements with a single free parameter, the primordial baryon-to-photon ratio.

In spite of these successes, the theory is clearly incomplete. It does not explain many things about Universe, but merely attributes these to appropriate initial conditions. A striking example is the fact that the temperature of the cosmic microwave background radiation is very nearly identical at antipodal points on the sky. How could this be?, we ask, when these points can have never communicated with each other—the light emitted from each is after all just reaching us, and we are in between them. So the initial conditions seem to be paradoxical and acausal. Similarly, despite the name, the Big Bang theory does not explain cosmic expansion; it just assumes that the initial conditions were rapidly expanding. A more accurate name would be the "big afterbang theory." The density and the expansion rate had to be incredibly uniform, and the geometry very flat to be consistent with the large smooth Universe we see today. Nevertheless the density could not have been exactly uniform. There must have been variations from place to place of the right magnitude (about a part in a hundred thousand) to seed the process of gravitational collapse which formed galaxies, stars ,and planets.

Inflation offers a solution to some of these puzzles.[1] An unusual form of matter, "scalar field potential energy," is assumed to have predominated over other forms of matter early on. Scalar field potential energy behaves like a cosmological constant, known since the 1930s to have a powerful repulsive gravitational field (see, e.g., Peebles[7]). This causes the Universe to expand exponentially, so that a region undergoing a relatively short period of inflation could nevertheless have expanded to a region much larger than the visible Universe today. This would resolve the paradox of the temperature's being the same in opposite directions, because those two regions were originally causally connected in the inflationary, pre-Big Bang epoch. Guth realized that the exponential expansion would also have a flattening effect akin to what happens when a lumpy deflated balloon is blown up.

Unlike a cosmological constant, scalar field potential energy can "switch itself off" after a period of exponential expansion, converting its energy into radiation and setting off the standard hot Big Bang. Thus inflation plays the role of setting up the initial conditions for the hot Big Bang, in a manner which preserves all the successes of the Big Bang theory.

Scalar fields were originally invoked for very different reasons in elementary particle physics, for the purpose of breaking symmetries and giving particles their mass. But unfortunately these scalar fields do not give interesting amounts of inflation, and special scalar fields have to be invoked, with potential energy functions, which are picked in an *ad hoc* manner. Because of this, inflationary models can at present only be viewed as a provisional.

The case for inflation nevertheless became much more convincing when it was realized that it automatically had a very beautiful (and originally unanticipated) side-effect, that of producing density inhomogeneities[5] similar to those needed to account for the formation of galaxies and other structures in the Universe.[6]

The primordial inhomogeneities are now being accurately probed by measurements of the cosmic microwave sky, providing us with detailed tests of inflationary models.[4] So inflationary models are certainly observationally testable. The last year has seen dramatic progress with the detection of the first "acoustic peak" in the angular power spectrum of fluctuations on the cosmic microwave sky. The existence of such a peak had been anticipated long before inflation was invented, but its location at the position expected for a flat Universe, with the density inhomogeneities expected in simple inflationary models" was rightly hailed as a success for inflationary theory. As someone who put a lot of effort into developing alternate theories of the origin of the inhomogeneities, I can testify that these observations have decisively disproved many alternate theories, thereby considerably strengthening the case for some simple inflationary model.

In spite of these major successes, inflation remains somewhat in a state of limbo because no convincing candidate for the needed scalar field has emerged. In fact, the most mathematically complete theories we have, string theory and supergravity, do not seem to yield fields with potential energy functions of the form needed without substantial special pleading.

ETERNAL INFLATION

The mechanism through which primordial inhomogeneities are produced in inflation is truly remarkable. Simple models of inflation involve a scalar field φ rolling down the hill provided by its assumed potential energy function $V(\varphi)$. As long as $V(\varphi)$ *is* positive, it acts like a positive cosmological constant which causes exponential expansion of the Universe. However, the field φ, like every other field, is subject to quantum mechanical fluctuations. As it rolls down the hill, in some regions φ fluctuates downwards and in others it fluctuates upwards. The former regions reach the bottom of the hill, and convert the energy stored in ϕ into hot radiation sooner: the latter regions do so later. Since the density of the radiation rapidly decreases as the universe expands, the regions undergoing heating later are left with a higher energy density than those undergoing heating sooner. This is a very beautiful idea, because it links gravity to quantum mechanics and in fact provides one of our few observational probes of quantum gravity. All the more remarkable is that it appears to be consistent with the latest observational data. *If this is the correct explanation of the origin of structure in the Universe, then quantum processes were crucial to the formation of the Universe we now see.*

The naive approach to inflation begs the question of why the scalar field started out up the hill. Of course that was necessary in order to get inflation going, but assuming that is really just assuming the desired answer. The mechanism of quantum fluctuations offers an intriguing way out of this impasse which has been seized upon by Guth and others. The idea is that those regions in which the field jumps uphill will not only undergo heating later, but will also actually undergo inflation for longer. Since inflation is exponentially powerful, regions undergoing inflation for longer expand to vastly greater size. The thought is therefore that in large parts of the Universe will have quantum-jumped uphill, and these parts will multiply. Within these exponentially growing regions, small "island Universes" will appear in which the field does roll down the hill and releases its energy into hot radiation. The picture is then something like a steady-state universe "in the large," where uphill quantum fluctuations continously replenish the inflating regions, so that the production of "island Universes" continues indefinitely.

We do not have the techniques to fully verify this behavior, not least because there is no consistent theory of quantum gravity we can use to perform the calculation. But in spite of this there are various approximate calculations we can perform and there is a fair amount of evidence that the basic physical picture is correct. Jumps of φ uphill lead to exponential expansion because they increase the potential energy $V(\varphi)$, increasing its repulsive effect.

Nevertheless I believe there is a serious flaw in the reasoning, which renders the eternal inflation effect irrelevant. The problem is that the argument neglects the constraint of causality.

We are all used to the fact that an event can only be said to be "caused" by something that precedes it. In relativity theory, causality is a stronger constraint because nothing travels faster than the speed of light. Thus something can only be "caused" by phenomena within a certain distance from it, roughly the speed of light times the time before the event.

This region of space-time is called the past light cone of the event. All the laws of physics we know (general relativity and quantum field theory) are consistent with causality in this form. Causality implies that all measurable quantities (more precisely, all correlators) at a set of spacetime points are fully determined by the complete set of measureable quantities (equivalently, correlators) evaluated on a space-like surface which fills the past light cones of all of the points in question.

FIGURES 1 and 2 show how causality works in an inflating spacetime. The solution of the field equations in general relativity with a positive cosmological constant is called de Sitter spacetime. It has the geometry of a four-dimensional hyperboloid embedded in five-dimensional Minkowski spacetime (FIG. 1). For example, if we start the Universe on a three-sphere which is static (i.e., not expanding) then it will expand due to the repulsive effect of the cosmological constant. The slice Σ of the spacetime (which is a three-sphere) grows exponentially, as we go forward or backwards in time due to this effect. The diagram also shows that de Sitter spacetime can be time-sliced as a closed Universe, a flat Universe, or an open Universe. This illustrates the

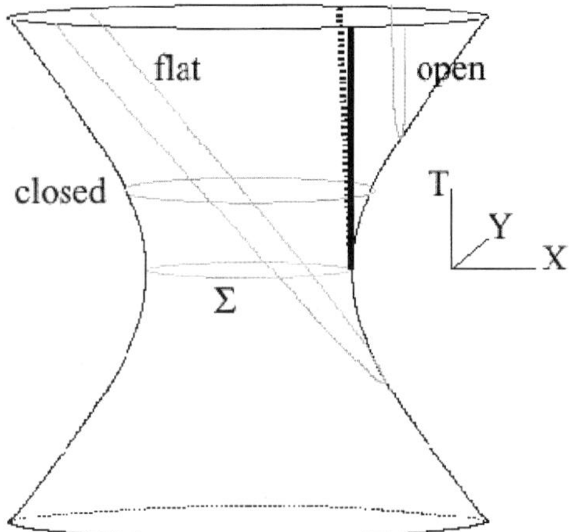

FIGURE 1. De Sitter spacetime is a four-dimensional hyperboloid embedded in five-dimensional Minkowski spacetime. A two-dimensional version is illustrated, embedded in three-dimensional Minkowski spacetime. It may be time-sliced on closed, flat or open slices as shown. In four-dimensional de Sitter spacetime the closed slices are three spheres and the open slices are three-dimensional hyperbolic spaces, or infinite open universes. The heavy vertical line (dotted on the rear side) represents the future light cone of a point on the equator Σ, and is the spacetime locus of a bubble wall nucleated there. Inside the bubble there is an infinite inflating open universe.

point that "three-volume," whilst an intuitive concept, is actually hard to define. If we speak of our initial inflating region expanding its volume by some amount, we need to specify which spatial slicing we use. By changing the slicing, we can completely change the inferred volume, by an infinite amount.

The second diagram (FIG. 2), shows the causal properties of de Sitter spacetime. In this diagram, physical length and time scales have been shrunk so that the whole of the spacetime is now a finite region. The infinite hyperboloid shown in FIGURE 1 is shrunk to a finite cylinder S^3 times an interval. In fact only the region to the future of the initial surface Σ is shown, for reasons which will be explained below. The important point, however, is that in the causal diagram (FIG. 2), light rays move at ± 45° to the vertical.

Now let us consider the production of island hot Big Bang Universes within the inflating spacetime. This description is relatively well understood in the case where the scalar field is trapped in a metastable minimum of its potential and out of which it quantum-tunnels. The tunneling causes bubbles to nucleate, and inside the bubbles the field rolls down the potential to the true minimum (assumed to be at $V = 0$). Outside the bubbles the field is nearly constant and the spacetime is near perfect de Sitter spacetime.

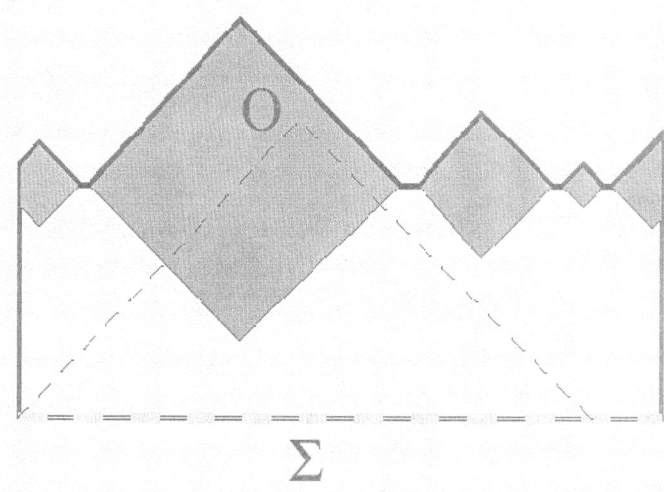

FIGURE 2. Causal diagram showing the formation of "bubble universes" within de Sitter space. de Sitter spacetime is represented as the surface of a two-dimensional cylinder (here seen from the side), where the circular cross sections are actually three spheres. The special three sphere Σ is the equator or initial condition surface for de Sitter space. The point *O* represents the location of an observer inside one of the bubble Universes. The past light cone does not intersect any of the other bubble Universes.

The interior of the bubble is the region inside the future lightcone of the nucleation point. For example, a bubble nucleating on the surface labeled Σ is bounded by a spherical bubble wall which grows out along the light cone indicated by the solid vertical line on the right of the diagram. It is a remarkable fact that the surfaces of constant field ϕ,, which are also the surfaces of constant energy density, turn out to be the *open* slices of de Sitter spacetime. These slices turn into the natural "constant time" slices in the ensuing hot Big Bang. Remarkably therefore, an infinite open universe is contained in the spacetime future of a perfectly finite bubble. Thus we see the first emergence of an infinity—*the spatial regions undergoing heating following inflation are in fact infinite open universes.*[12] As the lower diagram in FIGURE 2 indicates, a number of bubbles—in fact an infinite number—will nucleate in de Sitter spacetime, since the spacetime volume to the future of Σ is infinite. If the bubble nucleation rate is small, very few bubbles ever collide. So we have a picture of an external eternal inflating spacetime which contains an infinite number of infinite inflating "island Universes," which will heat up and produce independent hot Big Bangs.

The existence of these infinities of infinite Universes presents a real problem, which simply put is the question "Where are we?". It is not clear what the relative probability is to be in one part of an island Universe as opposed to another, or in one bubble or another. This lack of predictivity leads to what Vilenkin has termed the "predictability crisis" in inflation. Let me expand briefly on what this crisis is. We are only expecting to be able to calculate relative probabilities for different types of Universes, since that is all quantum theory allows us to consider. Relative probabilities are always well defined when the number of distinct possibilities is finite. For example if one takes a card from the top of a well-shuffled pack, one expects to get a heart or a club with equal probabilities. Likewise we could take many packs and keep more clubs than hearts to bias the probabilities, but as long as there were a finite total number of cards, there would always be a well-defined probability, given a random choice of card.

The situation becomes more difficult when there are an infinite number of possibilities. We cannot enumerate all the possible orderings and it is hard to say which are more probable than others. For example, take all the integers from one to infinity. If we put all the odd numbers first, it might appear that a number chosen at random would have to be odd. With the even numbers put first we would reach the opposite conclusion (this example has been used by Guth). We need to define the relative probabilities more carefully, and in such a way that the infinities are taken care of.

It seems to me that the original question of what an eternally inflating spacetime looks like is actually the wrong question. We should not ask "where are we?" in the infinite, quantum fluctuating spacetime. Rather we should note that causality implies *the only parts of the spacetime that ever influence what we see are the parts inside our past light cone*. An example is shown in FIGURE 2. An observer at O, inside a bubble, is causally disconnected from events in bubbles which never collide with that bubble. For an observer living inside a bubble, everything observable now or in the future should be fully determined by the physics inside her/his past light cone. It should be completely unnecessary to discuss the other bubbles. The constraint of causality is very important and in a stroke removes the twin problems of an infinite number of infinite open universes, leaving us with a much better-defined problem.

The question of what an inflating spacetime looks like should therefore be re-phrased. We should ask "Given that we have undergone a hot Big Bang, what is the most probable past within our past light cone?"

THE NO-BOUNDARY PROPOSAL

The question of our most probable past formulated above is of course only answerable within the context of some definite theory which tells you the probability measure on past spacetimes. The usual arguments for eternal inflation are vague on this point, and basically just assume that for some reason a region existed in which there were the right conditions for inflation to start. Hartle and Hawking proposed an ansatz[8] (which is a fancy way of saying they guessed!) formula for relative probabilities. One can easily imagine an infinite number of possible ansatzes for the initial conditions of the Universe. At the present time many of these might be perfectly consistent with observation. But Hartle and Hawking's proposal is appealing because it is based on general ideas which have a rationale beyond cosmology. In a strong sense I think it is the most conservative thing you could do. It may well fail precisely because it is too conservative. Space and time may be emergent rather than fundamental properties. Describing the Universe as a manifold may not be appropriate to its early moments. Nevertheless, precisely because the no-boundary proposal is *not* just cooked up to make inflation work, its failings and limitations may teach us something deeper about what is in fact required.

In physics we are now used to the fact that all we can ever predict in practice is probabilities. This was even true in classical physics, because nothing is ever measured perfectly and in consequence all predictions carry "error bars." But in quantum physics the issue is more fundamental, because the quantities we use to describe the world, and which we need in order to predict its future or to extrapolate back into the past (like position and velocity of a particle) are in principle impossible to measure beyond a certain limiting accuracy. Only certain questions are allowed, and these involve *correlations* between physical quantities. Paradoxically, in spite of the limitations quantum mechanics imposes on our ability to predict classical properties of the world, quantum physics is actually a far more complete theory than classical physics because it predicts the *correlations* with perfect accuracy.

Hartle and Hawking proposed to use the formulation of quantum mechanics defined by Dirac and Feynman, called the path integral formulation, to define all cosmological correlators. This formulation is widely accepted and is the basis for modern treatments of gauge field theories, well tested in the laboratory. But it usually does not claim to solve the initial conditions problem. Usually the path integral formula is used to give the quantum mechanical amplitude to be in a state B at time t_f, given that the system was in a state A at time t_i.

There is one situation, however, in which it becomes unnecessary to specify exactly what state the system was in. That is when the system is in thermal equilibrium. In this case, two beautiful things happen to the Dirac-Feynman formula. First, we identify the initial state A with the final state B, and *sum over all possible states*. Second, *real time is continued to imaginary time*, and the temperature of the system then places the role of the period of the system in imaginary time. In non-gravitational physics, one still needs to specify the temperature in order to specify all correlators. But in the context of gravity, the size of the system in imaginary time becomes a dynamical quantity to be determined by the theory itself. Thus Hartle and Hawking's proposal can actually become a completely self-referential prescription for the initial conditions, with the only input required being the dynamical laws of physics (the Lagrangian).

It may seem paradoxical that quantum physics yields better-defined probability measures than classical physics because in quantum physics certain quantities are inherently uncertain (Heisenberg's "uncertainty principle'). But, in fact, the two aspects are closely related. For example, the state of a particle in a box is described in classical theory by its location and velocity, which can in principle be measured with absolute certainty. If we limit the energy of the particle, there are still an infinity of possible states for it to be in (for example, differing infinitesimally in location). In quantum physics, however, if we limit the energy there are only a finite number of possible states. In all of these states both the location and velocity of the particle is to some extent uncertain.

The finiteness imposed by quantization is one aspect of obtaining a well-defined probability measure. But there is another aspect, namely the box! If the box is infinite, there are an infinite number of possibilities for the particle and again the probability measure is ill-defined. In many situations the precise nature of the box is irrelevant, as long as it is bigger than any relevant scale in the problem. But in some situations, it is all-important. For example, a gas particle in a box performs a random walk as it collides with other particles. For some questions, such as the average speed of the particle, the size of the box is irrelevant. But for others, such as "how far does it travel?," the walls of the box are all important. In cosmology, I want to argue that causality plays a role similar to the walls of the box.

Before proceeding let me list a few of the "technical difficulties" to be faced in implementing Hartle and Hawking's idea.

- Einstein gravity is non-renormalizable. This objection refers to the bad short-distance properties of the theory. These are certainly important but not central to the discussion here. Theories such as supergravity with improved ultra-violet properties are not conceptually different as far as the problems we are discussing.
- The Euclidean Einstein action is not positive definite, and therefore the Euclidean path integral is ill-defined. This is the "conformal factor" problem in Euclidean quantum gravity and I shall return to it below. Recent work has shown that at least for some choices of physical variables, and to quadratic order, this problem is overcome.[15–17] This means that, at least to the order of approximation to which calculations have been pursued, the problem is overcome.
- The sum over topologies in four dimensions is likely to diverge, as happens in string theory. Most likely one will need some formulation in which manifolds of differing topologies are treated together.

These problems are formidable. The main hope, it seems to me, is that our Universe appears to be astonishingly simple, well-described by a classical solution of great symmetry with small fluctuations present at a level of a part in a hundred thousand. This suggests that we may be able to accurately describe it using perturbation theory about classical solutions.

INSTANTONS

Hawking and I made progress with the Hartle–Hawking proposal when we realized there is a simple class of classical solutions to the field equations for gravity coupled to an inflationary scalar field, which yield a finite probability according to the Dirac–Feynman formula in imaginary time.[9] The solutions we studied yield a classical spacetime which is a real four-dimensional manifold in either real time t or imaginary time τ provided that the transition $t = -i\tau$ performed on the zero time surface Σ shown in FIGURE 3. The form of the solutions is generic for inflationary models with gently sloping potentials $V(\varphi)$, which are the simplest models.

Region I is an infinite inflating open "island Universe." Region II, an approximately de Sitter region ϕ, is nearly constant, bounded by a time-like singularity.

There are actually a one-parameter family of finite probability solutions, in which we have a one-to-one relation:

$$\phi_0 \leftrightarrow \Omega_0$$

where Ω_0 is the current density parameter, and ϕ_0 is the value of the scalar field at the regular pole of the instanton. For the solutions described above,

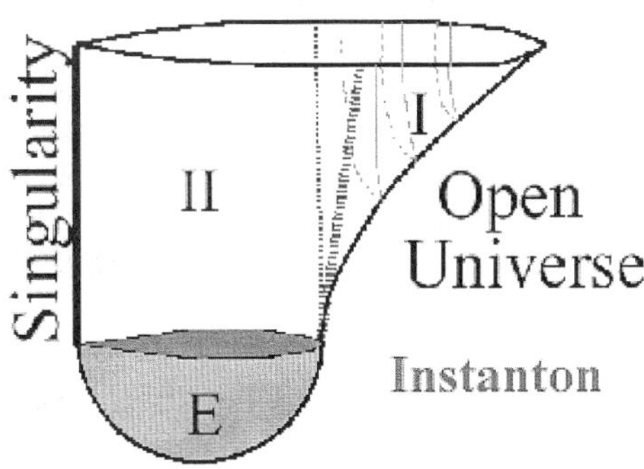

FIGURE 3. An infinite open Universe emerges from a "pea" instanton.

$\Omega_0 > 1$, but if one allows instantons which are singular at both poles, and symmetric about the maximum of b, these analytically continue to closed inflating Universes, and one can obtain any value of $\Omega_0 > 1$.

The classical instanton solutions could then be used to compute the quantum fluctuations about inflating spacetimes in a more precise manner than had hitherto been possible. In particular, previous calculations had to make more or less *ad hoc* choices for the initial quantum state. But in these new solutions the initial quantum state was initially and precisely defined from the no-boundary proposal.[10] I should also mention that doubts were raised about the Hawking–Turok solutions because they possess a singularity, albeit one mild enough that the quantum fluctuations about it are well-defined.[13,14] These doubts have since been dealt with by showing that by a suitable change of variables the singularity can actually be removed.[11] These developments are reviewed in Turok.[18] For the purposes of the present debate the precise predictions of the Hawking–Turok calculation are not relevant, though I must mention that work subsequent to this meeting has shown that one can actually calculate the volume factors appropriate to inflation in a slicing-independent way.[18] When one includes these volume factors, the most probable instantons are those in which the initial field ϕ_0 starts out large, and the scalar field potential energy density is at the Planck density, at which quantum gravitational effects become large. Therefore these calculations do seem to resolve the question of the most probable starting point for inflation, which is at the Planck density.

ETERNAL INFLATION ISN'T

The detailed predictions of the Hawking–Turok instantons are not crucial to the present discusssion. What is important is the claim of Guth and others that if such an instanton started the Universe, and started inflation, that there would be an infinite period of inflation between the instanton and ourselves. This is an assertion one can check by performing an analysis of the fluctuations about the instanton solution and the inflating spacetime which emerges from it. My conclusion from these calculations, at least within the context of the causal formulation given above, is that our past light cone contained only a brief episode of inflation, and if we follow it back, we are led after a very short time right back to the Planck density. For example in a theory with the $V = m^2\phi^2$ the duration of inflation turns out to be only of order $M_{Pl}m^{-2}$, or about 10^{12} Planck times, or about 10^{-31} seconds.

The reason my finding is so very different to that of Guth *et al.* is that I am asking a different question. Whereas Guth makes a comparison between the classical rolling of the field downhill and the quantum jumps uphill *in a single Hubble time*, I am asking for the integrated effect of the quantum jumps over the entire spacetime being considered. I view the classical solution as a first approximation to the entire classical spacetime, and the issue is then whether the effects of the quantum jumps, when integrated up over the past, significantly modify the classical solution. The main point is that the classical solution is *coherent*, so that the scalar field ϕ changes only in one direction as one tracks back in time: for a solution which starts at the Planck density (which is the case where the quantum effects are greatest) the scalar field monotonically increases as one goes back in time towards the Planck density.

However, the quantum jumps are *diffusive* in character. Because they are either positive or negative, their effect does not add up coherently. If the potential $V(\phi)$ were nearly constant, then the quantum diffusion would grow as the square root of the time. This lack of coherence means that the quantum jumps do not severely correct the classical solution *until one reaches the Planck density.* At the Planck density, all calculations break down and there is no point in arguing whether inflation is eternal or not at the Planck density because that really is the realm of the unknown.

What I would say, however, is that I think the study of inflating spacetimes is a challenging problem which forces us to carefully think through exactly what we should calculate. I think this is a very important and interesting exercise even if all it teaches us is about the limitations of existing theoretical frameworks.

ACKNOWLEDGEMENTS

I acknowledge my debt to many physicists for discussions of these issues, in particular Jim Hartle, A. Guth, A. Linde, V. Rubakov and A. Vilenkin. I also acknowledge my collaborators: S. Gratton, S. Hawking, T. Hertog, K. Kirklin, and T. Wiseman.

REFERENCES

1. Guth, A. *Phys. Rev.,* Vol. D23 (1981), p. 347 .
2. Linde, A.D., *Phys. Lett.,* Vol. 108B (1982), p. 389; Albrecht, A. and P. Steinhardt, *Phys. Rev. Lett.*, Vol. 48 (1982), p. 1220.
3. Vilenkin, A. *Phys. Rev.*, Vol. D27 (1983), p. 2848; Linde, A.D. *Phys. Lett.,* Vol. B175 (1986), p. 395; *Phys. Lett.*, Vol. B327 (1994), p. 208; Linde, A.D., D.A. Linde, and A. Mezhlumian. *Phys. Rev.*, Vol. D49 (1994), p. 1783; Vanchurin, V., A. Vilenkin, and S. Winitzki, gr-qc/9905097, *Phys. Rev.,* Vol. D61 (2000), 083507 .
4. Miller, A.D., *et al.*, *Astrophys.J.,* Vol. 524 (1999), p. L1; de Bernardis, P,. *et al.*, Nature, Vol. 404 (2000), p. 955; Hanany, S., *et al.,* astro-ph/0005123 (2000).

5. Hawking, S.W. *Phys. Lett.*, Vol. 115B (1982), p. 295; Guth, A.H. and S.-Y. Pi, *Phys. Rev. Lett.* Vol. 49 (1982), p. 1110; Starobinsky, A.A., *Phys. Lett.*, Vol. 117B (1982) p. 175; Bardeen, J., P. Steinhardt and M. Turner, *Phys. Rev.*, Vol. D28 (1983), p. 679.
6. Harrison, E.R., *Phys. Rev.*, Vol. D1 (1970), p. 2726; Zeldovich, Ya. B., *Monthly Notices of the Royal Astronomical Society,* Vol. 160 (1972), p. 1.
7. Peebles, P.J.E., *Principles of Physical Cosmology (*Princeton, NJ: Princeton University Press, 1993).
8. Hartle, J.B. and S.W. Hawking, *Phys. Rev.,* Vol. D28 (1983), p. 2960.
9. Hawking, S.W. and N. Turok, *Phys. Lett.*, Vol. B425 (1998), p. 25; Hawking, S.W. and N. Turok, *Phys. Lett.*, Vol. B432 (1998), p. 271.
10. Gratton, S.. and N. Turok, *Phys. Rev.*, Vol. D60 (1999), p. 123507; Hertog, T., and N. Turok, *Phys. Rev.,* Vol. D62 (2000), p. 083514; Hawking, S.W., T. Hertog, and N. Turok, *Phys. Rev.,* Vol. D62 (2000), p. 063502; Gratton, S., T. Hertog and N. Turok, *Phys. Rev.,* Vol. D62 (2000), p. 063501.
11. Kirklin, K., N. Turok, and T. Wiseman, hep-th/0005062, *Phys. Rev. D*, (2001), in press.
12. Bucher, M., A. Goldhaber, and N. Turok, *Phys. Rev.*, Vol. D52 (1995), p. 3314.
13. Linde, A., gr-qc/9802038, *Phys. Rev.* Vol. D58 (1998), 083514.
14. Vilenkin, A., hep-th/9803084, *Phys. Rev.,* Vol. D57 (1998) p. 7069; gr-qc/9804051, *Phys. Rev., Vol.* D58 (1998), 067301; gr-qc/9812027.
15. Khvedelidze, A., G. Lavrelashvili, and T. Tanaka, gr-qc/0001041.
16. Lavrelashvili, G., *Nucl. Phys. Proc.* Suppl. 88 (2000), p. 75.
17. Gratton, S. and N. Turok, hep-th/0008235 (2000).
18. Turok, N., *Before Inflation*, preprint (2000).

The Idea of a "Beginningless" World-Process

Perspectives from the Hindu Tradition

ANINDITA N. BALSLEV

Department of Philosophy, University of Copenhagen, Copenhagen, Denmark

ABSTRACT: This paper, while seeking to expose some of the basic ideas of Hindu cosmology, focuses on the philosophical and soteriological dimensions of the notions of "beginning" and "beginningless" in the discourses associated with dominant world-religions. It is hoped that a deeper grasp of these issues in a multi-religious context will help to bridge the distance between diverse traditions of thinking as well as facilitate the science–religion dialogue.

Keywords: Hindu cosmology; beginning; beginningless; Prakrti/Nature; time and cosmology

To participate in a project where the aspiration of the organizers is to gradually build bridges between scientific and religious discourses that are customarily kept apart by the hard boundaries of disciplines is an inspiring and a humbling task. This is especially so when the theme of the project is of such magnitude as expressed in the question: "Did the universe have a beginning?"

At the outset it may be observed that even if the word "beginning" (in sanskrit "ārambha") is derived from conventional language, the notion of beginning lends itself to a variety of interpretations in diverse contexts, to which the history of ideas bears witness. The significance of these conceptual formulations can be grasped only when the entire network of ideas that form a given discourse is taken into account, be that scientific, philosophical, or religious. We therefore need to proceed with caution. In the course of this presentation, I will briefly refer to examples from the Indian sources that demonstrate how different implications are read into the diverse philosophical formulations of this notion. An awareness of the absence of a general con-

Address for correspondence: Dr. Anindita Balslev, Department of Education, Philosophy, and Rhetoric, University of Copenhagen, Njalsgrade 80, DK-2300, Copenhagen, Denmark. Voice: 45 86 27 45 68; fax: 45 35 32 88 50.
anindita@hum.ku..dk

sensus about the conceptual content that is entailed in the notion of beginning along with all its ramifications, it seems to me, makes the complex theme of "cosmic questions" even more intriguing. It is indeed important for our present endeavor to uncover a conceptual space in which a dialogue between science and religion does not seek to ignore the questions that are intertwined with the soteriological dimensions of cosmological models or to stage an encounter of world religions in the process without the effort for an authentic comprehension of the concerns that prompt different religious traditions to answer in the affirmative or in the negative the question, "Did the universe have a beginning?"

With these preliminary remarks, let me directly focus on the Indian conceptual world. There are very early texts belonging to Hindu, Buddhist, and Jaina traditions in which one comes across a bewildering variety of speculations about the origin of the universe. However, there is a hymn in the Rigveda, one of humanity's oldest existing documents, that deserves special mention. This well-known hymn of creation expresses the query, "Whence all creation had its origin…," ponders over it, and then ends with the lines "… he, who surveys it all from the highest heaven, he knows—or maybe even he does not know."[1]

The early Upanishads, which have tremendous impact on the unfolding of subsequent Indian thought, not only contain records of an intense search for detecting a principle that regulates and controls all that underlies change and becoming, but also document the views of sceptics, agnostics, and naturalists who were against any explanation for occurrence of an event through causal operation. These views were carefully examined and rejected. What is of primary importance for our present discussion is to note that the idea highlighted in these early sources, starting from the Rigveda itself, is the notion of "Anādi Samsara/jagat"—the idea that the universe is beginningless. It is found not only in the Hindu tradition, with its roots in the Vedas, but also in the Buddhist and Jaina traditions, which are non-Vedic and non-theistic and yet advocate this idea. In other words, Anādi Samsara is a pan-Indian concept.

Before exposing other conceptual subtleties that are integral to philosophico-religious thinking regarding the large question about the nature and origin of the universe, let us note that by describing the world as Anādi, or beginningless, what is denied is the notion of an *absolute* beginning, that is, a beginning out of nothing. There has been ample discussion in the philosophical literature in favor of the idea that the occurrence of an event is inconceivable without invariable dependence on something else, that is, a cause. These also demonstrate at the same time the absurdities that will follow from holding a position in which a cause itself is taken to be of the nature of non-being. The idea that recurs in the tradition—in spite of the differences in metaphysical structures, epistemological theories, and the like, advanced by the proponents of the different schools—is that only that which is eternal (i.e., ever-present) or fictitious (i.e., never-present) can be said to be uncaused, whereas that to which a beginning or/and end can be ascribed (i.e., contingent) must have a cause, which accounts for rule and order in every case of occurrence and happening. Consequently, whether it is in cosmological speculations or for soteriological purposes or for the sake of forming a theological symbolism intended to accentuate the idea of an all-powerful, personal God, the Hindu traditions of thought invariably hold on to the basic tenet, as expressed in the Bhagavad Gītā, "nāsato vidyate bhāvo," that is, being cannot come out of nothing.[2]

Let me quickly remark in this connection that if no theistic school in the mainstream Hindu tradition had espoused a model similar to that of the "creatio ex nihilo" with a creator as the First Cause, it is on grounds that stem from ethico-religious considerations. This is quite a different issue, to which I will come back later.

One important feature of Hindu thought that relates cosmological concerns to those of ethics and religion is expressed in the notion of ṛta. This is an ancient idea connected with the initial perception that there prevails a cosmic order. However, the Hindus think that this order is present not only in nature, as expressed in the rhythmic recurrences of natural events, but that there is also a moral order. "Nothing is reaped which has not been sown" has this double connotation in the Hindu discourse.

It is also interesting to find that this idea is present in very early texts and that in some of the Vedic hymns the deity is described not only as an upholder of the physical, but also of the moral order. The idea that the universe is governed by laws that cannot be transgressed plays a significant role in Indian thought. At a later period, one comes across renditions where the theistic premise is discarded as superfluous; no executor is seen as necessary for the operation of this law. Examples of this position can be found not only in the Buddhist and the Jaina traditions, but also in the non-theistic Hindu schools of thought, where the moral order functions just like the natural order—an impersonal law governs it. This idea—that the moral, as the natural, is by no means a chaotic situation—is closely linked with the pan-Indian belief in karma.

However, for a deeper understanding of the variations in Hindu cosmological ideas, it is necessary to go to that stage of conceptual growth that saw the rise of distinct schools of thought. It is relevant, in this connection, to note that the sanskrit word "anādi" is constituted by adding the negative prefix "an" to the word "ādi." The word "ādi" can mean "beginning" or "first." This allows for interpreting the idea of a beginningless universe [anādi samsara], using two distinct cosmological models.

A non-theistic school, called Mīmāmsā, takes "anādi" to mean simply "beginningless." This school maintains that the world-process has no absolute beginning, but repudiates the more commonly held view that it is interrupted by periodic states of dissolution. The well-known phrase that highlights the Mimāmsā model of cosmology is that "this world has never been quite otherwise."

However, schools such as Vaiśesika or Sānkhya have adopted the second meaning of the word "ādi" and advocate the cosmological idea that every creation is a subsequent creation—not first. In other words, they employ the cosmological model that operates with the notion of repeated creation and dissolution. A perusal of Indian literature shows that many schools—theistic as well as non-theistic—have argued for and accepted this model. Support for this view is found in the Rigveda Samhitā itself in the statement that "The Lord created the sun and the moon like before...." A world-cycle is said to consist of billions of human years. The epic Mahābhārata makes much use of this notion and in the Bhagavad Gītā we find that a Kalpa (i.e., a world-cycle) is described as "a day of Brahmā," the creator. The night that follows is also said to be of equal length.

In other words, the alternatives that are before us are not whether the universe has a beginning, but rather whether it is to be taken as beginninglessly ongoing process or that this specific world-cycle has a beginning, but it can by no means be held to be the first creation—that is, this present world was preceded by a state of dissolution of a prior world, and so on and on. This latter position, serially viewed, is also a beginningless process, which is intermittently intercepted by states of collapse. In some sense, this latter model can be said to be the predominant cosmological model that wielded strongest influence on the Indian cultural soil.

The impact of this cosmological idea on the Indian mind is also to a large extent due to the Purānas, a body of mythological literature. Drawing on philosophical and astronomical sources, the Purānas make the most of these ideas in the construction of numerous myths and narratives that constantly inform one about the vastness of time, the immensity of the cosmos, and our place in it. These colorful stories, woven in the backdrop of a grandiose cosmological model where each world-cycle (Kalpa) is calculated in terms of billions of human years, tell us about the characteristics of the world in which we live. It is designated as "martya-loka" in sanskrit—that is, the mortal world, where all the inhabitants, without any exception, are subject to death. These stories seek to instill in us certain ethico-religious attitudes that can be used as guidelines during our transitory stay, while incessantly reminding us that this death-bound existence is, nevertheless, not without purpose and meaning.

At this point, it is pertinent to mention that the idea of cosmological cycles has often been confused with the notion of cyclic time. It is commonplace to maintain that the Indian notion of time is cyclic, as opposed to the Judeo-Christian understanding of linear time. Not only is this reading an oversimplification, but it has also created serious obstacles in the context of cross-cultural and inter-religious exchanges. The philosophical scenario in the Indian context is as can be expected in the case of any major philosophical tradition. Just as in the history of Western thought,[3] one encounters a wide range of positions, such as the view of absolute time, time as a relational concept, time as process, and time as appearance; similarly the different schools of Hindu, Buddhist, and Jaina traditions have put forward diverse views concerning time, such as the ideas of absolute and relative time, several variants of a discrete view of time, a view where space–time–matter are combined in the same principle, time as appearance, and so forth.[4]

In this connection, let me refer to the discussion that I earlier mentioned concerning the diverse manner in which the notion of "beginning" has been interpreted by various schools. Let me take the examples of Vaisesika and Sankhya to show how in each case these conceptualizations are in harmony with their over-all positions and with their respective theories of causality and views about time.

Sankhya, considered to be the oldest school of Indian philosophy, propounded a form of metaphysical dualism. The two principles are termed Puruṣa and Prakṛti, the former conceived as an unchanging principle of consciousness and the latter as ever-changing, ever-active Nature. It is in the notion of Prakṛti, the sanskrit word for Nature, that Sānkhya combines space, time, and matter in the same principle. An important cosmological idea that the school puts forward is the idea of cosmic evolution, claiming that there is a persistent tendency in Prakṛti to revert to its unmanifest state, that is, the state of cosmic dissolution. This is the pre-empirical aspect of Nature, when all heterogeneous manifestations cease, but it retains its primal dynamism.

Causal operation, this school maintains, makes manifest the effect. In other words, the effect is an actualization of that which was potentially present in the cause. However, in the frame of metaphysical pluralism propounded by the Vaiśeṣika school, the effect is conceived to be non-existent prior to causal operation. Thus, whereas for Sankhya a beginning refers only to an emergence of an effect that was pre-existent (that is, latent in the cause), there was no need to postulate any idea of an empty time; in Vaiśeṣika, an effect is seen as a new beginning, and, since it was absent before the causal operation took place, the idea of absolute and relative time came to play a significant role. What is of special interest for our present discussion is to note the Vaisesika insistence on the idea of "prior non-existence," which is evidently a temporal reference implying that a beginning is always an event-in-time.

Detailed argumentation in support and against all these positions, recorded in expository and polemical literature, are available. However, I would like to draw your attention here to the fact that to ignore all these varieties of views and describe the Indian conceptual experience of time simply as "cyclic" is a cliché that we must get rid of. Not only is this a distortion of the Indian philosophical scenario, but it also hampers cross-cultural exchanges. A perusal of literature shows many examples of misuse of time metaphors, but what is particularly disturbing is to see how cyclicity and linearity have ceased to be simple time metaphors that depict recurrence and irreversibility; rather, they have come to be associated with such concepts as that of history, progress, and even salvation. I have discussed these questions in greater detail elsewhere, as I have found that these improper metaphorical designations are used as a conceptual device to set major philosophico-theological traditions against each other.[5] It is precisely these that obstruct dialogue.

Cycles and arrows are major metaphors which not only form part and parcel of everyday discourse in various contexts, but also appear and reappear—sometimes assuming technical significance—in the frame of specific disciplines such as physics, cosmology, and theology. Recent writings of anthropologists and paleontologists[6] have shown what role these time metaphors actually play in various cultural traditions and why an exclusive emphasis on one leads to a distorted view. It seems to me that the crisis of the phenomenon of religious pluralism, in intra- and inter-cultural contexts, lies to a very large extent in our inability to open up a creative conversation which is informed and inspired by a spirit of critical understanding. Today, while discussing alternative modes of thinking about such multi-faceted problems as that of the question of the "beginning" of the universe, we need to be on our guard so that we avoid such pitfalls. What is vital, while considering such a large theme, is to be aware of the precise sense and the specific contexts in which a notion of "beginning" is forwarded or withheld, so that the religious meaning of these statements does not get lost in the process of relating it to any model proposed in scientific cosmologies, regardless of which one is in vogue or out of fashion in the current phase of the investigation.

It may be remarked here that occasionally in the current literature one comes across the view that a religious meaning with regard to the cosmos and of our place within it could be and has been attributed only within a scenario that had a narrow sense of space and time, where there were no ideas comparable to the modern conception of cosmos. This view seems to be at variance with the experience of the Indian conceptual world, in which a vast sense of space, time, and cosmos, instead of hampering the soteriological quest, is used in the religious discourse to motivate it. Thus, for example, the late nineteenth century saint Sri Ramakrishna compares countless worlds to innumerable crabs on a sandy beach in the rainy season in order to "invoke humility and inspire a religious quest in a visitor."[7]

In order to elaborate a bit more about religious concerns that have led to different usages of the idea of "beginning" in religious discourses, let me now go back to the question that I alluded to earlier: Why in the Indian Vedic tradition, in which schools and sub-schools have freely constructed alternative models of metaphysical structures and formulated different notions regarding being and non-being, space, time and causality—why has no theistic school dealt with the idea of creation "ex nihilo"?

Interestingly, a reference to such a conceptual possibility of a "first creation" by God is made in a well-known compendium of diverse views.[8] The position is called Iśvaravāda, that is, "Godism." This position, however, did not find much support and was repudiated by the theists themselves. Note, however, that the reason that is given against such a view is not as the Greek thinker Lucretius said—"Nothing is ever produced by divine power out of nothing"—but on ethico-religious grounds instead. The argument goes that had this been the case, the accountability for all the disparities and differences that are undeniably present even at the very moment of birth would lie in Almighty God. In other words, the theists were reluctant to maintain the idea of a first creation because it would imply—they thought—an unjust and a cruel creator who bestows favors on some and deprives others of the same. Focusing on ideas of divine compassion, mercy, and justice, the theists insisted that what lies at the core of differences is not the wielding of an arbitrary power by an all-powerful external agent. They claimed that the human condition is determined by human actions. Now we are back again to Karmavāda, which projects the view that actions are efficacious, that they are not without consequences. Note that this position seeks to avoid, on the one hand, fatalism [niyativāda] or predestination and, on the other hand, the view that events occur arbitrarily, capriciously, without rhyme or reason [yadṛcchavāda]. The idea of human responsibility and freedom as well as the inevitability of having to reap the fruits of one's actions are all woven together in this idea of karma. It is evident that, on the basis of such a reading, one cannot project a view of "first creation" with an absolute beginning. It is also obvious why in such a scenario the idea of rebirth is seen as a necessary sequel to the idea of karma. It may be noticed here that despite divergences on other matters, reflections on disparities and inequities of various sorts at birth itself have led not only the theistic schools but also non-theistic schools in the Hindu, Buddhist and Jaina traditions to adhere to a similar network of ideas. In short, the ideas of karma, rebirth, and that of a beginningless world-process remain pan-Indian concepts.

At this juncture, it is illuminating to probe the theological thinking that is at work for pronouncing that the world has a beginning. As is well known, while pondering over the statement in Genesis, "In the beginning Thou madest the Heaven and Earth," St. Augustine took seriously the question provoked by the philosophical temper of pre-Christian Greece, namely, "What was God doing BEFORE He made the Heaven and Earth?" Note the word "before." Augustine confronts the question and answers "Before Heaven and Earth there was no time…" and again "at no time then hadst Thou not made any thing, because time itself Thou madest." We need to dwell on this answer in order to note how the very idea of beginning acquires a theological dimension, transforming the idea of absolute beginning of the world as absolute dependence of the creation on the creator. I am inclined to remark that the notion that the world has a beginning in this context is not intended as a piece of empirical information about the world, but an unveiling of revealed truth which puts in relief the utter contingency of this world on a timeless Cause.

In fact, it is entirely on the basis of ethico-religious considerations that even earlier than Augustine, the Jewish thinker Philo of Alexandria had already argued against those in ancient Greece who found the idea of an absolute origination of the world to be an absurd proposition. Philo's theological argument was: "Those who assert that this world is unoriginated unconsciously eliminate that which of all incentives to piety is the most beneficial and the most indispensable, viz., providence. For it stands to reason that what has been brought into existence should be cared for by its Father and Maker."[9]

Reflecting on the motivations that prompt these arguments it is hard to conclude that the thesis—the world has a beginning—is meant to support a secular cosmologist; rather it is a declaration about the status of the world as it is revealed to the religious consciousness.

Indeed to take an overview of the cognitive situation where cosmological speculations and soteriological positions seem to have considerable bearing on each other, it calls for a careful disclosure of the concerns as these are expressed within their respective traditional frames of discourse. Thus, it may be observed that, generally speaking, in conventional usage as well as in theoretizations, a notion of beginning entails a reference to time. This is why the idea of the beginningless world-process can be rephrased by saying that there never was a time when the world was not. While reflecting on the idea of the beginning of the world, Kant formulated its conceptual content in the following words: "Since the beginning is an existence which is preceded by a time in which the thing is not, there must have been a previous time in which the world was not, i.e., an empty time."[10]

We have seen that such a reading could not have been of great use to a theologian as St. Augustine. For him the idea of beginning is impregnated with a profounder meaning that has theological import. This is also why it must not be taken to be providing empirical information about the world, but as a technical word that forms a part of a theological vocabulary.

It is now pertinent to observe that the notion of the world as temporal, as contingent, as being totally dependent on a timeless cause, described in personal or impersonal terms, recurs in diverse theistic or absolutistic soteriologies across cultures. In the philosophical literature of Advaita Vedānta, one comes across an in-depth analysis of such insights. In order to capture its intended meaning, note that the idea of cause acquires here a special significance. The conventional understanding of cause as temporally preceding an effect is no longer entertained at this level of the discourse. The idea of a "timeless cause" is conceived not in terms of temporal antecedence; there is no question here of any temporal sequence between the cause and the effect. This is why the school of Advaita Vedānta has preferred the coinage of the term "Adhisthāna ground" in order to indicate that Brahman is the indispensable ground for the appearance of the contingent world by introducing the doctrine of Māya. The notion that "the world is an appearance" is not an empirical judgment about the world. In the Advaita vocabulary, it has a precise meaning. The Advaita claim that the world has only empirical reality is not the same as saying that it is imaginary [alika] or that it is pure nothing [tucca]. The thrust of Advaita thinking is that Brahman alone has ontological reality: it is unsublatable in three times (i.e., it is that about which one cannot say that it is not, it was not, or it will not be). It is unchanging and unchangeable [Aparināii/Kuṭastha]. Thus, the status of the empirical world [Vyavahāra], which is the realm of change and multiplicity, is conceptually distinguished from that of Being par excellence [Paramārtha], on the one hand, and the absolute naught [Pratibhāsa], on the other.

In order to highlight this perception, let me quote the following lines by St. Augustine: "And I beheld the other things below Thee, and I perceived that they neither altogether are, nor altogether are not, for they are, since they are from Thee, but are not, because they are not, what Thou art. For that truly is, which remains unchangeably."[11]

Before concluding, let me emphasize that to say all this is far from claiming that there are not differences. Differences are inevitable as traditions attempt to weave diverse networks of ideas while confronting the wide range of issues and questions related to the cosmos and ourselves. In fact, the soteriological traditions are extraordinarily innovative in this respect. The aim of the dialogue is not to blur the differences, but to recognize these as they constitute the identities of diverse traditions. Let us, therefore, preserve and cherish these differences, but not by overlooking the focal points that override these differences.

There is a persistent note in the self-understanding of the Hindu tradition that the Reality that is grasped by our religious consciousness at its height is one, as the Upanashads put it, "Where all words come to a standstill"—it is inexpressible. This is also why the tradition has insisted on the idea that every expression that seeks to express the Inexpressible is one among many such possible expressions, thus echoing the seer in the Ṛgveda who exclaimed: "Ekam sat, vipra bahudha vadanti" [the Real is One, the sages call it by different names].

Gazing at the wide canvas where a range of ideas about the cosmos seem to be vying with each other to capture our attention, I would like to urge that in order to proceed with the dialogue between science and religion, it is necessary to take up the challenge of engaging in a project that seeks to incorporate into this dialogue the contents of religious consciousness, variously expressed in myth and symbols, philosophical conceptualizations and religious doctrines that are integral to the diverse world religions across cultures. In a discussion such as this, created with the special aim of removing the "hard" boundaries between scientific and religious thinking while contemplating such profound issues as "cosmic questions," we need to be on our guard so that we do not continue with the unfortunate practice that tends to play religious traditions against each other. There needs to be a will and readiness for fruitfully exploring the wisdom and insights that lie embedded in religious traditions—a dimension of knowledge about us and the cosmos that we often lose sight of at our peril. As we attempt in earnestness to comprehend these messages and their meaning, we will learn to discern the subtleties present in a given religious discourse where a word or a phrase derived from conventional language has to perform the symbolic function of pointing to that which is not of this world. Going through this process, one becomes more and more sensitive to the levels of thinking that operate in different genres of discourse that deal with these profound issues. Just as scientists with religious concerns or who have clear leanings toward specific religious traditions know that from science itself one cannot derive justification for any form of metaphysical materialism or reductionistic thinking that makes a religious quest appear futile, similarly scholars of religious traditions are aware that it is in vain that some seek to find support for any given religious tradition in scientific cosmology.

Progress in this matter calls for an openness that consciously avoids preconceived notions that prejudice the investigation from the start and thereby makes it possible to overcome the unwarranted antagonism between reason and faith, scientific observations and religious experience.

NOTES AND REFERENCES

1. This translation is from A.L. Basham, *The Wonder That Was India* (London: Sidgwick & Jackson, 1967).
2. *The Bhagavad Gita*, text & translation by S. Radhakrishnan (India: Blackie & Son) 4th reprint, 1976.
3. cf. Charles Sherover, *The Human Experience of Time* (New York: New York University Press, 1975).
4. cf. Anandita N. Balslev, *A Study of Time in Indian Philosophy* (Wiesbaden: Otto Harrassowitz, 1983).
5. cf. A. Balslev, "Time and the Hindu Experience" in *Religion and Time*, co-ed. A. Balslev & J.N. Mohanty (Leiden: E.J. Brill, 1993).
6. cf. Stephen Jay Gould, *Time's Cycles, Time's Arrows* (Cambridge, MA: Harvard University Press, 1987).
7. See *The Gospel of Sri Ramakrishna*, translated by Swami Nikhilananda (Madras: Ramakrishna Math, 1980.)
8. *The Sarvadarsanasamgraha of Madhavacharya*, Gaekwad Oriental Series, Baroda, 1924; translated by E.F. Cowell & E.E. Gough (London: Kegan Paul, 1904).

9. Philo I, translated by F.H. Colson and G.H. Whittaker, (London: 1929 [first printing]).
10. Immanuel Kant, *Critique of Pure Reason*, translated by N.K. Smith (London: Macmillan, 1970).
11. St. Augustine, *The Confessions*, translated by E.B. Pusseyi (New York: 1957) and *De Civitate Dei*, translated by P. Levine (Massachusetts, 1966).

Did God Create Our Universe?

Theological Reflections on the Big Bang, Inflation, and Quantum Cosmologies

ROBERT JOHN RUSSELL

Center for Theology and the Natural Sciences, Graduate Theological Union–Berkeley, Berkeley, California 94709, USA

Abstract: **The sciences and the humanities, including theology, form an epistemic hierarchy that ensures both constraint and irreducibility. At the same time, theological methodology is analogous to scientific methodology, though with several important differences. This model of interaction between science and theology can be seen illustrated in a consideration of the relation between contemporary cosmology (Big Bang cosmology, cosmic inflation, and quantum cosmology) and Christian systematic and natural theology. In light of developments in cosmology, the question of origins has become theologically less interesting than that of the cosmic evolution of a contingent universe.**

Keywords: **epistemic hierarchy; Big Bang cosmology; contingency; natural theology; inflation; quantum cosmology**

INTRODUCTION

Religious traditions shape, and are shaped by, culture as they seek to respond to fundamental questions about the meaning and purpose of life. Today more than ever, these questions are being catalyzed by the discoveries of science and the new scientific perspective on the universe. Fortunately, we live in a special period in which these great scientific discoveries and the enduring wisdom, truth, and values of the diverse religious traditions are together finding ways to enter into a new relationship of genuine respect and mutually critical dialogue. This conference proceedings offers a remarkable opportunity to strengthen this relationship and to advance the dialogue. In the process, it can make an important contribution to the public awareness that the options for science and religion are not restricted to unmitigated conflict or sterile isolation.

Arguments in support of this dialogue, though, are not widely known outside the specialized scholarly field of "science and religion," though they have been rigorously developed by four decades of such scholarship. It will

Address for correspondence: Dr. Robert John Russell, Graduate Theological Union, 2400 Ridge Road, Berkeley, CA 94709. Voice: 510-649-2485.
rrussell @ctns.org

be important, then, to include an initial section summarizing some of these key methodological arguments, and I will take this opportunity to offer a new approach that might facilitate the dialogue. The bulk of this short paper will then be a brief analysis of the extensive discussions of "God and cosmology" when the scientific model was the standard Big Bang cosmology. I will then turn to very recent discussions of this question in the context of its replacement, inflationary and quantum cosmologies. The closing section points to future directions for the dialogue. I would add that I am speaking from the perspective of a Protestant theologian with a background in physics, and I am grateful that other voices representing other religious traditions and perspectives included are included in this volume.

METHODOLOGY IN SCIENCE AND RELIGION

Before reflecting theologically on scientific cosmology, I first want to discuss the question of method: namely, how I intend to relate science and theology.[1] In essence, I will make a two-fold proposal: that the sciences and the humanities, including theology, form an epistemic hierarchy which ensures both constraint and irreducibility, and that theological methodology is analogous to scientific methodology, though with several important differences.

The idea of an epistemic hierarchy can be summarized here, but the methodological analogy will require further discussion below. In essence, the idea is that physics, for example, places constraints on biology: no biological theory should contradict physics, and so on up through the other sciences and humanities. On the other hand, the processes, properties, and laws of biology cannot be reduced without remainder to those of physics, and again on up through the other sciences and humanities. Though scholars differ on the precise ordering of the disciplines, and the role that cross-disciplinary fields like genetics plays in the scheme, the idea of an epistemic ordering like this is crucial both to warding off the philosophical claims of reductionism and a "dualistic" (or even more foliated) ontology of "levels."

To make my case, particularly in regard to the analogy between scientific and theological method, I am drawing directly on the pioneering writings of Ian Barbour,[2] as well as on those of Arthur Peacocke,[3] Nancey Murphy,[4] Philip Clayton,[5] John Polkinghorne,[6] and many others, each of whom has contributed to our growing understanding both of the hierarchy and the analogy. I will also suggest several ways we can expand on their work to portray the relationship as both genuinely asymmetrical and interactive. While these ways make the relationship explicitly asymmetrical, the underlying assumption of a hierarchy of knowledge in itself ensures an implicit asymmetry. Indeed, it will be the task of the paths to show that the asymmetry is not so entirely dominant that theological truth claims are irrelevant to those of science, as is implicit in most of the discussions of "theology and science." In short, it will be the task of the paths to show that there are viable ways from theology to science, without, of course, any sense of appeal to theological "authority."

Scientific Methodology

In the 1950s, Carl Hempel offered what has become a widely accepted description of how theories are constructed and tested in the natural sciences, drawing on arguments from the philosophy of science in the first half of this century.[7] Compared to the simpler idea of direct induction from data to theory as proposed by Bacon and Mill in the seventeenth century (FIG. 1A), Hempel portrayed scientific methodology in terms of a "hypothetical-deductive" path. One moves from data indirectly to the level of theory in a process that involves imagination, analogy, and models as well as logical inference. Then, as Karl Popper had shown earlier, theories are then open to falsification against the data by the predictions one deduces (FIG. 1B).

In the 1950s–60s, Thomas Kuhn,[8] Norwood Hanson,[9] Stephen Toulmin,[10] Imre Lakatos,[11] and others supplemented Hempel's account in a broader view which stresses the historical and contextual dimensions of scientific research. Barbour provides a particularly helpful overview of their work (FIG. 1C). According to Barbour, these philosophers showed that metaphysical concepts and assumptions pervade scientific theories and underlie scientific methodology. Data are "theory-laden," and theories influence the decisions as to which data are relevant. The testing of scientific theories is complicated, too, by the fact that *ad hoc* auxiliary hypotheses can always be constructed to ward off potential falsifiers. Networks of theories, and not just isolated concepts or equations, are tested as a whole. Finally, the criteria for choosing between rival theories goes far beyond predictive success to include coherence with other, accepted theories, explanatory scope, fertility in suggesting new domains for application, conceptual simplicity ("Occam's razor"), aesthetic qualities like beauty, and the avoidance of *ad hoc* moves. Since such criteria transcend the details of the particular theories being considered, they provide a framework for a rational choice between rivals.[12]

Theological Methodology as Analogous to Scientific Method

Barbour, Murphy, and Clayton claim that one can view theological method as analogous to scientific method (FIG. 2). I take their claim to be both a description of the way many theologians actually work and a prescription for progress in theological research. Here doctrines are seen as theories, working hypotheses held fallibly and constructed in light of the "data of theology"—for example, a combination of scripture, tradition, reason, personal and com-

A. Induction (Bacon and Mill):

B. Hypothetical-Deductive method / falsification (Carl Hempel, 1966; Karl Popper, 1932)

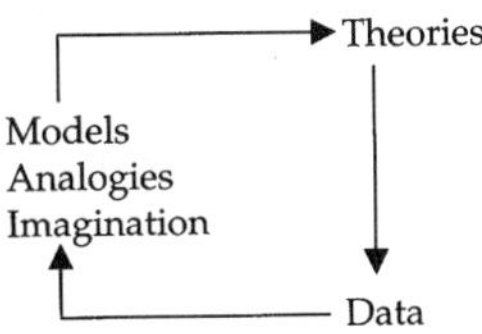

C. Contextual and historical analysis (Kuhn, Hanson, Feyerabend, Toulmin, Lakatos, 1960s)

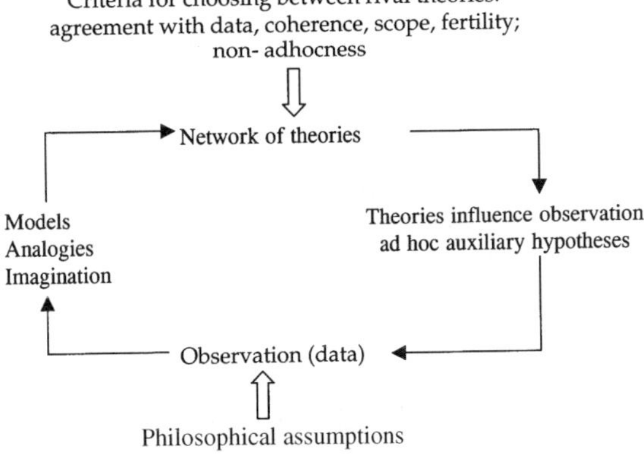

FIGURE 1. Three versions of scientific methodology. (Adapted from Barbour, 1990.)

munity experience, and the encounter with world cultures and with nature, including the discoveries and conclusions of the social, psychological and natural sciences. They are held seriously but tentatively, and they are open to being tested against such data. It is here in particular that the natural sciences are particularly germane: the theories and discoveries of cosmology, physics, evolutionary and molecular biology, anthropology, the neurosciences, and so on, should serve as crucial sources of data for theology, both inspiring new insights and challenging traditional, outmoded conceptions of nature.

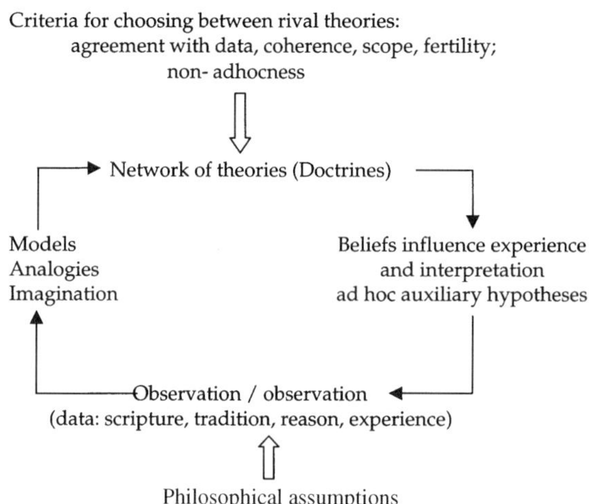

FIGURE 2. Theological methodology as analogous to scientific methodology. (Adapted from Barbour, 1990.)

There are, of course, important differences between the methods of theology and the natural sciences. One is that theologians lack criteria of theory choice which are agreed-upon in advance and which fully transcend the influences of the theories under dispute. Another difference involves the extent to which beliefs influence both the relevancy and the interpretation of data, and the power of imagination, analogy, and models in theory construction. A third difference is that, as in the social sciences, but unlike the natural sciences, many of the data for religious scholars come from subjects; in effect, religious scholars are typically seeking to interpret the interpretation of others—what Phil Clayton calls the problem of the "double hermeneutic." Murphy, drawing on Lakatos, has underscored the importance of "novel facts" in settling disputes and the avoidance of *ad hoc* arguments as a sign of epistemic progress in theology.

These similarities and differences make the appropriation of scientific methodology in theology both promising and challenging. My hope is that as theologians begin to shape their work in this way we will be able to decide whether such a move is genuinely fruitful.[3]

An Interaction Model of Theology and Science

Still, a major challenge exists for the constructive conversations between theology and science: Can theology and science be genuinely *inter*active, each offering something of intellectual value to the other, or is the only role

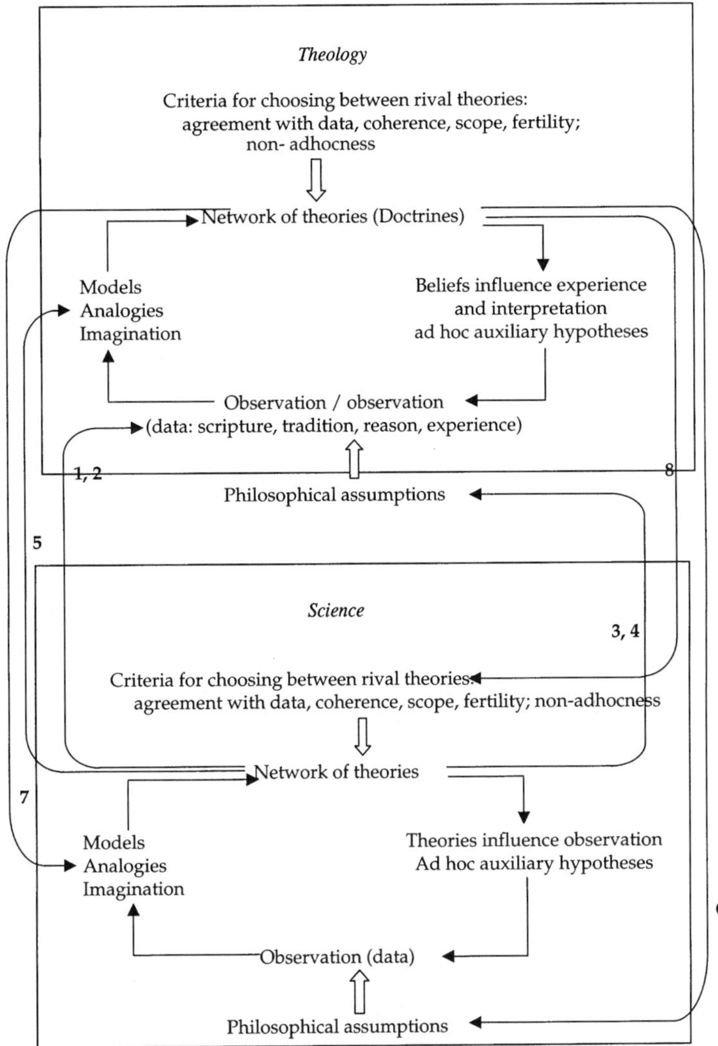

FIGURE 3. Method of creative mutual interaction.

for theology that of critically integrating the results of science into its own conceptual sphere? In order to answer this key question, let me suggest a new diagram which makes more explicit not only the ways in which science can influence theology *but also* the ways in which theology has been, and can now more explicitly be, an influence on science (FIG. 3). In one sense I am merely summarizing what has already been discussed by Barbour, Murphy, Clayton, and many others. In another sense I am offering a constructive proposal which could make the "theology and science" interaction much more explicit and, even more importantly, help us assess its true value to both communities.

The diagram consists of eight ways in which science might influence theology and theology, science. More ways could, and probably should, be added upon further reflection. Individual theologians or scientists typically use one path in particular, often without acknowledging the existence of the other paths. Some shift between them depending on the topic being addressed. My suggestion, though, is to consider what looking at the set of paths as a whole might tell us about the state of discussions in "theology and science" and what it might suggest for improving the conversations.

The eight paths divide into two sets: those now routine ones that describe the movement from science to theology, highlighting the differences in these ways, and those more controversial ones that describe the movement from theology to science, again highlighting their differences.

From Science to Theology

As FIGURE 3 suggests, there are at least five ways or "paths" by which the natural sciences can affect constructive theology. (I will focus on physics and cosmology for specificity, but my comments would apply to the other natural sciences as well.) In the first four, theories in physics, including the key empirical data they interpret, can act as *data for theology* both in a direct sense [(1) and (2)) and indirectly via philosophy ((3) and (4)]. (1) Theories in physics can act directly as data that place constraints on theology. So, for example, a theological theory about divine action should not violate special relativity. (2) Theories can act directly as data either to be "explained" by theology or as the basis for a theological constructive argument. Appropriate to our conference is the issue of $t = 0$ in standard Big Bang cosmology; $t = 0$ was often explained theologically via creation *ex nihilo*, as we shall see below. Note: the theological explanation should be considered a part of theology, and not as an explanation lying within the domain of science. (3) Theories in physics, after philosophical analysis, can act indirectly as data for theology. For example, the contingency of the Big Bang universe, as a philosophical claim based on science and given concrete expression by such issues as $t = 0$, can serve within natural theology as evidence for the existence of God. Similarly an indeterministic interpretation of quantum mechanics can function within theological anthropology as providing a precondition at the level of physics for the bodily enactment of free will. (4) Theories in physics can also act indirectly as the data for theology when they are incorporated into a fully articulated philosophy of nature (e.g., that of Alfred North Whitehead). Finally, (5) theories in physics can function heuristically in the theological context of discovery, by providing conceptual inspiration, experiential inspiration, practical/moral inspiration, or aesthetic inspiration. So Big Bang cosmology may inspire a sense of God's immanence in nature.

From Theology to Physics

To see the genuinely interactive, but asymmetrical, nature of the relations I am proposing, I will suggest at least three paths by which theology can influence science. First, though I want to stress at the outset that by influence I am in no way appealing to, or assuming that, theologians speak with some special kind of authority, whether based on the Bible, church dogma, magisterial pronouncements, or whatever. Quite the contrary; the overall context should be an open intellectual exchange between scholars based on mutual respect and the recognition of fallibility of hypotheses proposed by either side and based on scientific or theological evidence. Instead, the case I wish to make is that such influences have occurred historically and that they continue to occur in the contemporary scientific research. It is first of all, then, a descriptive claim, but it has a mildly prescriptive component as well: I believe a more intentional exploration of such influences could be fruitful for science as they have been for theology, and that they could be particularly fruitful for the "theology and science" interaction. That said, let's turn to three paths from theology to physics:

(6): As mentioned above, theological theories provide some of the philosophical assumptions that underlie scientific methodology. Historians and philosophers of science have shown in detail how the doctrine of creation *ex nihilo* played an important role in the rise of modern science by combining the Greek assumption of the rationality of the world with the theological assumption that the world is contingent. Together these helped give birth to the empirical method and the use of mathematics to represent natural processes.[14] Other assumptions grounded in the *ex nihilo* tradition, however, were not carried over into the scientific conception of nature, including goodness and purpose. It would be interesting to reopen the question of the value of these assumptions for contemporary science. Is there a sense, for example, in which neo-Darwinian evolutionary biology includes teleonomy? Do values have a partial, evolutionary grounding in nature? Would scientific theories which incorporate such ideas be more fruitful than those which do not, or are they hopeless ventures today?[15] (7) Theological theories can act as sources of inspiration in the scientific "context of discovery," that is, in the construction of new scientific theories. An interesting example can be found in the variety of theologies and philosophies which, to a varying degree, apparently influenced many of the pioneers of quantum theory in the period 1900–1930, including Vedanta for Schroedinger, Spinoza for Einstein, and Kierkegaard for Bohr.[16] Another example is the subtle influence of atheism on Hoyle's search for a "steady state" cosmology.[17] Still others include a Whiteheadian approach to science, in which experience or "prehension" is posited at every level of reality, including those treated by physics and biology.[18] Another example would be to search for temporal irreversibility in fundamental physics.[19] Finally (8), theological theories can lead to "selection rules" within the criteria of theory choice in physics.[20] For example, if one wants to consider a theological theory as true, then one can delineate what conditions must obtain within physics for the possibility of its being true. These conditions in turn can serve as reasons for an individual reseach scientist or group of colleagues to choose to pursue a particular scientific theory. The asymmetry between theology and science should now be quite apparent: theological theories do not act as data for science, placing constraints on which theories can be constructed as scientific theories do for theology. This, again, reflects the prior assumption that the sciences are structured in an epistemic hierarchy of constraints and irreducibility. It also safeguards science from any normative claims by theology. It does not, though, mean that theology cannot act to provide criteria for theory choice or inspiration for the construction of new scientific theories, as the older unidirectional relation between theology and science described (i.e., in which the sole task is the theological interpretation of scientific results).

Together these eight paths portray science and theology in a much more interactive, though still asymmetric, mode. I suggest calling this the method of creative mutual interaction. Given this method, we can begin to delineate the conditions needed for real progress in "theology and science." First, scholars in each field would need to find that such an interaction was fruitful according to the criteria of their own research field. So, would scientists feel that their research was more fruitful by having engaged with theology and philosophy in these ways? Would theologians consider their research to have benefited by engaging with science? Secondly, as major changes occur in one field and these changes are taken seriously by the other, would the corresponding effect of these changes be considered fruitful by scholars in that field? Ideally, a process such as this, once set in motion, could continue indefinitely. Finally, it might be possible to compare these results with those of scientists and theologians who have chosen not to engage in mutual interaction. It might also provide a useful typology for comparing and evaluating the ways that various scholars allow science to influence their theology—and vice versa! In any case, even accomplishing the first step would be a major event of enormous significance not only for theology and for science, but also, I believe, more generally for our contemporary culture, which is frequently skeptical (even bitter) towards religion (and sometimes towards science).

GOD, CREATION, AND SCIENCE

With the preceeding as background, we can now turn directly to the subject matter at hand: the relation of creation theology to scientific cosmology and, in specific, the question: "did God create our universe?" I will first give a brief analysis of the extensive discussions of this question when the scientific model was the standard Big Bang cosmology, and then turn to discussions of inflationary and quantum cosmologies.

Standard Big Bang and Creation ex Nihilo

There has been extensive discussion of the possible theological significance of standard Big Bang cosmology, particularly regarding the initial singularity, $t = 0$. Is $t = 0$ relevant to Christian theology, specifically to the claim that God created the universe? Questions of this sort represent what I am calling path (2), and we will touch on three kinds of responses to it.

One response is that it is directly relevant to theology: the scientific discovery of an absolute beginning of all things (including time) would provide empirical confirmation, perhaps even proof, of divine creation. This was the position taken by Pope Pius XII in 1951 in an address to the Pontifical Academy of Sciences.[21] In 1978 Robert Jastrow, then head of NASA's Goddard Institute for Space Studies, spoke metaphorically about scientists who, after climbing the arduous mountain of cosmology, came to the summit only to find theologians there already.[22] The idea that $t = 0$ provides strong, even convincing, support for belief in God is frequently advanced even today by conservative and evangelical Christians.[23]

Even before the address by Pius XII, Fred Hoyle, an outspoken atheist, together with colleagues Hermann Bondi and Thomas Gold, began to construct an alternative cosmology that would have no temporal beginning or end. Their "steady state cosmology" depicted the universe as eternally old and expanding exponentially forever. One reason Hoyle developed this model was his concern that Big Bang cosmology seemed, at least in the public mind, to support Christianity. In my view, Hoyle's concern suggests that he, too, at least implicitly assumed that $t = 0$ was directly relevant to Christian theology. What is crucial to note is that, whereas the previous examples represent a move from science to theology (path 2), Hoyle represents a move from theology[24] to science, proceeding along path (7) to construct an alternative cosmology to Big Bang cosmology. Moreover, the motivation to take path (7) implies that path (2) is implicitly accepted or there would be no theological reason to undertake it. This fact underscores the presence of relations between science and theology actually running in both directions! We'll see similar arguments when discussing Hawking's quantum cosmology below.

The diametrically opposite view is that $t = 0$ is completely irrelevant to theology. The strongest case for this comes from those who hold that theology and science as such are totally separate fields with no possible relationship between any elements in either area. Perhaps surprisingly, some of the strongest support for such a "two worlds" approach has come from the side of religion, particularly from Protestant liberal, existentialist, and neo-Orthodox theologies, and more recently from feminist, liberation and post-modern deconstructionist theologies—as well as from the National Academy of Sciences. What may seem more surprising is that several of the most important scholars in the theology and science interaction agree that when it comes to creation theology, *creatio ex nihilo* is an entirely philosophical argument for which empirical evidence is irrelevant. The contingency of the universe consists in its sheer existence, and is independent of the question of its temporal beginning. Arthur Peacocke, John Polkinghorne, and Bill Stoeger[25] take this position on the specific issue of $t = 0$. A final example would be process theology, which eschews *creatio ex nihilo* in general, and therefore the discussion of $t = 0$ in particular.

Finally, there are a variety of positions that one can take between the two extremes of direct relevancy and complete irrelevancy as developed by such scholars as Ian Barbour,[26] Phil Clayton,[27] Mark Worthing,[28] Howard van Till,[29] and Ted Peters.[30] I have sought to develop one which views $t = 0$ as indirectly relevant to theology by following path (3) instead of path (2).[31] Unlike those claiming that $t = 0$ is directly relevant to theology (path (2)), I claim that it is the philosophical interpretation of $t = 0$ which serves as data for theology. Unlike those who claim that it is only the strictly philosophical significance of cosmology that is relevant, and thus that $t = 0$ is irrelevant, I claim that scientific issues like $t = 0$ when interpreted philosophically are at least indirectly relevant to theology. Of course, such issues are highly nuanced by the theoretical framework of theoretical cosmology, with its penumbra of attendant assumptions about nature, and they may be forever beyond the possibility of direct empirical confirmation.[32] Nevertheless, when a philosophical analysis of nature is focused on such issues as these, it gives its theological appropriation at least some connection with the empirical world.

In my view, then, $t = 0$ gives the general philosophical category of contingency a modest empirical referent. In particular, $t = 0$ is an example of at least one form of past temporal finitude. Past temporal finitude, in turn, is a form of temporal finitude, temporal finitude is a form of finitude, and finitude, to complete the argument, is a form of contingency. In this sense $t = 0$ tends to confirm the much broader notion of the contingency of the universe, though the notion of contingency also includes the contingency of the laws and constants of nature and most fundamentally the contingency of its sheer existence. The contingency of the universe can then play a role in theology: it is a "prediction" of systematic theology (i.e., the datum of the world's contingency is "explained" in terms of the theology of creation) as well as a philosophical "datum" for natural theology (i.e., the datum of the world's contingency serves as a basis for an "argument for God"). $t = 0$ can thus play a helpful, if indirect, role in both systematic theology and natural theology, for it gives concrete empirical content to the much more diffuse philosophical meaning of contingency at work in these theologies. I believe this gives these theologies that take science seriously, but indirectly an advantage over the many other theologies present today that insulate themselves entirely from science. On the other hand, by embedding $t = 0$ within the philosophical form of theological data, the theological conversations about nature are not totally derailed when scientific cosmologies change, as they already have with the development of inflationary and quantum cosmologies (see below). I believe that this gives such theologies an advantage over those which tie their claims about the universe directly to specific results in science. Perhaps the best way to test my claims is to see what happens to the conversations between theology and science in mutual interaction when we move to inflationary and quantum cosmologies.

Before doing so, I want to close this section with a brief mention of the Anthropic Principle. As is well known, the issue here is the apparent "fine-tuning" of the universe that makes the biological evolution of life possible. As numerous scholars have argued in detail,[33] the physical conditions which make evolution possible impose an extremely narrow restriction on both the form of the fundamental laws of physics and the values of the constants of nature. If there is only one universe, as standard Big Bang depicts, it seems quite reasonable to ask why the laws of physics and the values of the constants of nature which characterize this universe happen to lie within these restrictions. To the theist, this stunning question can form the basis for a cosmological version of the "design" argument via path (2). Within the voluminous literature on the subject, a particularly striking recent argument has been given by cosmologist George Ellis and philosopher of religion Nancey Murphy. They argue that the most striking "fact" about the universe is not just the possibility of the evolution of life, but in particular of moral capacity, and it is this possibility which places severe restrictions on its "fine-tuning."[34]

Such a direct relation between cosmology and theology, though, is vulnerable to shifts in cosmologies just as we saw with the direct use of $t = 0$. In specific, if there are many universes, the "fine-tuning" of ours seems to dissolve in a "cosmic Darwinism," as we shall see below. However, there are indirect, ways to use "fine-tuning" within theology which draw out its implications for Big Bang cosmology while anticipating that the underlying scientific theories will change. They tend to follow path (3), embedding the Anthropic discussion within a philosophical analysis that then serves theological research.[35]

Inflation and Quantum Cosmology: The Shifting Conversations with Theology

Now let's turn to inflation and quantum cosmology. The inflationary Big Bang model was first proposed by Alan Guth and colleagues in the 1970s to overcome problems posed by the Big Bang model, including the horizon problem, the matter/antimatter ratio, and so on. The effect on $t = 0$ is fascinating: In some inflationary cosmologies, the Hawking-Penrose theorems don't apply. In these cosmologies we may never know whether or not an essential singularity exists, even if it does,[36] a situation that John Barrow calls "undecidable."[37] The move to inflationary models also provides a scientific explanation for why the initial conditions of the standard model seemed so "fine-tuned" for the evolution of life: such cosmologies can have countless domains in which the natural constants and even the specific laws of physics can vary.

More recently, attempts have been made to unify quantum physics and gravity and apply the results to cosmology. Proposals by Hawking and Hartle, Linde, Isham, Guth, Hawking and Turek, and others, are still in a speculative stage, but there are already some indications of what different quantum cosmologies might look like, including models with or without an initial singularity ("eternal inflation"), with open or closed domains embedded in an open or a closed mega-universe, and so on.[38]

What effect has the transition to inflationary and quantum cosmologies had on the conversations with theology? The answer is intriguing. Clearly the case for a direct relation between cosmology and theology in which $t = 0$ and the Anthropic Principle strongly support Christian theology is undercut, and the risk of making such as case would seem clearer than ever. Yet the story is more subtle. In 1988 Carl Sagan also seemed to assume that a direct relation exists between the issue of $t = 0$ and the existence of a creator God. In his introduction to Hawking's *Brief History* he claimed that Hawking's "no boundary" model, in which the universe has a finite past but no beginning point, leaves God with "nothing to do."[39] Sagan's argument apparently follows path (1) from science to theology and seeks to pose a constraint on what theology can claim: namely, if the universe has no beginning, then theologians cannot claim that God created it. Presumably the further, if implicit, constraint is that, since the universe is run entirely by natural laws, then all God could have done is to create it at the now non-existent beginning event; thus for all intents and purposes, God does not exist.

It is interesting to compare Sagan's move to that of Hoyle. Both apparently see a direct relation between a theological claim (e.g., the existence of God) and a scientific claim (e.g., the existence of $t = 0$), but they move in different directions: Hoyle's move is from atheism to science (path 7 and 8), while Sagan's is from science to theism.[40] Actually, Sagan's argument misses its mark since it is really aimed at a rather outmoded, Deistic conception of God in which all God is needed for is an initial creation. For theists of all varieties today, God is the ongoing Creator of the universe, even if there were no beginning, and the laws of nature, far from restricting God's interaction with and action in the world, are our meager attempts to describe one dimension of God's action as Creator.

I am suggesting instead that we look for indirect relations between cosmology and theology, ones in particular that involve a philosophical analysis of cosmological theories and their assumptions. In such an approach, the philosophical implications for theology of inflation and quantum cosmology may differ from those of Big Bang cosmology, but they are in no way eliminated. Instead, ontological, cosmological, and teleological issues resurface again as fertile sources of theological discussion.[41]

(a) Ontological: Whether or not there was a beginning, the very fact of existence drives us to ask why anything exists at all? Actually Hawking himself underscored the fundamental nature of this question, and the insight that science *per se* may not be able to answer it, when he wrote in the conclusions of *Brief History*, "What is it that breathes fire into the equations and makes a universe for them to describe?"[42]

(b) Cosmological: Why is the universe intelligible? Einstein returned to this fundamental question when he commented that "the eternal mystery of the universe is its comprehensibility."[43]

(c) Teleological: Why do the most general laws of physics, such as those underlying quantum gravity, and the natural constants used in quantum gravity, have the form and values they do? Could everything have been different such that no possible domain of the mega-universe would have been capable of life and mind? More generally, to what extent are inflation and quantum cosmologies relevant to theology? Are there specific features of inflationary and quantum cosmologies that deserve particular theological attention such as $t = 0$ received in the past with a theology of creation?

As we begin to probe questions like these, it will be exciting to see what results are found in the next few years.

Finally, the ongoing developments in quantum gravity/quantum cosmology provide a rich example of what I've called paths (7) and (8). Here in "real time" (and not just through such historical examples as that of Hoyle's steady state) we can study the ways non-scientific factors play a role in both the formation of new theories and the reasons for choosing between them. A nice example of (7) comes from a comparison between the work of Penrose and Hartle/Hawking. In Roger Penrose's approach, the universe arises through a small fluctuation in a quantum field in eternally existing superspace. According to Hawking, however, the universe as a whole arises from the dovetailing of three-geometries in a quantum superspace leading to the formation of the four-dimensional spacetime manifold with no initial singularity. How do we decide between them? According to Chris Isham, the Penrose approach runs into trouble by its arbitrariness: why should one point in an infinite and homogeneous superspace be the seed for the universe and not others?[44] In Isham's view, Hawking's model avoids this problem and is thus preferable. What is interesting here is the parallel Isham points out between his argument against the fluctuation model and Augustine's rejection of the Platonic demiurge model in which God creates the universe at a point in an eternally preexisting time. Instead, Augustine asserted that God creates time along with the universe[45]; it is this concept of time as arising with the universe that parallel's Augustine's conception of the creation of time by God.[46] Other scientists, including Hawking, have noticed this similarity, too.[47] As an example of path (8) we can consider the theological claim that the *imago Dei*, or image of God, includes the capacity for free will. But for us to be genuinely free, we must presuppose the possibility that we can enact our will by closing between alternative somatic dispositions, a possibility frequently seen as ruled out by Laplacian determinism, but regained through the indeterminism that Heisenberg's interpretation of quantum physics affords. It is reasonable, therefore, to extend this concern to the choice between an Einsteinian cosmology, with its inherent determinism, and a quantum cosmology. Clearly there are crucial scientific reasons for moving to the latter, but it is important to note that theology can be seen as offering intellectual reasons for such a move as well.[48]

PROSPECTUS FOR THE FUTURE DIALOGUE

Perhaps the most important result to emerge from the shifts in cosmology over the past decades is the emergence of the hot Big Bang as a "permanent" description of our universe from the Planck time some 12–15 billion years ago to the present. Gone is the time when Hoyle's steady-state model posed a serious challenge to the Big Bang, with its picture of a single, ever-expanding universe whose fundamental features were time-independent. Instead the domain of debate has shifted to the pre-Planck era and what might lie endlessly "before" the Big Bang in quantum superspace. We have witnessed what Joel Primack and Nancy Abrams call an "encompassing" revolution as distinguished from the kind of Kuhnian "replacing revolution" one usually thinks of when scientific paradigms change.[49] In such an encompassing revolution, the new paradigm (e.g., quantum cosmology) contains the old one, for example, Big Bang cosmology as a limit case (e.g., when quantum effects can be ignored). Said another way, we can have complete confidence in relying on the Big Bang scenario, since we know just where it fails: prior to the Planck time. In this sense the Big Bang is "here to stay."[50]

Given this perspective, the time is ripe for a renewed theological focus on the universe in which we have evolved, and a setting aside of what were interesting issues surrounding $t = 0$ but which are now becoming rapidly outmoded. Surely we would commit the "genetic fallacy" if we assumed that the most important clue to the universe we live in is found in its ancient origins. Instead we are poised, as never before, to focus research in theology and science on its 15-billion-year history and the evolution of life, at least on planet Earth and perhaps throughout countless galaxies.

Such a focus will lift up fundamental questions about the meaning of life and its relation to the universe in which it has evolved. If life is rare, does it reduce life to a meaningless surd? In my view, even if it turns out to be extremely scarce, it only renders it all the more precious.[51] But life may actually be abundant in the universe. If so, we may one day be able to decide on some deeply held questions about our own humanity. For example, does the evolution of intelligent life always include not only rationality but moral capacity as well, as it did on Earth? If so, will all such creatures experience moral failure, or is that tragedy limited to *Homo sapiens*? When life does experience moral failure, will those creatures, like us, claim to have an experience of transcendence and the offer of healing power? Will they speak in terms of God? The answer to these questions might serve both to illuminate the purposes of God in creating life in the universe and the question of our own meaning and purpose in the world. Hopefully the more interactive methodology between science and theology suggested in this chapter will enable scholars to form a more rigorous response to these fundamental questions.

NOTES AND REFERENCES

1. For the purposes of this paper I will use the term "theology" to signify the cognitive content of religious discourse. In specific I refer to those cognitive assertions about one's ultimate concern, including but not limited to assertions about the existence or non-existence of God. Theology arises through the critical reflection of individuals and communities on texts, traditions, culture, including philosophy and science, and personal and community experience. I do not intend to privilege the term "theology" to mean the specific cognitive claims of Christianity or any other religion, although it certainly includes them. Thus atheism is a form of theology in so far as the assertion that there is no God is a cognitive assertion bearing a truth claim about the reality of God.
2. Ian G. Barbour, *Religion in an Age of Science: The Gifford Lectures 1989–1991,* Volume 1 (San Francisco, CA: Harper & Row, 1990).
3. Arthur Peacocke, *Theology for a Scientific Age: Being and Becoming—Natural, Divine, and Human* (Minneapolis, MN: Fortress Press, 1993). See particularly Fig. 3, p. 217, and the accompanying text.
4. Nancey Murphy, *Theology in the Age of Scientific Reasoning* (Ithaca, NY: Cornell University Press, 1990).
5. Philip Clayton, *Explanation from Physics to Theology: An Essay in Rationality and Religion* (New Haven, CT: Yale University Press, 1989).
6. John Polkinghorne, *The Faith of a Physicist: Reflections of a Bottom-Up Thinker, The Gifford Lectures, 1993–4* (Princeton, NJ: Princeton University Press, 1994).
7. Carl Hempel, *Philosophy of Natural Science* (Englewood Cliffs, NJ: Prentice-Hall, 1966).
8. Thomas S. Kuhn, *The Structure of Scientific Revolutions* (Chicago, IL: University of Chicago Press, 1970).

9. Norwood Russell Hanson, *Patterns of Discovery* (Cambridge: Cambridge University Press, 1958).
10. Stephen Toulmin, *Foresight and Understanding: An Enquiry into the Aims of Science* (New York: Harper, 1961).
11. Imre Lakatos, *The Methodology of Scientific Research Programmes* (Cambridge: Cambridge University Press, 1978).
12. Two caveats are appropriate here. First, it is important to emphasize that these scholars differed in crucial ways about the philosophy of science. Barbour's point here is to stress what is shared by them and to represent it in a simplified, but instructive model (FIG. 1C). Secondly, it is also important to recognize that many of these ideas had been discussed previously. Hempel, for example, has underscored the influence of theory on observation. Still, neither he, nor Popper for that matter, incorporated it in the fundamental ways that Kuhn and Lakatos did.
13. For a thoroughgoing application of Lakatos's methodology in theology, see Philip Hefner, *The Human Factor: Evolution, Culture, and Religion* (Minneapolis, MN: Fortress Press, 1993).
14. For example, to view nature as created *ex nihilo* implies that the universe is contingent and rational, and these views provide two of the fundamental philosophical assumptions on which modern science is based. By the creation *ex nihilo* tradition I mean to include its long and complex development by Jewish, Muslim and Christian theologians and philosophers during what is often called the Patristic and Middle Ages. Of course, other sources of these assumptions were contributory, but it is important to remember that the doctrine of creation *ex nihilo*, has, in historical fact, served in this way. See for example Michael Foster, "The Christian Doctrine of Creation and the Rise of Modern Science," in *Creation: The Impact of an Idea*, Daniel O'Connor and Francis Oakley, eds. (New York: Charles Scribner's Sons, 1969); Eugene M. Klaaren, *Religious Origins of Modern Science: Belief in Creation in Seventeenth-Century Thought* (Grand Rapids, MI: William B. Eerdmans, 1977); David C. Lindberg and Ronald L. Numbers, *God & Nature: Historical Essays on the Encounter between Christianity and Science* (Berkeley, CA: University of California Press, 1986); Gary B. Deason, "Protestant Theology and the Rise of Modern Science: Criticism and Review of the Strong Thesis," in *The CTNS Bulletin,* Volume 6.4 (Autumn, 1986); Christopher Kaiser, *Creation and the History of Science* (London: Marshall Pickering, 1991).
15. For rarer, non-reductive views, see, for example, Francisco J. Ayala, "Darwin's Devolution: Design Without Designer," Wesley J. Wildman, "Evaluating the Teleological Argument for Divine Action," Charles Birch, "Neo-Darwinism, Self-organization and Divine Action in Evolution," and Ian G. Barbour, "Five Models of God and Evolution" in *Evolutionary and Molecular Biology: Scientific Perspectives on Divine Action* (Vatican City State: Vatican Observatory Publications, and Berkeley, CA: The Center for Theology and the Natural Sciences, 1998). Reductive views are frequently proposed by sociobiologists among others, and the literature is well known.
16. Historical work to date is suggestive though far from complete; a thorough study of this crucial period could help decide just how influential theology or philosophy was to each of the early quantum theorists.
17. For an extremely careful and recent account of the extra-scientific factors at play in cosmological debates in this century, including the implicit role of

religion, see Helge Kragh, *Cosmology and Controversy: The Historical Development of Two Theories of the Universe* (Princeton, NJ: Princeton University Press, 1996).

18. See John B. Cobb, Jr., and Charles Birch, *The Liberation of Life* (Cambridge: Cambridge University Press, 1981).
19. Ilya Prigogine, *From Being to Becoming: Time and Complexity in the Physical Sciences* (San Francisco, CA: W. H. Freeman, 1980).
20. In a similar way, John Barrow uses the Anthropic Principle, not as an argument for design, but as a way of allowing biology to place constraints on physics (i.e., conditions that are required if the evolution of life is to be possible), and these constraints lead Barrow to the discovery of new explanations of hitherto disparate phenomena in physics. Such explanations seem like prime examples of what Murphy, using Lakatos, would call "novel facts," suggesting Barrow's research program is progressive. See Barrow's paper in this volume. Note the unusual way in which "novelty" is used by Murphy /Latakos.
21. See the *Bulletin of the Atomic Scientists*, Volume 8 (1952), pp. 143–146, 165 for a translation of part of the Papal text. For an excellent discussion of it, see Ernan McMullin, "How should cosmology relate to theology?" in *The Sciences and Theology in the Twentieth Century*, A.R. Peacocke, ed. (Notre Dame, IN: University of Notre Dame, 1981), pp. 17–57. According to McMullin, the Pope later refrained from this claim after being cautioned by Georges Lemaitre. Though a Roman Catholic priest and one of the founders of Big Bang cosmology, Lemaitre's view was "two worlds": keep theology and science entirely separate.
22. Robert Jastrow, *God and the Astronomers* (New York: W.W. Norton, 1978), pp. 115–116.
23. For a scholarly argument, see W.L. Craig, *The Kalam Cosmological Argument* (London: Macmillan, 1979). For a popular account see Hugh Ross, *Creation and Time* (Colorado Springs, CO: NavPress, 1994).
24. Recall that I have defined theology broadly to include assertions about the existence or non-existence of God, and thus atheism.
25. See for example John Polkinghorne, *The Faith of a Physicist,* op. cit., chapt. 4, esp. p. 73; Arthur Peacocke, *Creation and the World of Science,* op. cit., p. 78–79; W.R. Stoeger, S.J., "Contemporary Cosmology and Its Implications for the Science-Religion Dialogue," in *Physics, Philosophy and Theology: A Common Quest for Understanding*, Robert John Russell, William R. Stoeger, S.J., and George V. Coyne, S.J., eds. (Vatican City State: Vatican Observatory, 1988) pp. 219–247, esp. p. 240, where Stoeger makes an interesting claim about the limitations of science regarding an absolute beginning.
26. Barbour's recent position has shifted from what it was in the 1960s, when in the context of two rival models (steady-state and Big Bang) he stressed the neutrality of theology to such specific aspects of cosmology as $t = 1$. In his 1990 Gifford Lectures he suggested that if a clear scientific consensus should emerge on the issue of $t = 0$, it would be relevant to theology. Compare Ian Barbour, *Issues in Science and Religion* (New York: Harper & Row, 1966), pp. 366–368; 377; 380; 414; 458 with *Religion in an Age of Science,* op. cit., 128–129. For a detailed analysis of Barbour's position, see Russell, "Finite Beginning...", op. cit.
27. Philip Clayton, *God and Contemporary Science* (Grand Rapids, MI: William B. Eerdmans, 1997).

28. Mark William Worthing, *God, Creation, and Contemporary Physics* (Minneapolis, MN: Fortress Press, 1996).
29. Howard J. van Till, "The Scientific Investigation of Cosmic History," in *Portraits of Creation: Biblical and Scientific Perspectives on the World's Formation* by Howard J. van Till, Robert E. Snow, John H. Stck, and Davis A. Young (Grand Rapids, MI: Eerdmans, 1990), pp. 82–125, see especially pp. 112–115.
30. Ted Peters, ed., *Cosmos as Creation: Theology and Science in Consonance* (Nashville, TN: Abingdon Press, 1989).
31. Robert John Russell, "Cosmology, Creation, and Contingency," in Peters, *Cosmos as Creation,* op. cit., pp. 177–209; Robert John Russell, "Finite Creation without a Beginning" in Russell, *et. al.*, *Quantum Cosmology and the Laws of Nature,* op. cit., pp. 293–329.
32. Bill Stoeger makes precisely this point in W.R. Stoeger, S.J., "Contemporary Cosmology and Its Implications for the Science-Religion Dialogue," in *Physics, Philosophy and Theology,* op. cit.,.
33. For references and discussion see Russell, *et. al.*, eds., *Quantum Cosmology and the Laws of Nature;* and John Leslie, *Universes* (London: Routledge, 1989).
34. Nancey Murphy and George F.R. Ellis, *On the Moral Nature of the Universe: Theology, Cosmology, and Ethics* (Minneapolis, MN: Fortress Press, 1996). Responses to their book can be found in the *CTNS Bulletin*, Volume 18.4 (Fall, 1999).
35. See Russell, "Cosmology, Creation, and Contingency," ibid.; R.J. Russell, "Contingency in Physics and Cosmology: A Critique of the Theology of Wolfhart Pannenberg" in *Zygon: Journal of Religion & Science,* Volume 23:1 (March, 1988), pp. 23–43.
36. In some inflationary scenarios, one of the key assumptions of the singularity theorems is violated during the inflationary epoch (i.e., the assumption that $+ 3p/c > 0$). Because of this we cannot infer that there must have been an initial singularity; there may have been, but it is also possible that these universes are eternally old. See Edward W. Kolb and Michael S. Turner, *The Early Universe* (Reading, MA: Addison-Wesley, 1990).
37. John D. Barrow, *Impossibility: The Limits of Science and the Science of Limits* (Oxford: Oxford University Press, 1998), p. 181.
38. See the papers from this conference. Some of the original papers include J.B. Hartle and S.W. Hawking, "Wave Function of the Universe," *Phys. Rev. D,* Volume 28 (1983), pp. 2960–2975, and A.D. Linde, "Particle Physics and Inflationary Cosmology," *Physics Today* Volume 40 (9) (1987), p. 61–68.
39. See the introduction by Carl Sagan in Stephen W. Hawking, *A Brief History of Time: From the Big Bang to Black Holes* (Toronto: Bantam, 1988), p. x.
40. It would be intriguing to know whether Hawking's work was triggered in part by motivations such as those of Hoyle.
41. For a detailed discussion, see Willem B. Drees, *Beyond the Big Bang: Quantum Cosmologies and God* (La Salle, IL: Open Court, 1990); Russell, *et. al., Quantum Cosmology and the Laws of Nature,* op. cit.
42. Hawking, *Brief History*, p. 174.
43. Albert Einstein, "Physics and Reality," *Ideas and Opinions* (New York: Dell, 1978), pp. 283–315. For an intriguing recent approach to this question see Michael Heller, "Chaos, Probability, and the Comprehensibility of the World," in Russell, *et. al., Chaos and Complexity,* op. cit., pp. 107–121.

44. C.J. Isham, "Creation of the Universe as a Quantum Process," in Russell, *et. al., Physics, Philosophy and Theology,* op. cit., pp. 375–408.
45. In Augustine's opinion, for God to wait, as it were, for a long time and then suddenly create the universe at some point in a pre-existing time would undercut the unchangeable character of God.
46. Philo of Alexandria took a similar position to that of Augustine.
47. For references to the literature on this point and further discussion, see Russell, "Finite Creation without a Beginning," in Russell, *et. al., Quantum Cosmology and the Laws of Nature,* op. cit., pp. 293–329, particularly pp. 318–320.
48. I hope that it no longer need be reiterated that the role of theology here is purely intellectual/academic, and in no way presupposes an appeal to religious "authority." Thus, while it could be seen as irrelevant, the voice of theology in such scientific issues would , it is hoped, not be seen as offensive.
49. References include Wolfhart Pannenberg, *Theology and the Philosophy of Science* (Philadelphia, PA: Westminster Press, 1976).
50. This is, of course, an overstatement. First of all, quantum gravity applies to the entire universe, not just its origins. If so, a careful philosophy of nature will have to take into consideration all the problems raised by such a theory, including those inherited from the philosophical problems of quantum mechanics. Second, the "demise" of steady-state cosmology may well be premature, since a number of cosmologists continue to construct models whose roots can be traced back to Hoyle's early work. See, for example, H.C. Arp, G. Burbidge, F. Hoyle, J.V. Narlikar & N.C. Wickramasinghe, "The extragalactic Universe: An Alternative View," *Nature,* Volume 346 (30 August, 1990), pp. 807–812.
51. If I was lost and thirsty in the trackless wastes of a desert and happened to see a palm tree on the horizon, I wouldn't say, "Oh well, since there's only one of them, it can't be important." Instead I would rejoice in the fact that even one tree exists, since it might mark an oasis and water.

The Meaning of "Design"

JOHN LESLIE

Professor Emeritus, Department of Philosophy, University of Guelph, Guelph, Ontario, Canada

ABSTRACT: **Our universe obeys elegant laws that permit living beings to evolve. This can suggest divine design. So can fine tuning of physical and cosmological parameters in ways that seem essential to life. Understanding the idea of design is, however, difficult for many reasons. For instance, could a designer be said to "fine tune" through choosing all-dictating laws very carefully? Again, would taking advantage of early quantum indeterminacies be a case of design, or would it be design-destroying interference? Can we speak of "design" if God is not a mind but an abstract Platonic principle? And what if, as Spinoza believed, the structure of our universe is just the structure of divine thinking? If such thinking extended to other universes which were lifeless, could those "exhibit design" simply through being orderly?**

KEYWORDS: **design; fine tuning; God; anthropic principle; universes; Plato; Spinoza**

I

"Design" means more than just order of some sort. No matter how you arrange books on a shelf, they will have some order or other, and the great philosopher Leibniz (1646–1716) noted that some formula or other could always be found to fit points scattered on paper randomly. Leibniz further remarked that some kinds of order may be interesting because they have what he called "richness"; they combine obedience to fairly simple laws with results which are complex without being merely untidy; but giving a more complete account of what "Leibnizian richness" means is a very hard task. You tend to end up with a collection of words such as "beauty" and "grandeur," which leaves you little the wiser. Luckily, there is no need for us to attempt the task. Instead, let us concentrate on the word "Design" as it appears in the name "The Argument from Design" or "The Design Argument for God's Existence."

Address for correspondence: Department of Philosophy, University of Guelph, Guelph, Ontario N1G 2W1, Canada. Voice: 519-821-2133; fax: 519-837-8634.
johnlesl@uoguelph.ca

The Argument from Design is really an argument *to* divine design from alleged signs of it. The idea is that the cosmos, as we can see by examining it, was selected for creation to serve a divine purpose, a purpose at least partially understandable because it is good. Leibnizian richness does enter into most people's thoughts about goodness; it is believed that a cosmos serving a divine purpose would have beauty, grandeur, *et cetera*; but more basic, typically, is the belief that the cosmos would be good *through containing intelligent living beings*. I think that makes excellent sense. Suppose, controversially, that God is to be understood as an immensely powerful person, and that any cosmos that this person created would be entirely outside him. Now, what if he had designed the cosmos so that no life would evolve in it? What if he had designed it just for its beauty and grandeur, which he alone could appreciate since nobody else would exist? Would that not make him rather a simpleton, somebody who would have actually to *create* his cosmos before he could properly appreciate the idea of cosmic beauty and grandeur, instead of just contemplating such beauty and grandeur in his mind's eye? Would he not have created something that was worthless *in itself*, since its only value would lie in something outside it, namely, his experience of looking at its structure? While the philosopher G. E. Moore wrote in his *Principia Ethica* (1903) that something could be intrinsically good merely through being beautiful, there being no need for anyone to exist to appreciate the beauty, this seems to me wrong, and in later writings Moore, too, came to think it wrong. He came to define the intrinsically good as what was worth having *in the sense in which an experience is had*. In a cosmos without living things clever enough to be worth calling *observers*, there would be nothing intrinsically good. So if a deity who is not rather a simpleton is to create a cosmos entirely outside himself, then that cosmos must contain intelligent life.

Living beings certainly *look as if* designed by somebody. Their parts come together to serve purposes in intricate ways. Hearts are fine mechanisms for pumping blood. Eyes are superbly constructed for collecting information. Still, we can accept this without accepting the Argument from Design. Darwin explained that the complex, elegant, useful arrangement of a living being's parts might well have come about without the action of a divine designer, through natural selection. When talking about hearts, a scientist of today could say "designed for pumping blood" without having to reject Darwin. All that would be meant would be that hearts were good at pumping blood, and had been produced by natural selection because of this.

It is impossible to prove firmly that Darwinian processes working on atoms which obeyed the laws of physics, and not supplemented by any Life Forces or miraculous acts of divine interference, would be enough to produce such structures as the human eye. Let me just say that any deity who supplemented laws of physics by Life Forces and acts of interference would have produced a disappointingly untidy universe. One would wonder why he had not simply decided to run the whole thing by magic. Naturally, we must avoid being narrow minded about what might count as laws of physics. My belief that everything obeys laws of physics is a hunch that events all conform to a fully unified set of laws, expressible by some reasonably short equation. The equation almost certainly leads to all kinds of phenomena which physicists have not yet dreamed of. The central point is merely that there are not three separate realms of Matter, Life, and Mind, each obeying basic laws peculiar to itself. It would, however, be absurd to try to prove this point firmly, which would involve knowing all the details of how the world works. Instead let me try to show that anyone accepting the Argument from Design could have plenty to offer as evidence without needing to speak of miracles or Life Forces.

We must not fancy that the only manner in which a designer could operate would be to take clay, so to speak, clay with properties beyond his control, and mould it into appropriate shapes. When the designer was God, he would have created his own clay with just the properties he wished. Divine design could be revealed by the fortunate nature of the physical laws which atoms and atomic particles obeyed. We could perhaps find evidence of design in the laws of special relativity, which permit living mechanisms to operate identically no matter how fast they move relative to one another. There is no problem of the forces inside some system acting particularly weakly in one direction, particularly strongly in another, just because the system is in rapid motion, absolutely, in the one direction rather than the other, for special relativity recognizes no such reality as being in rapid motion *absolutely.* Again, we might detect design in the laws of quantum physics that stop atoms from collapsing and that permit seemingly dissipated wave energy to be released in concentrated bursts so that it can do useful work. Let us pay special attention, though, to the marks of design that many have seen in the apparent fine tuning of our universe.

Recently, many physicists and cosmologists have argued that there is quite a problem in how our cosmic environment manages to be one in which Darwinian evolution can operate over long ages to produce living beings. The cosmic period known to us began with a Big Bang. It looks as if the early cosmic density, and the associated expansion speed, needed tuning with immense accuracy for there to be gas clouds able to condense into stars: tuning to perhaps one part in a trillion trillion trillion trillion trillion. Further, the strength of the nuclear weak force had to fall inside narrow limits for the Big Bang to generate any hydrogen (which was needed for making water and for long-lived, stable stars like the sun) and for the creation of all elements heavier than helium. Also the strength ratio between electromagnetism and gravity needed extremely accurate tuning, perhaps to one part in many trillion trillion, for there to be sun-like stars. Again, the existence of chemistry seemingly demanded very precise adjustment of the masses of the neutron, the proton, and the electron.

In a book of mine, *Universes* (1989), I made a long list of such claims about fine tuning. No doubt some of the claims will turn out to be wrong. For instance, it might be that the early cosmic expansion speed was more or less forced to be what it was, because of a process known as "inflation," and the people who think that inflation itself needed very precise tuning could be mistaken. What is impressive, I suggest, is not any particular one of the claims about fine tuning, but the large number of claims which seem plausible, and the consequent implausibility of thinking that every single claim is erroneous. Here, then, we might be thought to have evidence of design, provided we judged that such design would have been directed towards producing living beings in a non-miraculous fashion through making the world obey physical laws which led to the existence of stable stars, planets, and an environment with a rich chemistry in which life could evolve.

II

Do not imagine that dividing divine interference from natural physical processes is an easy affair. For one thing, it is standard theology to say that the cosmos would immediately vanish if God ceased to "conserve" it in existence from moment to moment, the theologians then adding that the laws of physics hold only because this is what God wills. Keeping everything in existence, and keeping it obedient to physical laws, are simply not counted by theologians as "interference" or "miracle." I see nothing wrong in this, but the point is a controversial one. It becomes particularly difficult to handle when you bear in mind two further points: first, that our universe can seem to obey laws of quantum physics which do not dictate precisely how events develop, and second, that quantum randomness, perhaps together with other types of randomness, may have had major effects on the general structure of the world which we see. It is often theorized that the strengths of various forces such as electromagnetism, gravity, and the nuclear weak and strong forces, and the masses of such particles as the neutron, the proton, and the electron, could all have been settled during early instants of the Big Bang *by random processes inside an initially very tiny domain which later grew large enough to include everything now visible to our telescopes.* Physicists speak, for example, of symmetry breaking by scalar fields whose values could have varied randomly from one very tiny domain to another. Against this background, how would things look to a theologian who believed *that God would have to choose at each instant precisely how the cosmos would be at the next instant when he "conserved" it in existence?*—when he preserved it, that is to say, but preserved it in a slightly changed form which led humans to speak of the action of physical forces. The laws of physics, as I noted, could fail to dictate exactly what would have to happen in order for them to be obeyed. They could be quantum-physical laws that left this up to God, in which case God might have chosen cunningly that events would in fact develop, at early instants of the Big Bang, in such a way that there would later be the sort of world which permitted the evolution of intelligent life because the strengths of its physical forces and the masses of its particles had been settled appropriately.

Having chosen cunningly how things would happen at early instants, God might also act rather similarly at various crucial later moments, ensuring that events which quantum physics allowed to develop along various different paths, most of them *not* leading to the evolution of intelligent life, in fact took one or other of the few paths leading to it. We might be unjustified in calling this type of thing "divine interference," as long as it did not happen on too large a scale. The distinction between designing a world's laws in a life-encouraging fashion and then leaving them to operate, and actually designing such things as eyes by, say, putting the optic nerves in the right places, is a sufficiently clear distinction—but in between there is a fuzzy area where what one person would call "messy divine interference" or "miracle" would be classified by another person as God just not choosing *perversely* to make events happen in life-excluding ways when life-encouraging ways were equally present among the possibilities allowed by physical laws, the possibilities among which God had to choose.

Another difficult point concerns whether we could say that divine design "used fine tuning" if the fundamental laws of physics were in fact all-dictating laws: laws with no free parameters. It could at first seem that there would then be just no way in which anything could be *tuned*. The strength ratio between electromagnetism and gravity, for instance, would have to be what it was, given that the fundamental laws were what they were, and so would the masses of the neutron, the proton, and the electron. Once the laws were in place, no divine designer could have been faced with a range of possibilities among which he could have chosen cunningly. All the same, there could be room for talk of "fine tuning," I suggest. For suppose that many slightly different systems of fundamental law, each dictating exactly how events would have developed, would all of them have led to the existence of a universe containing forces recognizable as gravity and electromagnetism, and particles recognizable as neutrons, protons, and electrons, but with the precise properties of those forces and particles differing in each case. I suggest that a divine designer could then be said to have "fine tuned" the properties by choosing the fundamental laws appropriately.

When even the distinction between "God's using physical laws" and "God's operating through miraculous acts of interference" can become fuzzy, this field is going to supply plenty of work for philosophers like me. But unfortunately it may not be work which settles anything of much importance. It may simply amount to recommending various ways of using words, on disappointingly arbitrary grounds. What does seem to me important, however, is that we distinguish firmly between *divine selection* and *observational selec-*

tion. Many scientists who describe our cosmic situation as "fine-tuned for life" believe that observational selection is at work here. They think that a gigantic cosmos includes hugely many domains worth calling "universes." The many universes might be widely separated in space, in a cosmos that had inflated enormously; or they might be successive oscillations of an oscillating cosmos; or they might spring into existence entirely independently. Now, several mechanisms have been proposed for making the various universes differ in the strengths of their forces, in the masses of their particles, and in other respects as well. Brandon Carter's "anthropic principle" then reminds us that only life-permitting conditions give rise to beings able to observe them.

Carter's *strong anthropic principle* says this about conditions in any cosmic region you decide to call "a universe," while his *weak anthropic principle* says the same thing about conditions in anything you prefer to call a spatiotemporal locality. Inevitably, though, one speaker's "large spatiotemporal locality" is another speaker's "universe," for there are no firm rules for using these words. The point to notice is that neither Carter's weak anthropic principle nor his strong anthropic principle has anything to do with divine design. These principles concern *observational selection effects*, period. When reminding us, with his strong anthropic principle, that the universe in which we find ourselves must (since we observers are in it, aren't we?) be a universe whose properties are not totally hostile to life and to intelligence, Carter has never meant that this universe *was forced to be* of a kind which would permit intelligent life to evolve, let alone that it had been positively compelled to contain intelligent living beings. He has always accepted that a great deal of randomness might enter into whether a universe developed life-permitting properties, and if it did, then whether living things, intelligent or otherwise, would actually evolve in it.

I ought to make clear that Carter has himself written so little about this area that he has now largely lost control of what the term "anthropic principle" means. I sometimes get the impression that most people use the phrase "believing in the anthropic principle" to mean something like "believing in divine design"; and, sure enough, when sufficiently many folk use words in a particular fashion, then that fashion can become *right*. Still, I recommend using the term "anthropic principle" in the way that Brandon Carter outlined.

Instead of confusing Carter's observational selection with divine selection, otherwise known as divine design, might we not *combine* these two things? Imagine God creating hugely many universes, the general properties of each universe being settled by random processes at early instants. Suppose that the likely outcome of such random processes would be that only a tiny proportion of the universes had properties permitting life to evolve. God could still be certain that life would arrive in many places if he created sufficiently many universes—perhaps infinitely many. And although it would now be observational selection, not divine selection, which guaranteed that intelligent beings found that their universes had properties of life-permitting kinds, God might still be counted not merely as a creator but also as a designer since he had at least ensured that the fundamental laws obeyed by all the various universes were laws leading living beings to evolve *in some of them.* Why not think along these lines?

I suspect that they would be unsatisfactory lines. Yes, a deity interested in producing good states of affairs might be expected to create infinitely many universes, for why be satisfied with creating only fifty-seven, or only thirty million? However, it could seem bizarre to imagine that this deity would create any universe *which he knew in advance would develop in a fashion totally hostile to intelligent life.* And he could of course know in advance whether a universe would become totally hostile if this depended on physical processes that were only partially controlled by fundamental laws since these laws, for instance ones of quantum physics, failed to dictate precisely what would happen, *so that the deity himself had to decide this* when exerting his power of conservation, of keeping things in existence while at the same time changing them slightly. Remember, divine conservation, the preservation of the existence of things, without which they would at once vanish, is very traditional theology. And theologians are not such fools as to fancy that "conservation" here means "preservation in a totally unaltered state" so that nothing ever changes.

III

If there exist hugely or infinitely many universes, then we could not expect our universe to be the very best of them. All the same, theologians might appear to face a severe difficulty in the fact that ours is a universe containing forest fires which burn animals alive, earthquakes which destroy buildings and the humans inside them, plagues, and so forth. Some people have concluded that anybody who had designed it could be interested only in producing intelligent life and not in good states of affairs. However, I suspect that a deity aiming to achieve good ends would not necessarily be in the business of saving living beings from all disasters. A universe designed by an all-powerful and benevolent being might still include many evils, for various reasons.

One possible reason is that it might, as the poet Keats suggested, be a universe designed not for its own goodness but as a "vale of soul-making." It might be a gymnasium for building up moral strength through often painful efforts. In the absence of strong moral fiber, heavenly bliss would not be deserved or perhaps could not even be *had*: the idea here is that only good souls would get pleasure from life in heaven. But a difficulty with Keats's theory is that it is not at all clear why God would not simply create souls complete with strong moral fiber. Would creating beings with pleasant personalities be dictatorial interference with freedom of the will? I cannot see that it would.

A better suggestion, I suspect, would be that any complex universe would be bound to include disasters if it obeyed causal laws. Think, here, of how it might well be impossible to create a universe in which every single coin of all the billions ever tossed was a coin which landed *heads*—assuming, that is to say, that the coins were governed by causal laws, not by magic. Now, it is not at all obvious that a universe designed for its own goodness would be better if it ran by magic, all such events as earthquakes being banned.

Note that a universe designed as a home for intelligent life would not necessarily include such life *from its earliest moments*. We need not picture God as forced to exist in solitary splendor until intelligent living beings had evolved in our universe. He could have created up to infinitely many earlier universes. What is more, those who agree with Einstein's views about time would say that even at our universe's earliest moments *it was true* that lives were being lived in it at later moments: moments "further along the fourth dimension." [Einstein tried to comfort the relatives of a dead friend by writing to them that he continued to be alive at earlier times. Many philosophers think this makes sense. They compare existing in the past or in the future to *existing on the left,* or *existing to the south.*]

Again, a universe designed as a home for intelligent life might still not be one in which any particular intelligent species, for example humankind, would be guaranteed to survive for long. Remember always that God may well have created a cosmos containing infinitely many universes, while even our own universe may, if the currently popular inflationary models are correct, stretch farther than our telescopes can probe by a factor of perhaps ten followed by a million zeroes. In this connection, think of Enrico Fermi's problem of why we have detected no extraterrestrials—a possible solution being that intelligent species almost always destroy themselves soon after inventing hydrogen bombs, germ warfare or highly polluting industrial processes. You can believe in a benevolent divine designer without rejecting this solution. In contrast, the solution that our own intelligent species happens to be *the very first of many thousands to evolve in our galaxy* could be judged preposterous.

If it does strike you as preposterous, then you might be interested in various themes of a book of mine published under the potentially alarming title *The End of the World.* Let me hurry to make clear that despite the title, plus the beautiful supernova exploding on the front cover of its new paperback version, I myself think that the human race has something approaching a half chance of spreading right across its galaxy. Still, I see considerable force in a point first noticed by Brandon Carter: that just as it could appear preposterous to view our intelligent species as the very first of many thousands, so it could appear preposterous to suppose that you and I were in a human race which was fairly certain to spread right across its galaxy, which would place us *among the earliest thousandth, the earliest millionth, or even the earliest billionth of all humans who will ever have lived.* It might well seem preferable to believe that humankind will become extinct in the not too distant future. Belief that divine benevolence designed our universe is compatible with thinking that Carter is right.

Also, belief in divine benevolence is compatible with recognizing that the laws of physics do permit the existence of hydrogen bombs—and may actually lead to *a vacuum metastability disaster* if physicists push their experiments beyond the energies that are generally considered to be safe, the energies that have already been reached in collisions between cosmic rays. There have been some (but not nearly enough!) discussions of this last point in the physics journals. In his book *Before the Beginning* (1997), Martin Rees, who is Britain's Astronomer Royal, draws firm attention to the calamity that might lie in wait for us here. If space is filled by a scalar field in a merely metastable condition, then a sufficiently powerful collision between particles might work like a pin pricking a balloon. As S. Coleman and F. De Luccia explained in 1980 in *Physical Review D*, a tiny bubble of new-strength scalar field might be formed, this at once expanding at almost the speed of light and destroying first the Earth, then the solar system, then our entire galaxy, *et cetera*. Divine design would not necessarily guarantee us against this. "Designed" need not be a word saying that the universe is always cozy, never threatening. If that were what it said, then the Design Argument for God's Existence would be utter rubbish.

IV

When all is said and done, is belief in God really any better than belief in magic spells? I think it is. Magic cannot be understood, but if God is real then a platonic approach might help us to understand why God is real. It could also have interesting things to suggest about the meaning of the words "God" and "divine design."

Let me first introduce the Platonism of one of this century's finest philosophers, A. C. Ewing. In his book *Value and Reality* (1973), Ewing suggested that God exists *simply because this is good.* What sense could we make of that idea? In *Universes*, and earlier in *Value and Existence* (1979), I commented that followers of Plato think it impossible, even in theory, to get rid of all realities. Even in a blank, an absence of all existing things, it would still, Platonists think, be a reality that two and two made four. It would still be real, in other words, that if there were ever to exist two groups of two things, then there would be four things. Similarly, it would be a reality that the blank was better than any world of people in agony that might replace it. It would be real that the absence of such a world of torment *was ethically required.* And likewise, the presence of a good world could be ethically required despite how there would be, in the blank, nobody to have a duty to produce such a world. The platonic suggestion is that ethical requirements can be real unconditionally, absolutely, eternally. And a further platonic suggestion is that when it is sufficiently weighty *an ethical requirement*—such as, perhaps, the requirement that there exist a supremely good divine person—*can be directly responsible for the actual existence of whatever it is that is required.* Asking Platonists to point to some mechanism which made any such requirement able to have this responsibility could be like asking them to point to *a mechanism which made misery an evil,* or to a mechanism which forced the experience of red to be nearer to that of orange than to that of yellow. For Platonists, these are not affairs which depend on mechanisms. Instead they are affairs of a sort which could explain why anything at all exists, and why any mechanism ever works: why, that is to say, there is a world that obeys causal laws which mechanisms can exploit.

Much more can be said about all this, but let us simply suppose that it does make some sense, as is accepted by John Polkinghorne in his recent book *The Faith of a Physicist* (1994). Like Ewing, Polkinghorne thinks Platonism could best be used to give us insight into why there exists a benevolent divine person who selects a world among all the worlds which are possible, and who wills that it shall exist. However, Ewing and Polkinghorne are little inclined to believe that this person selects anything in quite the way you and I do, with much hard effort to reach correct evaluations, noble struggles to direct acts of will towards good results, stiffening of arm muscles, and so forth. If God is indeed a person and a designer, then we must recognize that he is at least not a person quite like you and me and a designer quite like any architect apart, of course, from being smarter and more powerful. But many people, for instance Paul Tillich among recent theologians, have gone much further than recognizing that point. There is a long Neoplatonic tradition in which it seems to be argued (albeit obscurely) that "God" is just a name for the fact that an ethical need for the cosmos to exist is directly responsible for its existence. This tradition takes its inspiration from Plato's remark in the *Republic* that the Form of the Good "is itself not existence but far beyond it in dignity" since it is "what bestows existence upon things."

Here the idea of an omnipotent architect is entirely abandoned. We might still speak of divine design, but only on the grounds that good things were selected for existence by virtue of being good—the word "selected" being used, evidently, in an unusual sense because nobody would be doing the selecting. As the Neoplatonist Plotinus expresses the matter in his *Third Ennead*, the cosmos exists "not as a result of a judgement recognising its desirability, but by sheer necessity"; "effort and search" play no part in the creative process; yet the outcome, "even had it resulted from a considered plan, would not have disgraced its maker."

A compromise between this Neoplatonic explanation for the cosmos and belief in a divine designer can be found in Spinoza's world-picture, for which Einstein expressed admiration. In my understanding of Spinoza's difficult writings, Spinoza believes that the cosmos exists because this is ethically required, which provides a reason for calling the cosmos "God." However, it is also true that God is an immensely knowledgeable mind and that there exists nothing outside this mind. How can that be so ? The answer is that the divine mind contemplates everything worth knowing, including what a universe would be like if obedient to the laws which our universe obeys, and how it would feel to be each of the conscious beings in such a universe. Now, says Spinoza, the divine mind's contemplation of this *just is* the reality of our universe and of every conscious being in it. Your own knowledge of precisely what it feels like to be you is simply God's contemplating exactly how it must feel to be somebody with precisely your properties—such as, perhaps, the property of not believing a word Spinoza says.

Spinoza seems to have viewed the cosmos as obedient throughout to a single set of laws. This strikes me as unfortunate. If the divine mind really did contemplate everything worth knowing, then presumably it would contemplate all the details of many beautiful, grand universes obeying laws that were very different from those of our universe, even to the extent of being laws incompatible with the evolution of life of any kind. Perhaps infinitely many universes would exist in the divine thought (which is, remember, where Spinoza thinks that you and I and all our surroundings exist). Yet even so, there could be limits to how far the divine thought ranged. The divine mind might not be cluttered with thoughts about absolutely all facts, including facts concerning all the messy forms which universes could take if they obeyed no laws whatever. We might regard all the universes that God thought about as universes *selected* for being thought about because each obeyed laws of some sort. In view of their being in this way selected, we might even speak of their law-controlled structures as "instances of divine design." It would, however, be Brandon Carter's *observational selection* which then ensured that the universe studied by human physicists was a universe whose laws permitted the evolution of intelligent living beings.

BIBLIOGRAPHY

1. S. Coleman and F. De Luccia, "Gravitational effects on and of vacuum decay," *Phys. Rev.* D., Vol. 21 (1980), No. 12., pp. 3305–3315.
2. J. Leslie, *Value and Existence* (Oxford: Blackwell, 1979); *Universes* (London and New York: Routledge, 1989); *The End of the World: The Science and Ethics of Human Extinction* (London and New York: Routledge, 1996).
3. G.E. Moore, *Principia Ethica* (Cambridge: Cambridge University Press, 1903).
4. J. Polkinghorne, *The Faith of a Physicist* (Princeton, NJ: Princeton University Press, 1994).
5. M. Rees, *Before the Beginning: Our Universe and Others* (Reading, MA: Addison-Wesley, 1997).

Cosmology, Life, and the Anthropic Principle

JOHN D. BARROW

Centre for Mathematical Sciences, Cambridge University CB3 0 WA, England

ABSTRACT: We discuss some ways in which the age, size, and structure of the Universe, together with the values of its defining constants, satisfy the necessary conditions for life. The implications for teleology and the existence of other universes are also discussed.

KEYWORDS: cosmology; life; design; anthropic principles

COSMOLOGY, STARS, AND LIFE

Prior to the discovery of the expansion of the Universe there was little that cosmology could contribute to the question of extraterrestrial life aside from probabilities and prejudices. After our discovery of the expansion and evolution of the Universe, the situation changed significantly. The entire cosmic environment was recognized as undergoing steady change. The history of the Universe took on the complexion of an unfolding drama in many acts, with the formations first of atoms and molecules, then galaxies and stars, and most recently, of planets and life. The most important impact of the discovery of the Universe's expansion is that it gives the Universe a changing history and thereby links the cosmos to the local conditions needed for the existence of life.

In the 1930s, the distinguished biologist J.B.S. Haldane took an interest in Milne's proposal[1] that there might exist two different timescales governing the rates of change of physical processes in the Universe: one, t, for "atomic" changes and another, τ, for "gravitational changes," where $\tau = \ln(t/t_0)$ with t_0 constant. Haldane explored how changing from one timescale to the other could alter one's picture of when conditions in the Universe would become suitable for the evolution of biochemical life.[2,3] In particular, he argued that it would be possible for radioactive decays to occur with a decay rate that was constant on the t timescale but which grew in proportion to t when evaluated on the τ scale. The biochemical processes associated with energy derived from the breakdown of adenosine triphosphoric acid would

Address for correspondence: Prof. John D. Barrow, Centre for Mathematical Sciences, Wilberforce Road, Cambridge CB3 0WA, England, U.K.

yield energies which, while constant on the t scale, would grow as t^2 on the τ scale. Thus there would be an epoch of cosmic history on the t scale before which life was impossible but after which it would become increasingly likely. Milne's theory subsequently fell into abeyance, although the interest in gravitation theories with a varying Newtonian "constant" of gravitation led to detailed scrutiny of the paleontological and biological consequences of such hypothetical changes for the past history of the Earth.[3] Ultimately, this led to the formulation of the collection of ideas now known as the Anthropic Principles.[4,5]

Another interface between the problem of the origin of life and cosmology has been the perennial problem of dealing with finite probabilities in situations where an infinite number of potential trials seem to be available. For example, in a universe that is infinite in spatial volume (as would be expected for the case for an expanding open universe with non-compact topology), any event that has a finite probability of occurring should occur not just once, but infinitely often with 100 percent probability if the spatial structure of the Universe is exhaustively random.[6] In particular, in an infinite universe we conclude that there should exist an infinite number of sites where life has progressed to our stage of development. In the case of the steady-state universe, it is possible to apply this type of argument to the history of the universe as well as to its geography because the universe is assumed to be infinitely old. Every past-directed world line should encounter a living civilization. As a result, it has been argued that the steady-state universe makes the awkward prediction that the universe should now be teeming with life along every line of sight.[3]

The key ingredient that modern cosmology introduces into considerations of biology is that of *time*. The observable universe is expanding and not in a steady state. The density and temperature are steadily falling as the expansion proceeds. This means that the average ambient conditions in the universe are linked to its age. Roughly, in all expanding universes, dimensional analysis tells us that the density of matter, ρ, is related to the age t measured in comoving proper time and Newton's gravitation constant, G, by means of a relation of the form

$$\rho \approx \frac{1}{Gt^2} \tag{1}$$

The expanding universe creates an interval of cosmic history during which biochemical observers, like ourselves, can expect to be examining the Universe. Chemical complexity requires basic atomic building blocks which are heavier than the elements of hydrogen and helium that emerge from the hot early stages of the universe. Heavier elements, like carbon, nitrogen, and oxygen, are made in the stars as a result of nuclear reactions that take billions of years to complete. Then they are dispersed through space by supernovae, after which they find their way into grains, planets, and ultimately, into people. This process takes billions of years to complete and allows the expansion to produce a universe that is billions of light-years in size. Thus we see why it is inevitable that the universe is seen to be so large. A universe that is billions of years old and hence billions of light-years in size is a necessary prerequisite for observers, on the basis of chemical complexity. Biochemists believe that chemical life of this sort, and the form based upon carbon in particular, is likely to be the only sort able to evolve spontaneously. Other forms of living complexity (for example that being sought by means of silicon physics) almost certainly can exist, but this artificial "life" is being developed with carbon-based life-forms as a catalyst rather than by spontaneous evolution.

The inevitability of universes that are big and old as habitats for life also leads us to conclude that they must be rather cold on the average because significant expansion to large size reduces the average temperature inversely in proportion to the size of the universe. They must also be sparse, with a low average density of matter and large distances between different stars and galaxies. This combination of low temperature and density also ensures that the sky is dark at night (the so-called Olbers' Paradox, first noted by Halley[7]) because there is too little energy available in space to provide significant apparent luminosity from all the stars. We conclude that many aspects of our Universe that superficially appear hostile to the evolution of life are necessary prerequisites for the existence of any form of biological complexity in the Universe. Habitable universes need to be big and old, dark and cold.

Life needs to evolve on a timescale that is intermediate between the typical timescale that it takes for stars to reach a state of stable hydrogen burning, the so called main-sequence lifetime, and the timescale on which stars exhaust their nuclear fuel and gravitationally collapse. This timescale t_*, is determined by a combination of fundamental constants of Nature

$$t_* \approx \left(\frac{Gm_N^2}{hc}\right)^{-1} \bullet \frac{h}{m_N c^2} \approx 10^9 \text{ yrs} \quad \textbf{(2)}$$

where m_N is the proton mass, h is Planck's constant, and c is the velocity of light.[3,8]

In expanding universes of the Big Bang type the reciprocal of the observed expansion rate of the universe, Hubble's constant $H_0 \approx 70$ Km·s^{-1}Mpc^{-1}, is closely related to the expansion age of the universe, t_0, by a relation of the form

$$t_0 \approx \frac{1}{H_0} \quad \textbf{(3)}$$

The fact that the age $t_0 \approx 10^{10}$ yr deduced from observations of H_0 in this way is a little larger than the main-sequence lifetime, t_*, is entirely natural in the Big Bang theory; that is, we observe a little later than the time when the Sun forms. However, the now defunct steady-state theory, in which there is no relation between the age of the universe (which is infinite) and the measured value of H_0, would have had to regard the closeness in value of H_0^{-1} and t_* as a complete coincidence.[9]

BIOLOGY AND STARS: IS THERE A LINK?

It is interesting to explore how the apparent coincidence between the time required for stars to burn hydrogen and the time needed for biological evolution to produce advanced lifeforms bears on the question of the relative uniqueness or profusion of life in the universe around.

Evidently, in our solar system life first evolved quite soon after the formation of a hospitable terrestrial environment. Suppose the typical time that it takes for life to evolve is denoted by some timescale t_{bio}, then from the evidence presented by the solar system, which is about 4.6×10^9 years old, it is seems that

$$t_* \approx t_{bio},$$

At first sight we might assume that the microscopic biochemical processes and local environmental conditions that combine to determine the magnitude of t_{bio} are *independent* of the nuclear astrophysical and gravitational processes that determine the typical stellar main-sequence lifetime, t_{ms}. However, this assumption leads to the striking conclusion that we should expect extraterrestrial forms of life to be exceptionally rare.[3,10,11] The argument, in its simplest form, is as follows. If t_{bio} and t_* are independent, then the time that life takes to arise is random with respect to the stellar timescale t_*. Thus it is most likely that either $t_{bio} \gg t_*$ or that $t_{bio} \ll t_*$. Now if $t_{bio} \ll t_*$, we must ask why it is that the first observed inhabited solar system (that is, us) has $t_{bio} \approx t_*$. This would be extraordinarily unlikely. On the other hand, if $t_{bio} \gg t_*$, then the first observed inhabited solar system (us) is most likely to have $t_{bio} \approx t_*$ since systems with $t_{bio} \gg t_*$ have yet to evolve. Thus we are a rarity, one of the first living systems to arrive on the scene. In this case, we are led to a conclusion, an extremely pessimistic one for the SETI enterprise, that $t_{bio} \gg t_*$.

In order to escape from this conclusion we have to undermine one of the assumptions underlying the argument that leads to it. For example, if we suppose that t_{bio} is not independent of t_*, then things look different. If t_{bio}/t_* is a rising function of t_*, then it is actually likely that we will find $t_{bio} \approx t_*$. Livio[12] has given a simple model of how it could be that t_{bio} and t_* are related by a relation of this general form. He takes a very simple model of the evolution of a life-supporting planetary atmosphere like the Earth's to have two key phases that lead to its oxygen content:

Phase 1: Oxygen is released by the photodissociation of water vapor. On Earth this took 2.4×10^9 yr and led to an atmospheric O_2 build-up to about 10^{-3} of its present value.

Phase 2: Oxygen and ozone levels grow to about 0.1 of their present levels. This is sufficient to shield the Earth's surface from lethal levels of ultra-violet radiation in the 2000–3000Å band (note that nucleic acid and protein absorption of ultra-violet radiation peaks in the 2600–2700 Å and 2700–2900 Å bands, respectively). On Earth this phase took about 1.6×10^9 yr.

Now the length of Phase 1 might be expected to be inversely proportional to the intensity of radiation in the wavelength interval 1000–2000 Å, where the key molecular levels for H_2O absorption lie. Studies of stellar evolution allow us to determine this time interval and provide a rough numerical estimate of the resulting link between the biological evolution time (assuming it to be determined closely by the photodissociation time) and the main sequence stellar lifetime, with[12]

$$\frac{t_{bio}}{t_*} \approx 0.4\left(\frac{t_*}{t_{sun}}\right)^{1.7},$$

where t_{sun} is the age of the Sun.

This model indicates a possible route to establishing a link between the biochemical timescales for the evolution of life and the astrophysical timescales that determine the time required to create an environment supported by a stable hydrogen burning star. There are obvious weak links in the argument. It provides a necessary condition for life to evolve, but not a sufficient one. We know that there are many other events that need to occur before life can evolve in a planetary system. We could imagine being able to derive an expression for the probability of planet formation around a star. This would involve many other factors which would determine the amount of material available for the formation of solid planets with atmospheres at distances that permit the presence of liquid water and stable surface conditions. Unfortunately, we know that there were many "accidents" of the planetary formation process in the solar system which have subsequently played a major role in the existence of long-lived stable conditions on Earth.[13] For example, the presence of resonances between the precession rates of rotating planets and the gravitational perturbations they feel from all other bodies in their solar system can easily produce chaotic evolution of the tilt of a planet's rotation axis with respect to the orbital plane of the planets over times much shorter than the age of the system.[13,14] The planet's surface temperature variations, insolation levels, and sea levels are sensitive to this angle of tilt. It determines the climatic differences between what we call "the seasons." In the case of the Earth, the modest angle of tilt (approximately 23°) would have experienced this erratic evolution had it not been for the presence of the Moon.[13,15] The Moon is large enough for its gravitational effects to dominate the resonances that occur between the Earth's precessional rotation and the frequency of external gravitational perturbations from the other planets. As a result the Earth's tilt wobbles only by a fraction of a degree around 23° over hundreds of thousands of years—enough perhaps to cause some climatic change, but not catastrophic for the evolution of life. Some argue that life will always find a way to overcome local climatic diversities in order to become complex. However, the evidence on Earth does not really support that optimistic view. There are many continents on which higher apes did not evolve. There are many vacant terrestrial niches that remain unfilled by complex organisms. Living things are overwhelmingly tiny.

This shows how the causal link between stellar lifetimes and biological evolution times may be rather a minor factor in the chain of fortuitous circumstances that must occur if habitable planets are to form and sustain viable conditions for the evolution of life over long periods of time. The problem remains to determine whether the other decisive astronomical factors in planet formation are functionally linked to the surface conditions needed for biochemical processes.

HABITABLE UNIVERSES

We know that several of the distinctive features of the large-scale structure of the visible universe play a role in meeting the conditions needed for the evolution of biochemical complexity within it.

The first example is the proximity of the expansion dynamics to the "critical" state that separates an ever-expanding future from one of eventual contraction to better than 10 percent. Universes that expanded far faster than this would be unable to form galaxies and stars and hence the building blocks of biochemistry would be absent. The rapid expansion would prevent islands of material separating out from the global expansion and becoming bound by their own self-gravitation. By contrast, if the expansion rate were far below that characterizing the critical rate, then the material in the universe would have condensed into dense structures and black holes long before stars could form.[3,16–19]

The second example is that of the uniformity of the universe. The non-uniformity level on the largest scales is very small, $\Delta \approx 10^{-5}$. This is a measure of the average relative fluctuations in the gravitational potential on all scales. If Δ were significantly larger, then galaxies would have rapidly degenerated into dense structures within which planetary orbits would be disrupted by tidal forces, and black holes would form rapidly before life-supporting environments could be established. If Δ were significantly smaller, then the nonuniformities in the density would be gravitationally too feeble to collapse into galaxies and no stars would form. Again, the universe would be bereft of the biochemical building blocks of life.[20]

In recent years the most popular theory of the very early evolution of the universe has provided a possible explanation as to why the universe expands so close to the critical life-supporting divide and why the fluctuation level has the value observed. This theory is called "inflation." It proposes that during a short interval of time when the temperature was very high (say ~ 10^{25} K), the expansion of the universe *accelerated*. This in turn requires the material content of the universe to be temporarily dominated by forms of matter which effectively antigravitate for that period of time.[21] This in turn requires that their density ρ and pressure, p, satisfy the inequality[19]

$$\rho + \frac{3p}{c^2} < 0 \qquad (4)$$

The inflation is envisaged to end because the matter fields responsible decay into other forms of matter, like radiation, which do not satisfy this inequality. After this occurs, the expansion resumes the state of decelerating expansion that it possessed before its inflationary episode began.

If inflation occurs, it offers the possibility that the whole of the visible part of the universe (roughly 15 billion light years in extent today) has expanded from a region that was small enough to be causally linked by light signals at the very high temperatures and early times when inflation occurred. If inflation does not occur, then the visible universe would have expanded from a region that is far larger than the distance that light can circumnavigate at these early times and so its smoothness today is a mystery. If inflation occurs, it will transform the irreducible quantum statistical fluctuations in space into distinctive patterns of fluctuations in the microwave background radiation. Future satellite observations will be able to detect them if they were of an intensity sufficient to have produced the observed galaxies and clusters by the process of gravitational instability.

As the inflationary universe scenario has been explored in greater depth it has been found to possess a number of unexpected properties which, if they are realized, would considerably increase the complexity of the global cosmological problem and create new perspectives on the existence of life in the universe.[19,22,23]

It is possible for inflation to occur in different ways in different places in the early universe. The effect is rather like the random expansion of a foam of bubbles. Some inflate considerably, while others hardly inflate at all. This is termed "chaotic inflation." Of course, we have to find ourselves in one of the regions that underwent sufficient inflation so that the expansion lasted for longer than t_* and stars could produce biological elements. In such a scenario the global structure of the Universe is predicted to be highly inhomogeneous. Our observations of the microwave background temperature structure will only be able to tell us whether the region that expanded to encompass our visible part of the universe underwent inflation in its past. An important aspect of this theory is that for the first time it has provided us with a positive reason to expect that the observable universe is not typical of the structure of the universe beyond our visible horizon, 15 billion light-years away.

It has subsequently been discovered that under fairly general conditions inflation can be self-reproducing. That is, quantum fluctuations within each inflating bubble will necessarily create conditions for further inflation of microscopic regions to occur. This process of "eternal inflation" appears to have no end and may not have had a beginning. Thus life will be possible only in bubbles with properties that allow self-organized complexity to evolve and persist.

It has been found that there is further scope for random variations in these chaotic and eternal inflationary scenarios. In the standard picture we have just sketched, properties like the expansion rate and temperature of each inflated bubble can vary randomly from region to region. However, it is also possible for the strengths and number of low-energy forces of Nature to vary. It is even possible for the number of dimensions of space which have expanded to large size to be different from region to region. We know that we cannot produce the known varieties of organized biochemical complexity if the strengths of forces change by relatively small amounts or in dimensions other than three because of the impossibility of creating chemical or gravitational bound states.[3,24–27]

The possibility of these random variations arises because inflation is ended by the decay of some matter field satisfying Equation (4). This corresponds to the field evolving to a minimum in its self-interaction potential. If that potential has a single minimum, then the characteristic physics that results from that ground state will be the same everywhere. But if the potential has many minima (for example like a sine function), then each minimum will have different low-energy physics, and different parts of the universe can emerge from inflation in different minima and with different effective laws of interaction for elementary particles. In general, we expect the symmetry breaking [which chooses the minima in different regions] to be independent and random.

CHANGING CONSTANTS

We have so far been entertaining the possibility that the distribution of matter, or the rate of expansion of the universe, might differ from place to place or on the average. This does not involve ditching any cherished notions of physics. But it is interesting to consider the consequences of a world in which the constants of physics are slightly changed—either in a hypothetical universe with different true constants or a world (which might be ours) in which they are not all truly constants, but either vary from place to place very slowly or have their origins in processes with quasi-random aspects that could have fallen out differently. The first detailed considerations of this sort were made by Hoyle in 1953, when he recognized that the presence of a significant level of carbon in the universe hinges upon a fine coincidence of physical constants taking values that just allow the carbon nucleus to possess a resonance for the production of carbon from helium, yet just fails to possess a resonance for the reaction that would then burn it all away into oxygen. Later, Freeman Dyson pointed out the significance of the non-existence of the diproton, helium-2, in Nature. If it did exist, then very rapid hydrogen-burning would allow stars to race through their evolutionary history, producing black holes and dead relics long before planets could form or life could evolve.

Considerations like these, together with the light that superstring theories have shed upon the origins of the constants of Nature, mean that we should assess how narrowly defined the existing constants of Nature need to be in order to permit biochemical complexity to exist in the Universe.[3,28] For example, if we were to allow the ratio of the electron and proton masses $\beta = m_e/m_N$ and the fine structure constant α to change their values (assuming no other aspects of physics are changed by this assumption—which is clearly going to be false!), then the allowed variations are very constraining. Increase β too much, and there can be no ordered molecular structures because the small value of β ensures that electrons occupy well-defined positions in the Coulomb field created by the protons in the nucleus. If β exceeds about $5 \times 10^{-3}\,\alpha^2$, then there would be no stars. If modern grand unified gauge theories are correct, then α must lie in the narrow range between about 1/180 and 1/85 in order that protons not decay too rapidly and a fundamental unification of non-gravitational forces can occur. If, instead, we consider the allowed variations in the strength of the strong nuclear force, α_s, and α then roughly $\alpha_s < 0.3\alpha^{1/2}$ is required for the stability of biologically useful elements like carbon. If we increase α_s by 4 percent, there is disaster because the helium-2 isotope can exist (it just fails to be bound by about 70 KeV in practice) and allows very fast direct proton + proton $\rightarrow$ helium-2 fusion. Stars would rapidly exhaust their fuel and collapse to degenerate states or black holes. In contrast, if α_s were decreased by about 10 percent, then the deuterium nucleus would cease to be bound, and the nuclear astrophysical pathways to the build up of biological elements would be blocked. Again, the conclusion is that there is a rather small region of parameter space in which the basic building blocks of chemical complexity can exist.

We should stress that conclusions regarding the fragility of living systems with respect to variations in the values of the constants of Nature are not fully rigorous in all cases. The values of the constants are simply assumed to take different constant values from those that they are observed to take and the consequences of changing them one at a time are examined. However, if the different constants are fully linked together, as we might expect for many of them if a unified Theory of Everything exists, then many of these independent variations may not be possible. The consequences of a small change in one constant would have further necessary ramifications for the allowed values of other constants. One would expect the overall effect to be more constraining on the allowed variations that are life-supporting. For examples of such coupled variations in string theories, see Refs. 29, 30, and 31, and see also Ref. 38.

These considerations are likely to have a bearing on interpreting any future quantum cosmological theory. Such a theory, by its quantum nature, will make probabilistic predictions. It will predict that it is "most probable" that we find the universe (or its forces and constants) to take particular values. This presents an interpretational problem because it is not clear that we should expect the most probable values to be the ones that we observe. Since only a narrow range of the allowed values for, say, the fine-structure constant will permit observers to exist in the Universe, we must find ourselves in the narrow range of possibilities that permit them, no matter how improbable they may be.[19,32] This means that in order to fully test the predictions of future Theories of Everything we must have a thorough understanding of all the ways in which the possible existence of observers is constrained by variations in the structure of the universe, in the values of the constants that define its properties, and in the number of dimensions it possesses.

DESIGN

If we look back at the history of natural theology we find that there is an ancient tradition of trying to draw teleological conclusions about the nature of the Universe from the structure of the Universe.[3] Despite the superficial diversity of content and sophistication in these "design arguments"—some arguing for the existence of a Designer, some for the anthropocentric purpose of the Universe, some for optimality of the Universe in different senses—they can be neatly classified by means of a distinction between laws and outcomes.[33] The oldest and most common design arguments are about fortuitous coincidences in the world of outcomes: the fact that animal needs seem to be so well met by their physiologies and habitats, that the geology and motion of the Earth is conducive to the presence of life, that the eye is so remarkable an optical instrument, and so on. These arguments were extremely commonplace in the eighteenth and early nineteenth centuries and are particularly simple to state. As a result they were extremely persuasive and their seductive power is still in evidence today. The flaws in the arguments based upon them are logically subtle and rarely persuasive to non-scientists (a state of affairs that some biologists find exasperating[34]). However, in the living world the discovery of the power of evolution by natural selection provided a simple alternative explanation for the many remarkable examples of apparent design in Nature that the proponents of this form of the design argument had so carefully documented. Indeed, this collection of examples of adaptations had played an important role in stimulating Darwin to find an explanation for them. The other stream of design arguments, which became fashionable in Newton's time, were based upon the nice structure of the laws of Nature. Sometimes they were called *eutaxiological* design arguments.[3,35] Newton's discovery of laws of motion and gravity provided the impetus and scientific basis for this type of design argument. Richard Bentley's Boyle Lecture provided the first public platform for argument of this sort. They appealed to the rationality, symmetry, specifivity, and simplicity of Nature's laws as evidence for a Designer. This type of design argument is much more sophisticated than the first. You have to know about mathematics and physics to appreciate its force. You need to be able to work out the chains of consequences for human life of altering some aspect of the laws of Nature (for example changing Newton's law of gravity). As a result it was less persuasive to non-scientists than the design argument based on the fortuitous relationships between outcomes. However, one can see that (unless the laws of Nature themselves evolve in some way that we do not suspect) natural selection does not affect this form of the argument. We note that many modern discussions in "God and the New Physics" focus upon this side of the design argument. This reflects the extent to which physicists who work on elementary particle physics and gravitation are Platonic in outlook. They see the underlying laws, symmetries, and mathematical structure of the physical world as the primary source of wonder and inquiry in their scientific work. I suspect that when it comes to this aspect of the Universe almost all physicists would say that the Universe is obviously "designed," in the sense that it possesses order, it is not random, it is not a muddle of half-baked structures and unreliable laws. Where they would differ is on the issue of what the cause of such "design" is. Why does it require a Designer? If it does, what link can we make between this Designer and traditional concepts of God? These questions draw us off from the observational evidence along various directions of metaphysical speculation. However, we must be very careful when drawing metaphysical conclusions from the nature of physics. As if the intrinsic uncertainties in getting the physics right were not large enough, there is also an alarming non-uniqueness associated with these extrapolations into the metaphysical realm. Let me give one example of how deductions from the supposed nature of the laws can also be philosophically ambiguous. There can exist representations of physical laws that are equivalent in mathematical content and in the observational predictions they make, yet which diverge philosophically when some meaning is ascribed to them. A simple example is provided by laws of motion, like Newton's. These are commonly presented as causal (non-teleological) laws: the application of the law to a present state determines the future. Seen like this there can be no teleological aspect: there is no final state that is fixing the trajectory of motion by means of some target that is to be reached. However, we know that causal laws like these can be replaced by the requirement that some quantity (the action) be minimized when considered for all the possible trajectories that the motion might take between points A and B. This minimization principle chooses the same path from A to B as is dictated by the causal law of motion. However, the action principle formulation has a teleological aspect. The path is fixed by initial and final conditions being specified. Thus any interpretation of the form of the laws is fraught with ambiguity in this case.

POSSIBLE RESPONSES TO COINCIDENCES

The modern approach to the issues that used to be the raw material of design arguments is characterized by the cosmological approach to anthropic coincidences. We know that there are many aspects of the Universe's global and local structure, and of its laws and other defining constants, which appear crucial for the existence of life as we know it. Rather than try to use this state of affairs to produce a modern version of the design argument, we ask the anthropic question: What are the necessary conditions for the existence and persistence of complexity (life) in a Universe? The absence of a real definition of life prevents us being more specific, but the relaxation to complexity allows us to avoid being rigidly anthropocentric. We would like to know how special these necessary conditions are in some space of possibilities. If, as is increasingly seen to be the case, there are aspects of the early evolution of the Universe that introduce random variations in the structure of the Universe (for example, to the distribution of matter density, temperature, matter–antimatter balance, cosmological expansion rate, etc.) or to the quantities that we call constants of physics through symmetry breakings or quantum gravitational corrections, then we must introduce the anthropic question in order to evaluate the relevant probabilities when asking how likely it is that the Universe possesses some particular property. At present we believe that Nature is pointing us towards a description of her laws which are far more unified and economical in their schema than we have so far been able to see. At present we are able to contemplate the consequences of altering the values of different fundamental constants of physics or aspects of the Universe's large-scale structure independently. One suspects that in the future these different defining features of the Universal laws will turn out to be linked together and that there will be very few (perhaps only one, or zero ??) independent changes that are self-consistently possible.

The chaotic and eternal inflationary universes mean that we must take seriously the possibility that the global structure of the Universe displays significant variations. The different self-reproducing regions of the chaotic and eternal inflationary universe pictures should display different densities and fluctuation levels. In some versions they will display different fundamental physical constants and even different numbers of large dimensions of space. Subsequently, the symmetry breakings that occur at phase transitions during the early history of the universe can introduce further random variation in cosmological quantities like the matter–anti-matter asymmetry. With regard to the existence of apparent coincidences between the values of constants of physics and the large-scale structure of the Universe which make our existence possible, we have several options. We can shrug our shoulders and say there is one and only one possible universe and it is just good luck that we are here. This option includes the traditional teleological interpretations that appeal to theistic design.

A second option is that some (unknown) physical process creates the fine tunings. That is, they are not disconnected. Attempts to provide partial explanations of this sort can be found in the speculations of Edward Harrison[36] and Lee Smolin.[37] Theories of this sort need to harness the possibility of producing sequences of universes, in the way suggested by eternal inflation, in order to create a population of possible universes upon which some selection process can operate. In Harrison's case that selection process is artificial selection imposed by intelligent beings who are aware of the need for certain structural properties of the universe to be tuned from cycle to cycle in order to optimize the conditions for the subsequent evolution of life. In Smolin's case the selection is for the maximal production of black holes. Both scenarios have their problems. Harrison's has to get going in the first place. Smolin's is undermined by the fact that there appear to be small changes (like those that lead to the binding of the di-proton) that will increase the black hole production. Also the values of the constants that maximize black hole production may not allow life to exist. Moreover, such maxima may not exist for all conceivable variations of the constants (especially for the variation of G).

A third option is that all possibilities exist, either in some quantum cosmological ensemble or actually in an infinite space and time of the sort suggested by the chaotic and eternally inflating universes. In this case the anthropic coincidences just tell us something about the size of the region of life-supporting universes in the collection of all possibilities. In this picture life evolves in some (or even all) of the places where it is possible for it to do so no matter how improbable that sequence of events might be. A problem with this perspective is knowing how large to make the ensemble of variations. It could include structural properties of the universe and its constants, but it might also include the underlying mathematical and logical structures upon which our description of it is based.

A fourth possibility is that life is a good deal easier to evolve than we have concluded from the particular situation that is on view to us. It may be that there are many other routes to biological complexity which would also be possible if the constants and laws of physics were markedly different. In this case the fine tunings that we see are illusory, a consequence of our limited understanding of life and the conditions required for its emergence. These four options alone show how difficult it is to draw firm theological or philosophical conclusions about any aspect of the Universe from cosmology. The four options we have listed, like that of teleological Design, are all consistent with the evidence.

ACKNOWLEDGMENT

The author was supported by a PPARC Senior Fellowship.

REFERENCES

1. E.A. Milne, *Relativity, Gravitation, and World Structure (*Oxford: Clarendon Press, 1935).
2. J.B.S. Haldane, *Nature,* Vol. 139(1937), p. 1002 ; *Nature,* Vol. 158 (1944), p. 555 .
3. J.D Barrow and F.J. Tipler, *The Anthropic Cosmological Principle (*Oxford: Oxford University Press,1986).
4. B. Carter in *Confrontation of Cosmological Theories with Observation*, ed. M. Longair (Dordrecht, the Netherlands: Reidel, 1974).
5. J.D. Barrow, "The Mysterious Lore of Large Numbers," in *Modern Cosmology in Retrospect*, eds. B. Bertotti *et al.* (Cambridge: Cambridge University Press, 1990).
6. G.F.R. Ellis and G.B. Bundrit, *Quart. J. Roy. Astron. Soc*. Vol. 20 (1979), p. 37.
7. E.R. Harrison, *Darkness at Night* (Cambridge, MA: Harvard University Press, 1987).
8. R.H. Dicke, *Rev. Mod. Phys.* Vol. 29(1957), pp. 355, 363; and *Nature* Vol. 192 (1961), p. 440.
9. M.J. Rees, "Comments on Astrophysics," *Space Phys*., Vol. 4 (1972), p. 182 .
10. B. Carter, *Phil. Trans. Roy. Soc*. A, Vol. 310 (1983),p. 347.
11. J. Leslie, *Universes* (London: Routledge, 1989).
12. M. Livio, *Ap. J.*, Vol. 511 (1998), p. 429.
13. J.D. Barrow, *The Artful Universe* (Oxford: Oxford University Press, 1995).
14. J. Laskar and P. Robutel, *Nature*, Vol. 361 (1993), p. 608.
15. J. Laskar, F. Joutel, and F. Boudin, *Nature*, Vol. 361(1993), p. 615.
16. C.B. Collins and S.W. Hawking, *Ap. J.,* Vol. 180 (1973), p. 317.
17. J.D. Barrow, *Quart. J. Roy. Astron. Soc.,* Vol. 23 (1982), p. 344 .
18. M.J. Rees, *Before the Beginning* (New York: Simon and Schuster, 1997).

19. J.D. Barrow, *The Origin of the Universe* (London: Orion, 1994).
20. M. Tegmark and M.J. Rees, *Ap. J.,* Vol. 499 (1998), p. 526.
21. A. Guth, *The Inflationary Universe* (London: Jonathan Cape, 1997).
22. A. Linde, *Physics Today* , Vol. 40 (9)(1987), p. 61.
23. A. Vilenkin, *Phys. Rev. Lett.*, Vol. 74 (1995), p. 846.
24. P. Ehrenfest, *Ann. Physik,* Vol. 61(1920), p. 440.
25. G. Whitrow, *Brit. J. Phil. Sci.*, Vol. 6(1955), p. 13.
26. F.R. Tangherlini, *Nuovo Cim.*, Vol. 27 (1963), p. 636.
27. M. Tegmark, *Class Q. Grav.*, Vol. 14 (1997),p. L69.
28. B.J. Carr and M.J. Rees, *Nature,* Vol. 278 (1979), p. 605.
29. W. Marciano, *Phys. Rev. Lett.*, Vol. 52 (1984), p. 489.
30. J.D. Barrow, *Phys. Rev.* D. Vol. 35 (1987), p. 1805.
31. M.J. Drinkwater, J.K. Webb, J.D. Barrow, and V.V. Flambaum, *Mon. Not. Roy. astron. Soc.*, Vol. 295 (1998), p. 457.
32. V. Rubakov and M.E. Shaposhnikov, *Mod. Phys. Lett.* A 4 (1989), p. 107.
33. J.D. Barrow, *Theories of Everything* (Oxford and NY: Oxford University Press, 1991).
34. R. Dawkins, *The Blind Watchmaker* (Harlow: Longman, 1986); G. Williams, *Plan and Purpose in Nature* (London: Weidenfeld,1996).
35. L.E. Hicks, *A Critique of Design Arguments* (New York: Scribners,1883).
36. E.R. Harrison, *Quart. J. Roy. Astron Soc.*, Vol. 36 (1995), p. 193.
37. L. Smolin, *The Life of the Cosmos (*London: Weidenfeld, 1997).
38. H. Sandvik, J.D. Barrow, and J. Magueijo, "A Simple Varying-Alpha Cosmology," astro-ph/0107512 (2001).

The Argument from Design

What Is At Stake Theologically?

ANNA CASE-WINTERS

McCormick Theological Seminary, Chicago, Illinois 60637, USA

ABSTRACT: This article offers a brief overview of the argument for God's existence grounded in the evidence of design. It pays particular attention to the way the argument has evolved over time and in relation to changing scientific perspectives. The argument from design has in fact been formulated and reformulated in response to the discoveries and challenges it has encountered from the field of science. The conclusion of the article explores the theological importance of this argument with respect to its extent and its limits

KEYNOTE: Thomas Aquinas; arguments for existence of God; Aristotle; Karl Barth; Charles Darwin; emergence of life; evidence of design; David Hume; intelligibility of the universe; Immanuel Kant; meanings of design; Isaac Newton; William Paley; problem of evil

INTRODUCTION

I have been asked to survey the history of the *argument from design*. This presents something of a challenge since the argument from design has a long and winding road with many interesting turns and occasional "dead end "signs along the way. In this brief time we will have to content ourselves with an aerial survey of the landscape it has traveled. But perhaps even that will be instructive for our purposes. I will move chronologically through its formulations, challenges, and reformulations up to its contemporary forms. Concluding comments will ask what is at stake theologically in this whole effort.

The argument from design should be distinguished from its close relative, the cosmological argument: Why is there something and not nothing? The existence of the cosmos as a whole is contingent; it is not self-explanatory; and it does not by careful examination reveal to us its own necessity. An argument for the existence of God may be posed on the ground that something exists.

Address for correspondence: Dr. Anna Case-Winters, McCormick Theological Seminary, 5555 South Woodlawn Avenue, Chicago, IL 60637. Voice: 773-947-6321; fax: 773-288-2612. acase-winters@mccormick.edu

NOTE: This article was originally published in slightly different form in *Zygon* (March 2000)

The argument from design works from what exists. The world evidences order, adaptation, directionality—design. Therefore it is argued that an intelligent designer must have brought it into being. This argument gets the name *teleological*[2] from the Greek word *telos,* which means "end" or "goal." Teleological order entails the notion that processes or structures are fitted to bring about certain results—and in that sense are "designed." [3]

EARLY GREEK PHILOSOPHY AND THE EARLY CHURCH

Forms of the argument go way back in Western classical tradition. Perhaps we should begin where it all begins—with the early Greeks. The pre-Christian Stoics believed that the order and harmony of the cosmos demanded explanation.[4] In 45 BCE, the Roman lawyer Cicero, in his book "The Nature of the Gods,"[5] presented both sides of the argument. Speaking for the Stoics, who favored a teleological view, he posed the question, "When we see a mechanism such as a planetary model or a clock, do we doubt that it is the work of a conscious intelligence? So how can we doubt that the world is the work of the divine intelligence?"

The Atomists (who were in the Epicurean camp) disagreed. Cicero presents their view as well, "The world is made by a natural process, without any need of a creator.... Atoms come together and are held by mutual attraction." No intelligent designer need be postulated. And if there were an intelligent designer, the atomist Lucretius adds, the world in some respects is really badly designed.[6] When we read of these two contesting points of view from all the way back in 45 BCE, today's conversations feel like *déjà vu*—all over again!

The early church eagerly took up the idea of nature as a witness to God, and Tertullian even spoke in terms of a double revelation in "God's two books" the book of nature and the Bible.[7] Nature's design—as seen in the order and beauty of the heavens, the anatomy and physiology of living creatures, and the suitability of the environment to support life—became and has continued to be for Christian theology a pointer to God.

THE MIDDLE AGES: CLASSIC FORMULATION

After the fall of the Roman Empire interest in the natural world dwindled and with it the pursuit of science and natural theology.[8] It was not until the 13th century that long-lost classical philosophy and science were rediscovered. With this turn the argument from design reemerged and received its classic formulation. Aristotelian physics, with its emphasis on causality, became widely influential. Purely physical processes were frequently explained in terms of "ends."

You will recall that for Aristotle there were four distinguishable types of cause:

Final cause is at the level of the maker of an object.

Formal cause is the design or blueprint according to which it is made.

Material cause is the raw material from which it is made.

Efficient cause:is the effort applied in actually making the object.

The point of our exploration seems to be to discuss whether there is a formal cause (a design), and the theological argument proceeds from there to final cause. If there is a design there must be a designer.

Thomas Aquinas, like most good theologians,[9] was conversant with the science and philosophy of his day. Aristotelian physics shaped his theology. The assumption that an effect cannot be greater than the cause and that something can be known of the cause by observing the effect—these became building blocks of his formulation of the argument from design.

Thomas's arguments for the existence of God work *a posteriori* from some observed facts of existence—effects—to their ultimate cause. The most famous of his arguments are the "five ways"[10]:

(1) The First Way begins with the point that things in the world are always changing or moving. Yet they lack the consciousness to be self-moving/changing and therefore must be moved and changed by another. This all has to stop somewhere. Thomas concludes the existence of one, Unmoved Mover.

(2) The Second Way argues from the observation that nothing is self-caused else it would have had to precede itself. But again the series of causes must stop somewhere, thus the need for a First Cause,[11]

(3) The Third Way reasons from the contingent character of things in the world (none of this has to be) to the existence of a totally different kind of being, a Necessary Being.

(4) The Fourth Way argues from the gradations of goodness, truth, and nobility in the things to the existence of a being that is most true, most good, and most noble—one who is the cause of these things in others. (Similarly fire—the hottest thing—is the cause of the heat in things that are hot).

(5) The Fifth Way, perhaps the closest to our present concern, starts from the orderly character of mundane events. Things meet their goals, even things that lack consciousness. Yet nothing that lacks awareness can tend toward a goal without direction from something that has awareness. As an arrow requires an archer to reach its goal, so also universal order points to the existence of an intelligent Orderer of all things.

At the end of each "way," Thomas simply comments "and this is what everybody understands by God."[12]

In each case he is arguing from what is evident in the world (as an effect) to what must be the case in the way of a Cause to bring about such an effect. Thomas seems to have favored the first form of the argument presenting God as the "Unmoved Mover." He treats this one most extensively. Remember that in the science of his day, 13th century physics and astronomy, the four basic elements were thought to be under dynamic influence of the stars, and lower celestial bodies were considered to be moved about by those at greater distance from the earth. Everything that moved did so because it was moved by something else. God is the Unmoved Mover behind all the motion.

THE SCIENTIFIC REVOLUTION: CHALLENGES AND NEW FORMS

When Newton began working out the physical laws of nature, he in a sense demolished this form of the argument, for he gave an explanation of the motion of bodies according to fundamental mechanical physical laws. No appeal to direct divine intervention to move things around in space was needed.

But in another sense he only reformulated the argument, for he assumed God was the architect of these physical laws he had discovered. Science could explain matter and motion without recourse to supernatural forces, but these mechanical secondary forces were simply the out-working of structural conditions given by God at the creation.

As the scientific revolution made many new discoveries, there was in fact more to work with theologically—from God's book of nature. But there came to be greater ambivalence about the place of natural theology. Some scientists were concerned that appeal to final causes might usurp attention to physical causes. Science needed to preserve its integrity and not be a "quarry" mined for theological arguments. And some orthodox theologians, on the other hand, were concerned that natural theology might usurp revelation.[13]

Nevertheless most theologians, philosophers and scientists (people like Francis Bacon, Robert Boyle, René Descartes, and Isaac Newton) assumed the legitimacy of natural theology. Francis Bacon, founder of the new scientific approach, adopted Tertullian's view and wrote[14] "God's two books are... first the Scriptures, revealing the will of God, and then the creatures expressing his power; whereof the latter is a key unto the former."[13]

EIGHTEENTH AND NINETEENTH CENTURIES: NEW FORM AND NEW CHALLENGES

New Form

In the eighteenth century, philosopher William Paley reformulated the argument from design by attending to specific instances of design.[15] He took the eye as a case in point and the "ways in which the various parts of the eye cooperate in a complex way to produce sight." To explain this adaptation of means to ends, he claimed that we need to postulate an intelligent designer, much as we would if we found a watch while "crossing a heath" (rather than assume it had come together by chance, we would assume tht an intelligent designer put it together).

For the record I just want to note the title of Paley's book *Natural Theology: Or, Evidences of the Existence and Attributes of the Deity, Collected from the Appearances of Nature.* Those were more confident days indeed!

New Challenges

David Hume in *Dialogues Concerning Natural Religion*[16] attacked Paley's position for privileging the model of human design of artifacts. This he claimed skews the argument. Why not use another model, for instance the model of biological generation, which does not require intentional design? One could as easily say the universe is like an organism and therefore there must be a cosmic womb.

Paley's argument is an analogy, it is not a proof. Of course much of our working knowledge depends on analogies,thought constructions rather than direct access to reality—things in themselves. The question is whether a chosen analogy is a good one, bearing a useful resemblance to reality—always a contestable point.

Paley had his defenders, who preferred his analogy to Hume's. They observed that in biological generation creatures reproduce themselves rather than producing new and various things. When we query why a rabbit has organs that are so well adapted to meet its needs, we are not helped by the answer that this is because it springs from other rabbits that were similarly adapted. It only pushes the question further back. [17]

Immanuel Kant in his *Critique of Practical Reason* also put forward objections to the argument from design. He thought that science and religion should be completely separated and natural theology was for him a contradiction in terms.[18] Nevertheless he said of himself, "Two things fill my mind with wonder and awe…the starry sky above and the moral law within."[19] Still it was the latter—the moral law within—and not the former that he took to be the clearer pointer to God and God's goodness. He constructed his own ethical argument for the existence of God. Something must account for the "moral law within." There must be a highest good, a coincidence of virtue and happiness. God must exist as the guarantee of the triumph of the good, for we do not see it in this life.

With the publication of Charles Darwin's *The Origin of Species*[20] in 1859 the argument from design met a truly formidable challenge to its credibility. In the theory of evolution there came to the fore a genuine alternative explanation for apparent design in organisms. One was not left with mere chance on the one hand or intelligent design on the other. Organic structures come to be what they are by development from simpler forms through purely natural processes of mutation and natural selection over an extended period of time. No intelligent designer is needed to design the eye for sight.

TWENTIETH CENTURY: NEW FORMS AND NEW CHALLENGES

New Forms

One might think that Darwin had dealt arguments from design the decisive blow, but the argument arises with new vitality and reemerges in the twentieth century. Now the shape is no longer examination of the particular instances of design but general principles behind apparent design. In a manner parallel to what happened with Newton's discovery of physical laws, with Darwin's discovery of principles of natural selection the theological interest shifts from particular divine interventions to the wider divine design: What makes mutation and natural selection work in the way that it does? How did material existence come to be self-organizing in the way that it is?

You see this taking shape in the work of F.R. Tennant[21] in the 1920s. He presents a fresh discussion of the teleological argument pointing to six kinds of adaptation that seem to evidence design and when taken together to point toward a theistic interpretation.[22]

(1) The intelligibility of the world.
(2) The adaptation of living organisms to their environment.
(3) The ways in which inorganic life is conducive to the emergence and maintenance of life.
(4) The way in which the natural environment nurtures moral development in human beings through coping with hardships.
(5) The over-all progressiveness of the evolutionary process.
(6) The aesthetic value of nature.

Here we have in rudimentary form elements of what will become the argument from design in the contemporary discussion—the intelligibility of the universe and its suitability for life.

But not so fast! We must first encounter the new challenges of the twentieth century. Two of the challenges I will name are explicitly theological challenges.

New Challenges

1. Neo-orthodoxy

With the theology of Karl Barth[23] and the advent of neo-orthodoxy, the twentieth century experienced a theological disillusionment with natural theology—the idea that there is a point of contact whereby we may easily perceive who God is by studying the natural world. The risk of natural theology is that what we discover will not be God, but our own reflection, which we then name as God. It is too easy to find God in our race, our culture, our interests. Barth's context was, as you may recall, Hitler's Germany and the rise of the Third Reich and the failure of Protestant liberalism to issue a prophetic challenge. Barth insisted on the prophetic distance of revelation—over against the culture Christianity of his day. So the early Barth said *no* (or *Nein*!) to natural theology and cautioned that God is "wholly other."

2. Evil in the Twentieth Century

A second challenge that arises in the twentieth century is the problem of evil. That is not exactly a new challenge. It is one to which any form of the argument from design has to give a thoughtful response. But it has lately been sharpened in new ways. The optimism of the Enlightenment and the nineteenth century—that every day and every way things are getting better and better—has been severely chastened in our time. Two world wars, the Holocaust, ethnic cleansing—evil has proven too pervasive and too heinous in the twentieth century for it to be dismissed as a brief passage on the way to God's good ends, the necessary dark shades in God's beautiful painting.

Any argument for the existence of God working from design has some hard questions to answer here. Some forms of it simply will not stand the test of evil. If by "design" we mean that whatever comes to be in world process can unequivocally be identified as happening by God's design, according to God's will, then the theodicy problem rears its ugly head. Unless we are willing to sacrifice the theological affirmation of divine goodness, we will need to think again about what we mean by "God's design."

The mixture of good and evil offers evidence sufficiently ambiguous not to require belief in a good and all-powerful creator. Many theologians today, particularly process theologians, are giving a more careful accounting of the nature of God's power. The exercise of divine power must be sufficiently subtle to allow for real freedom in world process. Both the problem of evil and the scientific picture of how the world works invite us to a reconstruction in this direction. To see God at work in setting initial conditions conducive to flourishing of life and working in concert with natural processes is a role more theologically arguable and more consonant with what we know from science than one that presents God as overriding natural processes and controlling events so as to unilaterally determine every state of affairs.

Minimally, if the shape of things can be admitted to be conducive to the realization of valuable ends credence is lent to the hypothesis that the universe is designed for good purposes.

3. Chaos theory/quantum mechanics

The last challenge I will name comes out of quantum mechanics and chaos theory. The reintroduction of the role of chance and contingency in the way the world works has for many, challenged notions of design. Biologist Jacques Monod in *Chance and Necessity* has expressed the conclusion of some: "The ancient covenant is in pieces: man at last knows that he is alone in the unfeeling immensity of the universe, out of which he has emerged only by chance. Neither his destiny nor his duty have been written down."[24]

Some renderings of the teleological argument for the existence of God do assume that by design we must mean that there is "a detailed preexisting blueprint in the mind of God."[25] This plan is foreordained and is working itself out in all its detail. Such a view of design is antithetical to chance.

But must design be understood in such a constraining mode? What if it is part of the "design" that some things happen by necessity, others by chance, and yet others in open interplay of relative freedom? A design might include a whole range on the spectrum: contingency as well as regularity, chaos as well as order, novelty as well as continuity.

Ian Stewart, in his mathematics of chaos, has noted that with the advent of quantum mechanics the clockwork universe of Newton's day has become a cosmic lottery.[26] As he notes, "The very distinction…between the randomness of chance and the determinism of law, is called into question. Perhaps God can play dice, and create a universe of complete law and order, in the same breath."[27] As we learn more about chaos theory the question becomes "not so much whether God plays dice, but how God plays dice."[27]

Contemporary theologians who wish to uphold design are responding variously to the observations of science that much of what occurs in the universe is random activity, pure chance. Ian Barbour has offered a helpful typology that I think accurately reflects the basic theological options on the horizon today[25]:

(a) One way of responding is to claim that what appears to be random is only apparently so. Einstein was himself persuaded of this position.[26,27] We simply cannot see the causal activity behind it. Some theologians see God in control of even the subatomic indeterminacies. If such a view is taken, there are questions to be answered regarding all the blind alleys, waste, suffering, and evil that have attended this process so carefully designed and closely controlled by God.

(b) Another way of responding—more common among theologians—is to rethink the meaning of design as a general directionality and not as a detailed blueprint. Design might be the systematic conditions that

make life and consciousness possible. This view is more conducive to evolutionary understandings and has the capacity to incorporate into "design" elements of chance as well as necessity. This has profound implications for the way in which God and God's relation to the world are viewed. As Polkinghorne expressed it, this view is "consistent with the will of a patient and subtle Creator, content to achieve his purposes through the unfolding of process and accepting thereby a measure of the vulnerability and precariousness which always characterize the gift of freedom by love."[28]

(*c*) The third view is very much like the last one, except that it wants to extend the role of God in the process. The second view has God setting the conditions conducive to life and then not interfering with the system. An option from process theology would envision a more interactive role for God. God's purposes are expressed not only in the unchanging structural conditions, but also in the novel possibilities introduced. Divine creativity works within order and chaos persuading toward good ends. It works with and does not coerce the self-creating activity of creatures.

There is generally speaking a willingness to reconstruct our theology as illumined by what we learn from science about the way the world works. So you see these revisions attending the discoveries of quantum mechanics and chaos theory.

CONTEMPORARY FORMS: INTELLIGIBILITY AND SUITABILITY FOR THE EMERGENCE OF LIFE

Today we are seeing forms of the argument from design that seem actually to be generated within the field of science rather than theology. This happens where scientists reflect upon their discoveries and begin to ask the mega-questions—these questions are not the sole province of philosophers and theologians. Two forms of the argument from design are grounded in the intelligibility of the universe and its suitability for the emergence of life. Remarkably there are attributes of the universe that make it amenable to our rational understanding and to life as we know it:

Intelligibility[29]

Mathematician and physicist Paul Davies has observed, "The success of the scientific method at unlocking the secrets of nature is so dazzling it can blind us to the greatest scientific miracle of all: science works. Scientists themselves normally take it for granted that we live in a rational, ordered cosmos subject to precise laws that can be uncovered by human reasoning. Yet why this is so remains a tantalizing mystery. [31]

Why is the universe intelligible?[32] Why do mathematical principles apply? Why does our science work? "Einstein said that the only thing that is incomprehensible about the world is that it is comprehensible."[33]

Our universe manifests order, unity, and coherence such that "laws of physics discovered in the laboratory apply equally well to the atoms of a distant galaxy."[34] It is not only orderly; it manifests a very particular kind of order; poised, as Davies has noted, between the twin extremes of simple regimented orderliness and random complexity. Organized variety is what we see.

Suitability for the Emergence of Life

Moreover this organization was not built into the universe at its origin. It has emerged from primeval chaos in a sequence of self-organizing processes that have progressively enriched and complexified the evolving universe in a more or less unidirectional manner."[35,36]

Nature seems to operate with a kind of "optimization principle whereby the universe evolves to create maximum richness and diversity. The fact that this rich and complex variety emerges from the featureless inferno of the Big Bang, and does so as a consequence of laws of stunning simplicity and generality, indicates some sort of matching of means to ends that has a distinct teleological flavor to it."[37,38]

Theoretical physicist Steven Weinberg at the end of his book, *The First Three Minutes*, makes the statement, "the more the universe seems comprehensible, the more it also seems pointless."[39] Analysis of cosmos does not for him yield clear and evident purpose. But advocates of the anthropic principle, John Barrow and Frank Tipler (also theoretical physicists), make a rather different interpretation. The very laws that Weinberg takes to be indifferent to human beings seem to them to suggest the presence of an intelligence that "wanted" beings like us to evolve.[40]

Biological systems do have some very particular requirements and these requirements are in fact met by nature. There are cosmic coincidences of striking proportions. The odds against this special set of physical conditions and natural laws that make our lives possible are astronomical.

Stephen Hawking has said, "The odds against a universe like ours emerging out of something like the Big Bang are enormous. I think there are clearly religious implications."[41]

Detractors will say that we could only observe a universe that is consistent with our existence—and surely that is a truism. And there is a possibility that there are other universes. Perhaps if there were a near infinite number of universes the probability would increase that somewhere this special set of conditions would obtain. It is also possible that other forms of life vastly different from our own have emerged elsewhere under different initial conditions and physical laws. So far we do not know of any. For now this must remain an open question.[42]

"If it is the case that the existence of life requires the laws of physics and the initial conditions to be fine-tuned to high precision, and that fine-tuning does in fact obtain, then the suggestion of design seems compelling."[43] It is at least not a more extravagant metaphysical claim than the claim for infinite random universes. In fact some would argue that the hypothesis that there exists an intelligent designer serves as a simpler and therefore better explanation (applying the criterion of Occam's razor).[44]

CONCLUSION: WHAT IS AT STAKE THEOLOGICALLY?

We have in the intelligibility of the universe and in its suitability for life arguments from design that are emerging from within the scientific community. From this scientific picture of the universe, theologians make the interpretive leap to the existence of an intelligent designer—a Creator with an investment in life, and an even, apparently, intelligent life.

Do we see design in this highly improbable "unified system of mutually adjusted and mutually supporting adaptive structures"?[43] Is it reasonable from this to suppose that an intelligent being created the universe? If we do see design, it is hard not to make the leap to thoughts of an intelligent designer. It is a Cheshire cat sort of thing. While we may imagine a designer without a design, a design without a designer would be a surprising thing indeed!

Even if we grant that this is a reasonable inference, it is still a bit of an interpretive leap and not something all impartial observers would automatically conclude. The evidence of design does not coerce a conclusion that there is a designer. But if it is a reasonable inference, theologically it gives us something.

But it does not give us everything. Natural theology can take us so far and no further. Evidence of design gives us a designer but not yet "God" in the sense of the creator of all things visible and invisible, infinite in goodness, wisdom, and power.

If I were to answer my own question posed in the subtitle, "what is at stake here theologically?," I would have to say not as much as we might imagine. In the argument from design we have a pointer toward God, not a proof for God.

Whether one believes or does not believe is a question of interpretation. Any conclusion we reach is "underdetermined by the data." But what do we make of the fact that design is everywhere apparent?

For believers it feels like a substantial confirmation of our belief in God. There is a consonance between what we see here and what we believe. There is a reason to believe that it is not unreasonable to believe.

For the person who does not believe in God the evidence of design in the universe is a source of fascination and wonder—not unlike the experience to which believers refer when they talk about encounter with the *mysterium tremendum.*

Does the universe as a whole have an "end" in the sense of a *telos*, a purpose? In actualizing maximal value? In the evolution of conscious being capable for relation and moral development? In the glory of God? What is a suitable candidate for "in Tennyson's words, the 'far off divine event, toward which the whole creation moves'"?[46] And how would we show that the process manifests progress toward this end? Many questions remain here. I look forward to our exploring them further.

NOTES AND REFERENCES

1. There are three other classical arguments from the existence of God. The teleological argument should be seen in their company: (*a*) The Ontological Argument: From Anselm, that God must exist by definition. For if by definition God is "that than which nothing greater can be conceived," and it is better to exist than not to exist, then existence must be attributed to God. (*b*) The Ethical Argument: From Kant, that something must account for the "moral law within." There must be a highest good, a coincidence of virtue and happiness. God must exist as the guarantee of the triumph of the good, for we do not see it in this life. (*c*) The Cosmological Argument: Why is there something and not nothing? The existence of the cosmos as a whole is contingent; it is not self-explanatory; it does not by careful examination reveal to us its own necessity. An argument for the existence of God may be posed as a response to the question of why is there anything at all.
2. The concept of teleological ordering should be distinguished from simple causal ordering. To say that the wind is fitted to circulate dust in the air is an example of causal ordering, but to say the eye is fitted for sight is an example of teleological ordering. It pertains to the adjustment of means to (presumably valuable) ends.
3. William P. Alston, "Teleological argument for the existence of God," in *Encyclopedia of Philosophy* (New York: Macmillan Company, 1967), p. 84.
4. Norma Emerton, "The Argument from Design in Early Modern Theology," *Science and Christian Belief*, Vol. 1 (2): 129–147 (1989).
5. Cicero, *The Nature of the Gods,* 2.97.
6. Emerton, p. 130.
7. Emerton, p. 131.
8. Emerton, p. 132.
9. For Thomas faith (*fides*) is midway between opinion and knowledge (*scientia*).
10. These are not entirely original with Thomas, but rely upon Plato, Aristotle, Avicenna, Maimonides, and Augustine. The section in which they are found takes the form of a question, "Is there a God?" and begins with the objections that there must not be a God because there is evil in the world and because natural effects can be explained by natural causes.

11. Another observation regarding causality is in order here. For Aquinas, all causes acting in the physical universe were instrumental and had to be used, as it were, by a primary agent. To assume that all this causation is self-explanatory is like expecting that a bed will be constructed if only one puts the tools and materials together "without a carpenter to use them." Aquinas then images God on the model of a craftsman.
12. Vernon J. Bourke, "St. Thomas Aquinas," in *Encyclopedia of Philosophy (*New York: Macmillan Company, 1967).
13. Emerton, p. 133
14. Francis Bacon, *The Advancement of Learning* (1605), 1.6.16.
15. William Paley, *Natural Theology: Or, Evidences of the Existence and Attributes of the Deity, Collected from the Appearances of Nature* (New York: McGraw-Hill, 1976).
16. David Hume, *Dialogues Concerning Natural Religion*
17. We are not out of the woods with this response, however. Hume countered that if the best answer is that there is an intelligent designer, then we still have to give an account for why the designer has a mind that is so well fitted for designing. If the design comes from the designer, where does the designer come from? Either way we end up with an infinite regression.
18. Emerton, p. 145.
19. Immanuel Kant, "Conclusion" in *Critique of Practical Reason* (1788).
20. Charles Darwin, *The Origin of Species* (1859).
21. F.R. Tennant, *Philosophical Theology* (192?).
22. Alston, p. 86.
23. Karl Barth and Emil Brunner, *Natural Theology*, trans. Peter Frankel (London: The Centenary Press, 1946).
24. Jacques Monod, *Chance and Necessity*, trans. A Wainhouse (London: Collins, 1972).
25. Ian Barbour, *Religion in an Age of Science* (San Francisco, CA: Harper Collins, 1990).
26. Ian Stewart, *Does God Play Dice?: The Mathematics of Chaos* (Oxford: Basil Blackwell, 1989).
27. Stewart, p. 2.
28. John Polkinghorne, *One World: The Interaction of Science and Theology* (Princeton: Princeton University Press, 1987), p. 69.
29. "Human beings have always been struck by the complex harmony and intricate organization of the physical world. The movement of the heavenly bodies across the sky, they rhythms of the seasons, the pattern of a snowflake, the myriads of living creatures so well adapted to their environment—all these things seem too well arranged to be a mindless accident. It was only natural that our ancestors attributed the elaborate order of the universe to the purposeful workings of a deity"[30] (p. 44). But with the increased understanding that science has brought we no longer need explicit theological explanations for these phenomena. We know that the laws of nature are such that "matter and energy can organize themselves into complex forms and systems"[30] (p. 44). The questions that remain concern why the universe is lawful and coherent and unified in this way. Why is it intelligible?
30. Paul Davies, "The Unreasonable Effectiveness of Science" in *Evidence of Purpose*, ed. John Marks Templeton (New York: Continuum, 1994).

31. Paul Davies, *The Mind of God: The Scientific Basis for a Rational World* (New York: Simon & Schuster, 1992).
32. "This cosmic order is underpinned by definite mathematical laws that interweave each other to form a subtle and harmonious unity. These laws are possessed of an elegant simplicity, and have often commended themselves to scientists on grounds of beauty alone. Yet these same simple laws permit matter and energy to self-organize into an enormous variety of complex states, including those that order that has produced them"[31] (p. 21).
33. Barbour, p. 141.
34. Davies, 1994, p. 47.
35. Davies, 1994, p. 45.
36. The evolutionary process seems to reflect at the very least a directionality, "a general trend toward greater complexity, responsiveness, and awareness"[25] (p. 172). Even Jacques Monod (studiously avoiding the term teleology) will admit in this a "teleonomy" that attends all evolutionary biology in reproductive invariance and structural teleonomy.
37. Davies, 1994, p. 46.
38. Sir John Eccles has helpfully shown that whatever may be the case about a designer having purposes for world process, it is certainly the case that there is evidence of "purposes" internal to the system, purposes that he breaks down into various differentiated levels: apparent purpose, living purpose, conscious purpose, and self-conscious purpose. These roughly correspond to levels of the prebiotic, the reptile, the mammal, and the human being.
39. Steven Weinberg, *The First Three Minutes* (London: Andre Deutsch, 1977).
40. In Polkinghorne's (1987) discussion.
41. Stephen Hawking, *Stephen Hawking's Universe*. (New York: William Morrow, 1985), p. 121.
42. Analysis of the laws of nature reveals a finely tuned system conducive to the emergence of life. Even a small change in the physical constants would result in an uninhabitable universe. For example, the inverse square laws that apply to gravitational, electric, and magnetic forces are essential to the stability of the atoms and solar systems. Even a small change in the force–distance relation would jeopardize life as we know it. There are countless other instances of "remarkable coincidences"[25] (p. 136). As Paul Davies observes, "attempts to explain this 'too good to be true' arrangement by invoking an infinity of random universes require metaphysical assumptions at least as questionable as those of design"[30] (p. 56).
43. Davies, 1994, p. 51.
44. "In the most general terms, we claim that the relation between fine-tuning and the theory of design is hypothetico-deductive, if there is a designer, this fact explains the fine-tuning and is thereby confirmed. More specifically, our claim is that given a theological research program that includes the theory that the universe was created (designed) by a God whose aim was personal relations with sentient beings, the fine tuning of the universe can be seen to provide novel confirmation, in Imre Lakota's terms,"[45] (p. 63).
45. Nancey Murphy and George Ellis, eds., *On the Moral Nature of the Universe: Theology, Cosmology, and Ethics* (Minneapolis: Fortress, 1996).
46. Alston, p. 46.

ADDITIONAL REFERENCES NOT CITED IN TEXT

1. Templeton, John Marks, ed. *Evidence of Purpose* (New York: Continuum, 1994.)
2. Thomas Aquinas *Summa Theologica*, Part I, Question 2, Article 3.
3. Wilcox, David. "How Blind is the Watchmaker?" in *Evidence of Purpose* ed. John Marks Templeton (New York: Continuum, 1994).

A Universe with No Designer

STEVEN WEINBERG

University of Texas–Austin, Austin, Texas, USA

ABSTRACT: Does the universe shows signs of having been designed by a deity like those of traditional monotheistic religions? Physics is in a better position than religion to give a partly satisfying explanation of the world. Recent developments in cosmology help explain why the measured values of the cosmological constant and other physical constants are favorable for the appearance of intelligent life. The presence of evil and misery disturbs those who believe in a benevolent and omnipotent God. It is not necessary to argue that evil in the world proves that the universe is not designed, but only that there are no signs of benevolence that might have shown the hand of a designer.

KEYWORDS: fine-tuning; anthropic principle; chaotic inflation; theodicy

I have been asked to comment on whether the universe shows signs of having been designed. I don't see how it's possible to talk about this without having at least some vague idea of what a designer would be like. Any possible universe could be explained as the work of some sort of designer. Even a universe that is completely chaotic, without any laws or regularities at all, could be supposed to have been designed by an idiot, as Macbeth suggested.

The question that seems to me to be worth answering, and perhaps not impossible to answer, is whether the universe shows signs of having been designed by a deity more or less like those of traditional monotheistic religions—not necessarily a figure from the ceiling of the Sistine Chapel, but at least some sort of personality, some intelligence, who created the universe and has some special concern with life, in particular with human life. I expect that this is not the idea of a designer held by many here. You may tell me that you are thinking of something much more abstract, some cosmic spirit of order and harmony, as Einstein did. You are certainly free to think that way, but then I don't know why you use words like "designer" or "God," except perhaps as a form of protective coloration. It would not surprise me to find that John Polkinghorne and I agree about what are the interesting questions; where we do disagree is in the answers.

Address for correspondence: Dr. Steven Weinberg, Department of Physics, University of Texas–Austin, Austin, TX78712-1081. Voice: 512-471-4394; fax: 512-471-4888.
weinberg@physics.utexas.edu

NOTE: An extended version of this talk has been published in the *New York Review of Books* (Oct. 21, 1999, pp. 46–48).

It used to be obvious that the world was designed by some sort of intelligence. What else could account for fire and rain and lightning and earthquakes? Above all, the wonderful capabilities of living things seemed to point to a creator who had a special interest in life. Today we understand most of these things in terms of physical forces acting under impersonal laws. We don't yet know the most fundamental laws, and we can't work out all the consequences of the laws we do know. The human mind remains extraordinarily difficult to understand, but so is the weather. We can't predict whether it will rain one month from today, but we do know the rules that govern the rain, even though we can't always calculate their consequences. I see nothing about the human mind any more than about the weather that stands out as beyond the hope of understanding as a consequence of impersonal laws acting over billions of years.

There do not seem to be any exceptions to this natural order, any miracles. I have the impression that these days most theologians are embarrassed by talk of miracles, but the great monotheistic faiths are founded on miracle stories—the burning bush. the empty tomb, an angel dictating the Koran to Mohammed—and some of these faiths teach that miracles continue at the present day. The evidence for all these miracles seems to me to be considerably weaker than the evidence for cold fusion, and I don't believe in cold fusion. Above all, today we understand that even human beings are the result of natural selection acting over millions of years of breedings and eatings.

I'd guess that if we were to see the hand of the designer anywhere, it would be in the fundamental principles, the final laws of nature, the book of rules that govern all natural phenomena. We don't know the final laws yet, but as far as we have been able to see, they are utterly impersonal, and quite without any special role for life. Henri Bergson and Obi-Wan Kanobi were wrong: there is no life force. As Richard Feynman has said, when you look at the universe and understand its laws, "the theory that it is all arranged as a stage for God to watch man's struggle for good and evil seems inadequate."

True, when quantum mechanics was new, some physicists thought that it put humans back into the picture, because the principles of quantum mechanics describe what *observers* would find under various conditions. But, starting with the work of Hugh Everett 40 years ago, there has been a reinterpretation of quantum mechanics as the objective (and deterministic) unfolding of a wave function that describes the observer as well as the system being observed. This work is not completed, so I can't say that we have a satisfactory completely objective formulation of quantum mechanics, but I think we will have.

I have to admit that even physicists going as far as they can go, when they have a final theory, will not have a completely satisfying picture of the world, because we will still be left with the question "Why?" Why *this* theory, rather than some other theory? For example, why is the world described by quantum mechanics? Quantum mechanics is the one part of our present physics that is likely to survive in any future theory, but there is nothing logically inevitable about quantum mechanics; I can imagine a universe governed by Newtonian mechanics instead. So there seems to be an irreducible mystery that science will not eliminate.

But religious theories of design have the same problem. Either you mean something definite by a God, a designer, or you don't. If you don't, then what are we talking about? If you do mean something definite by "God" or "design," or if for, instance, you believe in a God who is jealous, or loving, or intelligent, or whimsical, then you still must confront the question "Why?" Your faith can leave you with no explanation why the universe is governed by *that* sort of God, rather than some other sort of God.

In this respect, it seems to me that physics is in a better position to give us a partly satisfying explanation of the world than religion can ever be, because although physicists won't be able to explain why the laws of nature are what they are and not something completely different, at least we may be able to explain why they are not *slightly* different. For instance, no one has been able to think of a logically consistent alternative to quantum mechanics that is only slightly different. Once you start trying to make small changes in quantum mechanics, you get into theories with negative probabilities or other logical absurdities. When you combine quantum mechanics with relativity, its logical fragility increases. You find that unless you arrange the theory in just the right way you get nonsense, like effects preceding causes, or infinite probabilities. Religious theories, on the other hand, seem to be infinitely flexible, with nothing to prevent the invention of deities of any conceivable sort.

Now, it doesn't settle the matter for me to say that we cannot see the hand of a designer in what we know about the fundamental principles of science. It might be that, although these principles do not refer explicitly to life, much less human life, they are nevertheless craftily designed to bring it about.

Some physicists have argued that certain constants of nature have values that seem to have been mysteriously fine-tuned to just the values that allow for the possibility of life, in a way that could only be explained by the intervention of a designer with some special concern for life. I am not impressed with these supposed instances of fine-tuning. For instance, one of the most frequently quoted examples of fine-tuning has to do with the energy of a certain excited state of the carbon nucleus. The build-up in stars of elements necessary for life, like carbon and oxygen, depends on the carbon nucleus' having an excited state at an energy within a narrow range, where in fact just such a state is found. The reason that it has to have this energy is to provide a way for carbon nuclei be formed in stars in collisions of helium nuclei with unstable beryllium nuclei, which is a necessary step in the build-up of all elements heavier than helium. But recent calculations show that, as has been long expected, without any fine-tuning of the constants of nature one would in any case expect the carbon nucleus to have a state like an unstable molecule, consisting of a helium nucleus and a beryllium nucleus, which would naturally have an energy close to the values necessary for the synthesis of carbon and heavier elements.[1]

There *is* one constant whose value does seem remarkably well adjusted in our favor. It is the energy density of empty space, also known as the cosmological constant. It could have any value, but from first principles one would guess that this constant should be very large—much too large to allow matter to clump together in the early universe, which is the first step in forming galaxies and stars and planets and people. It's too early to tell whether this is a real problem, or whether there is some fundamental principle that explains why the cosmological constant must be this small.

But even if there is no such principle, recent developments in cosmology offer the possibility of an explanation why the measured values of the cosmological constant and other physical constants are favorable for the appearance of intelligent life. Sidney Coleman has shown how quantum mechanical effects can lead to a picture of the wave function of the universe in which the wave function is the sum of many different terms, each term corresponding to a big (or little) bang in which what we call the constants of nature take all possible values. Also, as you have heard here from Alan Guth, in the "chaotic inflation" theories of Andre Linde and others our Big Bang is supposed to be just one episode in a much larger universe in which big bangs go off all the time, each with different values of the fundamental constants.

In any such picture, in which the universe contains many parts with different values for what we call the constants of nature, there would be no difficulty in understanding why these constants take values favorable to intelligent life. There would be a vast number of big bangs in which the constants of nature take values unfavorable for life, and much fewer where life is possible. You don't have to invoke a benevolent designer to explain why we are in one of the parts of the universe where life is possible. In all the other parts of the universe there is no one to raise the question.

To conclude that the constants of nature have been fine-tuned by a benevolent designer is like saying "Isn't it wonderful that God put us here on earth, where there's water and air and the surface gravity and temperature are so comfortable, rather than some horrid place, like Mercury or Pluto." Where else in the solar system but on earth could we have evolved?

Reasoning like this is called "anthropic." Sometimes it just amounts to an assertion that the laws of nature are what they are so that we can exist, without further explanation. This seems to me to be little more than mystical mumbo-jumbo. On the other hand, if there really is a large number of worlds in which some constant takes different values, then the anthropic explanation of why in our world it takes values favorable for life is just common sense, like explaining why we live on the earth rather than Mercury or Pluto. The actual value of the cosmological constant, recently measured by observations of the motion of distant supernovas, is about what you would expect from this sort of argument; it is just about small enough to prevent it from interfering with the formation of galaxies and so on. But we don't yet know enough about physics to tell whether there are different parts of the universe in which what are usually called the constants of physics really do take different values. This is not a hopeless question; we will be able to answer it when we know more about the quantum theory of gravitation than we do now.

It would be evidence for a benevolent designer if life were better than could be expected on other grounds, including anthropic grounds. A certain capacity for joy would naturally evolve through natural selection, as an incentive to animals who need to eat and breed in order to pass on their genes. It may not be likely that evolution would produce animals who are fortunate enough to have the leisure and the ability to do science and think abstractly, but our sample of what is produced by evolution is very biased, by the fact that it is only in these fortunate cases that there is anyone thinking about cosmic design. Astronomers call this a selection effect. (Astronomers like Sandra Faber have to worry continually about selection effects of one sort or another.) The universe is very large, and it should be no surprise that, among the enormous number of planets that support only unintelligent life and the still vaster number that cannot support life at all, there is some tiny fraction on which there are living beings who are capable of thinking about the universe, as we are doing here. The real question is whether life is better than would be expected from what we know about natural selection, taking into account the bias introduced by the fact that we are thinking about the problem.

This is a question that everyone will have to answer for him- or herself. Being a physicist is no help with questions like this, so I have to speak from my own experience. My life has been remarkably happy, probably in the upper 99.99 percentile of human happiness, but even I have seen a mother die painfully of cancer, a father's personality destroyed by Alzheimer's disease, and scores of second- and third-cousins murdered in the Holocaust. Signs of a benevolent designer are pretty well hidden.

The prevalence of evil and misery has always bothered those who believe in a benevolent and omnipotent God. Sometimes God is excused by pointing to the need for free will. Milton gives God this argument in *Paradise Lost:*

> I formed them free, and free they must remain
> Till they enthral themselves: I else must change
> Their nature, and revoke the high decree
> Unchangeable, eternal, which ordained
> Their freedom; they themselves ordained their fall.

It seems a bit unfair to my relatives to be murdered in order to provide an opportunity for free will for Germans, but even putting that aside, how does free will account for cancer? Is it an opportunity of free will for tumors?

It is not necessary for me to argue here that the evil in the world proves that the universe is not designed, but only that there are no signs of benevolence that might have shown the hand of a designer. But, in fact, the perception that God cannot be benevolent is very old. Plays by Aeschylus and Euripides make a quite explicit statement that the gods are selfish and cruel, though they expect better behavior from humans. God in the Old Testament demands of us that we be willing to sacrifice our children's lives at His orders, and the God of traditional Christianity and Islam damns us for eternity if we do not worship Him in the right manner. Is this a nice way to behave? I know, I know, we are not supposed to judge God according to human standards, but you see the problem here: if we are not yet convinced of His existence, and are looking for signs of his benevolence, then what other standards *can* we use?

In an e-mail message from the American Association for the Advancement of Science, I learned that the aim of this conference is to have a constructive dialogue between science and religion. I am all in favor of a dialogue between science and religion, but not a constructive dialogue. Religion has done some good in the world, but on balance its effects on our lives have been awful. This is much too big a question to be argued here, so I'll just have to state my own opinion: with or without religion, good people would tend to behave well and bad people would do evil things, but the peculiar contribution of religion throughout history has been to allow good people do evil things. One of the great achievements of science has been, not to make it impossible for intelligent people to be religious—the example of John Polkinghorne shows that this is not impossible—but at least to make it possible for them not to be religious. We should not retreat from this accomplishment.

NOTE

1. This excited state of the carbon nucleus is 7.65 million electron volts (MeV) above the energy of the carbon nucleus in its normal state, the state of lowest energy. Calculations [Livio, M., D. Hollowell, A. Weiss, and J. W. Truran, *Nature,* Volume 340 (1989), p. 281] show that it would be necessary to increase the energy of the excited state by considerably more than 0.06 MeV in order significantly to reduce the amount of carbon and heavier elements produced in stars. Since 0.06 MeV is less than 1 percent of 7.65 MeV, this may look like some sort of fine-tuning is at work. But, as Livio *et al.* point out, if we think of this excited state of the carbon nucleus as an unstable state of a beryllium nucleus and a helium nucleus, then we should compare 0.06 MeV with the energy of the excited state relative to the total energy of a beryllium nucleus and a helium nucleus, which is only 0.281 MeV, rather than with the energy of the normal state. Since 0.06 MeV is 21 percent of 0.281 MeV, this is not a very impressive example of fine-tuning. Recent calculations [Hong, S.H., and S. J. Lee, nucl-th/9903001, to be published] show that in fact there would be expected to be an unstable state of a beryllium nucleus and a helium nucleus at about this energy, with no fine-tuning needed. For a contrary view, see H. Oberhummer, P. Richler, and A. Csoto, preprint nucl-th/9810057.

Understanding the Universe

JOHN POLKINGHORNE

Queens' College, Cambridge University, Cambridge, U.K.

ABSTRACT: Physics constrains metaphysics but does not determine it. The central metaphysical question is that of scope; a true theory of everything must take personal experience as seriously as the impersonal. Science raises questions of intelligibility and anthropic fine-tuning that go beyond its own self-limited power to answer. Theism provides a coherent response, as well as furnishing a foundation for the human encounter with value. Evolutionary insight affords some insight in relation to problems of physical evil, but eventual cosmic futility can only find an answer in the faithfulness of a Creator whose purposes are not frustrated by death.

KEYWORDS: anthropic principle; contingency; death; evil; evolution; intelligibility; metaphysics; metaquestions; self-consciousness; value

Is the universe designed? If it is, we shall not learn about it by looking for items trademarked "The Heavenly Construction Company," any more than, if it is not designed, we shall learn that by finding objects stamped "Blind Chance Rules." Science by itself will not tell us the answer to this question, either in the positive or in the negative. The reason for this is simple. The question of design is a metaphysical question, a question that goes "beyond" (*meta*) physics, and metaphysical questions must receive metaphysical answers, given for metaphysical reasons. Physics—or science generally—constrains metaphysics, but it does not determine it, just as the foundations of a house constrain what can be built on them, but they do not determine the actual form of the edifice. You can get the idea by thinking about another metaphysical issue: the nature of causality. Take non-relativistic quantum mechanics. Is it an indeterministic theory or not? Niels Bohr says Yes; David Bohm says No. Their interpretations are completely contrasting, but their radically different theories both lead to the same physical consequences. There is no empirical scientific test that can settle the matter between them. Whether they are aware of it or not, the 99.9% of physicists (among whom I number myself) who take the conventional view of an indeterministic quan-

Address for correspondence: Rev. Dr. John Polkinghorne, 74 Hurst Park Avenue, Cambridge CB4 2AF, U.K.Voice: 44-1223-360743; fax: 44-1223-35522.

tum mechanics, do so for metascientific reasons. Chief among these is the belief that Bohm's very clever ideas are, in fact, too clever by half, for they have a metaphysically unattractive air of contrivance about them.

That judgment illustrates the sort of considerations that are relevant to assessing metaphysical proposals. The criteria include a variety of desirable properties, many of which we also associate with a successful fundamental physical theory: economy, elegance, naturalness, and adequacy to experience. As we know in the case of science itself, these characteristics are not always easy to define in a watertight way, but it is often not difficult to find agreement in a truth-seeking community about when they have been fulfilled. There is greater possibility for disagreement in metaphysics than in science, and one reason for that is that a very significant metaphysical criterion is that of *scope*. How wide-embracing should be the understanding that the theory will afford us? What range of experience should it take into account? I shall be arguing today for a generously conceived metaphysic that takes personal experience as seriously as impersonal, and I shall be rejecting what I see as a narrow scientism. Some of the disagreements among us will certainly stem from this question of the breadth of phenomena we are trying to understand.

Although we are rightly impressed by the many things that science can account for satisfactorily, we should also recognize that this great success has been purchased by a degree of modesty of ambition. Science limits itself to considering only certain kinds of experience. Broadly speaking, its concern is with the impersonal dimension of reality. Galileo had the brilliant idea, followed so strictly by successive generations of physicists, of confining attention to the primary quantities of matter and motion, and to set aside what he called the secondary characteristics of human perception, such as color. This neglect of what the philosophers call *qualia* (that is to say, feelings such as seeing red or judging someone to be trustworthy) was an immensely successful technique of investigation. It would, however, be a very bad mistake to equate Galileo's methodological strategy with an act of ontological judgment —that is to say, a verdict on the nature of reality. Such a confusion would, in my view, result in a woefully inadequate metaphysics. Physics may tell us that music is vibrations in the air, and neurophysiology may describe the consequent patterns of neuron excitation that result from those airwaves impinging on the eardrum, but to suppose that this discourse is adequate to the phenomenon of music would be totally misleading. The mystery and reality of music slips through the wide meshes of the scientific net.

Metaphysics cannot tolerate such an impoverished scientism, for its grand aim is truly to be a Theory of Everything, obtained, not by Procrustean truncation of experience until it has been reduced to a scale so limited that it can be condensed into a formula that can be written on a T-shirt, but by taking absolutely seriously the many-layered richness of the reality in which we live. It will not grant an automatic priority of the objective over the subjective, of the impersonal over the personal, of the repeatable over the unique.

Interestingly enough, some of these wider issues that a true metaphysics must consider relate to questions that arise from our experience as scientists, but that go beyond the merely scientific. They center round two great metaquestions: Why is science possible at all? Why is the universe so special?

Those of us privileged to be scientists are so excited by the quest to understand the workings of the physical world that we seldom stop to ask ourselves why we are so fortunate. Human powers of rational comprehension vastly exceed anything that could be simply an evolutionary necessity for survival, or plausibly construed as some sort of collateral spin-off from such a necessity. How could that kind of argument possibly relate to our amazing ability to understand the strange and counterintuitive quantum world of subatomic physics, or to comprehend the cosmic structure of curved space? The point is reinforced by considering what the Nobel Prize-winning physicist, Eugene Wigner, called "the unreasonable effectiveness of mathematics." Wigner's brother-in-law, Paul Dirac, one of the founding fathers of quantum mechanics, said that his fundamental belief was that the laws of physics are expressed in "beautiful equations." The relentless and highly successful pursuit of a beautiful equation was how Dirac discovered the relativistic equation of the electron, and consequently antimatter. Einstein discovered the equations of general relativity in a similar fashion.

Mathematics is abstract human thinking. When this most austere of subjects proves to be the key to unlock the secrets of the physical universe, something very unexpected is happening. The unreasonable effectiveness of mathematics is a phenomenon that the mathematicians, in their modest way of speaking, would call "non-trivial." Non-trivial is a mathematical word meaning "highly significant." This raises the metaquestion of why this is the case.

In dealing with a question of that kind, I want first of all to say two things by way of preliminaries. One is that it is a question that should be pressed. My instinct as a scientist is to seek understanding through and through and it seems to me that it would be intolerably intellectually lazy just to shrug one's shoulders and say "That's just the way it is—and a bit of good luck for you people who are good at mathematics." The second thing I want to say is that deep metaphysical questions of this kind are too profound to have simple knockdown answers to them. When we enter the realm of metaphysical enquiry, we are in a domain where no one has access to absolute rational certainty. For that reason, I could not use so curt and categorical a title for my talk as that chosen by Steve Weinberg for his. This character of metaphysical argument does not mean that we shall not have reasons for the answers that we propose, for stating them will involve invoking the metaphysically desirable properties I have already discussed, particularly that of scope. However, none of us can pretend that our answers are logically coercive in a $2 + 2 = 4$ way. I am going to propose theistic responses to the questions we are concerned with. I think I have good reasons for my beliefs, but I do not for a moment suppose that my atheistic friends are simply stupid not to see it my way. I do believe, however, that religious belief can explain more than unbelief can do.

Back then to the metaquestion of why science is possible at all in the deep way that it is. I have described a physical world whose rational transparency makes theoretical physics possible and whose rational beauty guides and rewards those who inquire into its structure. In a phrase, it is a world shot through "with signs of mind." I believe that it is an attractive, coherent, and intellectually satisfying explanation of this fact that there is indeed a divine Mind behind the scientifically discerned rational order of the universe. In fact, I believe that science is possible because the physical world is a creation and we are, to use an ancient and powerful phrase, creatures "made in the image" of the Creator (Genesis 1:26). I regard this insight as the primary ground for believing that the universe is designed. I make no apology for speaking in theistic terms, for if the universe is designed, who could be its designer other than a Creator-God?

A second reason for that belief can be found in the answer to my second metaquestion: Why is the universe so special? Of course, here I am referring to the findings of the Anthropic Principle. Other contributors to this volume have already outlined the many considerations that lead us to conclude that the laws of nature as we observe them in our universe are precisely those that permit the development of carbon-based life, in the sense that even very small changes in intrinsic force strengths would have broken links in the long, delicate and beautiful chain of consequences linking the early universe to the existence of life today here on Earth. I agree with John Leslie's analysis, presented in his book *Universes*, that suggests, firstly that it would be irrational just to shrug this off as a happy accident, and secondly that there are two broad categories of possible explanation: either many universes with a vast variety of different natural laws instantiated in them, of which ours is the one that by chance has allowed us to appear within its history; or a single universe that is the way it is because it is not "any old world," but a creation that has been endowed by its Creator with just the circumstances that will allow it to have a fruitful history. I simply want to make two comments on this analysis.

The first is to emphasize that *both* proposals are metaphysical in character. That is clear enough in the case of creation, but it is also true of a many-universes proposal that is wide enough in scope actually to serve as an explanation. Of course, an inflation-expanded structure, containing many domains with differing consequences of spontaneous symmetry breaking, could give vast regions in which effective force constants differed and in one of which they might take the anthropically desirable values, but that would still require that the overall Grand Unified Theory was constrained in its character in order to permit this to happen. Something specific and requiring explanation would remain. I regard quantum cosmology and baby universes as being too precarious and speculative in character at present to rely on. In any case, the underlying theory would again have to take an appropriate form. One might comment that quantum theory, general relativity and suitable matter fields do not "come for free," so to speak. It is important to recognize that that anthropic fruitfulness requires the right kinds of laws as well as the right values for the parameters appearing in those laws. It is very difficult to see what could ensure this right character of fundamental physical law, realizing the necessary combination of flexibility and stability required to make fruitful evolution possible, other than a strictly metaphysical proposal.

The second point is to agree with Leslie that, in relation to the Anthropic Principle, it is a metaphysically even-handed choice between many universes and creation. It seems to be six of one and half a dozen of the other, neatly illustrating the point I made that these kinds of argument are not knockdown in their character. However, I believe that, while the many-universes hypothesis seems to have only one explanatory piece of work that it can do, there are other kinds of explanation that the thesis of theism can afford, such as granting an understanding of the intelligibility of the universe, and also providing the ground for the widely attested phenomena of religious experience. (Of course, the history of religions is a tangled tale. Weinberg is right to draw our attention to the sad fact that religion can cause good people to do bad things, but we should also recognize that religious conversion has often led to bad people becoming able to do good things.) My conclusion is to prefer the explanation of anthropic fruitfulness in terms of a Creator, which strikes me as being more economic and forming part of a cumulative case for theism.

There are a number of other considerations that can be held to be relevant to the discussion of whether the universe is designed. One concerns what seems to me to be the most astonishing (and most significant) event known to us that has happened since the Big Bang. I refer, of course, to the dawning of self-consciousness here on Earth (and perhaps elsewhere). In us, the universe has become aware of itself. You remember that Blaise Pascal said that human beings are thinking *reeds*, so insubstantial on the grand scale of the cosmos, but we are *thinking* reeds, and so greater than all the stars, for we know them and ourselves and they know nothing. With, for example, Paul Davies in his book *The Mind of God*, I cannot regard this dawning of consciousness as being just a fortunate accident in the course of an essentially meaningless cosmic history. We know from the Anthropic Principle that the potentiality for this happening was present in the ground rules of the universe from the beginning. I see this striking emergence as a signal of meaningfulness, amounting to an intimation of intrinsic design. Something is going on in what is happening in cosmic process. Of course, I am talking in terms of a propensity to great fruitfulness. There is much that is contingent in the way that this has been realized. I am not claiming that it was laid down from all eternity that *Homo sapiens* should have five fingers! I will return later to the matter of contingency.

Meanwhile, let us recognize that our human self-consciousness enables us to look through many different windows onto reality. We are not confined to the impersonal scientific perspective of Galileo's primary quantities, but we have access to those personal qualities that Galileo set aside. I take with the greatest seriousness our human encounters with beauty and with moral imperative. I see them as affording us windows onto the reality within which we live and not, as I think Steven Weinberg does, as being internally constructed human attitudes through which we defy an intrinsically meaningless and hostile universe.

I have already drawn attention to science's inadequacy in relation to music. Its reductionist strategy can never do justice to a work of art, for a Leonardo painting is much more than a collection of specks of paint of known chemical composition. It would be a disastrous mistake to throw away the insights of aesthetics, for they must find their proper place in a true Theory of Everything.

I believe that the same is true of our ethical intuitions. I know something about what the anthropologists tell us about the cultural tricks of perspective that different societies impose upon their discernment of moral issues. Of course, we must pay attention to these matters but, when all is said and done, I personally cannot believe that my conviction that torturing children is wrong is just a convention of my society. It is a fact about reality, the way things are. We have access to moral knowledge, which is knowledge of a totally different kind from scientific knowledge, for ethical insights are more than disguised genetic survival strategies. If that were not so, what would be the grounds on which Richard Dawkins could, on the last page of *The Selfish Gene*, urge us to rebel against their influence?

A true cosmology—an adequate account of reality—will have to take these issues into account. The physical world that science describes is the carrier of beauty and the arena of moral decision, value-laden aspects of reality that science does not describe. One of the attractions of theism is that it offers a way of tying these different levels of experience together, which otherwise might seem so unconnected with each other. Just as we can understand the rational order of the world that science discovers, as being a reflection of the mind of its creator, so we can understand our aesthetic experience as being a sharing in the Creator's joy in creation, and our moral intuitions as being intimations of God's good and perfect will. And, I would want to add that dimension of human experience that testifies to a meeting with the sacred as being encounter with the divine presence.

But is there not a fatal flaw in all this argument? I refer to the greatest difficulty for theism, namely the problem of the evil and suffering so manifestly present in the world. I think that this problem holds more people back from religious belief than any other, and those of us who are believers can never be unaware of it, or untroubled by the challenge it presents. Could one really claim that so apparently dysfunctional a universe was one that exhibited design? Does not our very sense of value, to which I have appealed, make us rebel against a strange and bitter "creation"?

The questions proliferate. Is not evolutionary history a tale of struggle and competition, of death as the necessary cost of life, of the blind alleys of extinction that have dealt death blows to 99.9% of the species that have ever lived? Is not the role of chance, in the evolutionary interplay of chance and necessity, the conclusive sign that the universe's history is, as Macbeth said, "a tale told by an idiot, full of sound and fury, signifying nothing"?

I do not wish to deny that these are serious questions. But, in an unexpected way, science's insights have been moderately helpful to theology in its wrestling with the problem. The key concept, in fact, is evolution itself. It is historically ignorant to suppose, as the modern myth does, that Darwin was opposed by solid ranks of obscurantist clergymen when *The Origin of Species* was published in 1859. In both Britain and the United States there were Christians, like Charles Kingsley and Asa Gray, who welcomed his insights. Kingsley coined the phrase that, in a nutshell, sums up a theological understanding of evolution. He said that God had done something cleverer than producing a ready-made creation, for God had created a world "that could make itself." If there is a God who is the God of love, then creation could never be just the divine puppet theatre. The gift of love is always the gift of a due independence, as wise parents know in relation to their children. The God of love must be one who allows creatures to be themselves, and to make themselves by exploring the endowment of potentiality given to creation. "Chance" simply means historical contingency—this happens rather than that. It is not automatically to be given the tendentious adjective "blind," as if it were an unambiguous sign of meaninglessness. Rather, it may be seen as signifying the shuffling exploration and realization of fertile possibilities, by which creation makes itself. This due independence of process is a good gift, but it has a necessary cost attached to it. Raggednesses and blind alleys, a well as fruitful outcomes, are inescapable accompaniments of this evolving self-realization. Biology even helps theology a little with the deep question of theodicy, the problem of the evil and suffering of the world. Exactly the same biochemical processes that enable some cells to mutate and produce new forms of life—in other words, the very engine that has driven the stupendous four billion year history of life on Earth —these same processes will inevitably allow other cells to mutate and become malignant. In a non-magic world, it could not be different, and the world is not magic because its Creator is not a capricious Magician. I do not pretend for a moment that this insight removes all the perplexities posed by the sufferings of creation. Yet it affords some mild help, in that it suggests that the existence of cancer is not gratuitous, as if it were due to the Creator's callousness or incompetence. We all tend to think that if we had been in charge of creation we would have made a better job of it. We would have kept the nice things (flowers and sunsets) and got rid of the nasty (disease and disaster). The more science helps us to understand the process of the universe, the more, it seems to me, to cohere into a single "package deal." The light and the dark are two sides of the same coin.

One final threat to any claim that the universe is designed, as the expression of a divine purpose, needs to be considered. In a sense, it is the ultimate difficulty, for it concerns how things will end. As cosmologists peer into the universe's future, they tell us that the only honest answer is "badly." There are the two possible alternative scenarios of collapse or decay, but either way carbon-based life will prove to have been a transient cosmic episode, and eventually all is condemned to futility. We know about built-in obsolescence in car design, but surely Creatorly design should be able to do better than that? Can one wonder that Steven Weinberg notoriously said that the more he understood the universe, the more it seemed pointless to him? We have to consider what response theology can make to this.

I do not think that the problem posed by cosmic death on a time-scale of tens of billions of years is altogether different from the problem posed by the even more certain knowledge of our own deaths on a time-scale of tens of years. In each case, what seems to be put in question is the genuineness of the Creator's concern for creatures. Do they—do we—matter to God only transiently? Those of us who believe in the steadfast faithfulness of God must reply that creatures matter to God for ever. What cosmic and human death remind us of is that an evolutionary optimism, based on total fulfillment in terms of unfolding present process, is an illusion. Death is a real end, but it is not the ultimate end, because only God is ultimate. As a Christian speaking in this season of Easter, I affirm my belief in a destiny beyond death. Such a destiny could never arise naturally, but it can only be the outcome of a great divine redeeming act. To trust that that is God's ultimate purpose is an exciting and mysterious belief, which I cannot find time to explain and defend on this occasion, though I have sought to do so in some of my writings (*The Faith of a Physicist*).

I agree with Steve Weinberg that the issue of apparent cosmic futility is one of paramount importance. In reality, it is the question of whether the universe makes *total* sense or not. We disagree about the answer to that. Such disagreement is possible because, as I have repeatedly acknowledged in this paper, none of us has access to logical certainty in metaphysical matters. What I have sought to show is that religious believers who see a divine Mind and Purpose behind the universe are not shutting their eyes and irrationally believing impossible things. We have reasons for our beliefs. They have come to us through that search for motivated understanding that is so congenial to the scientist. That search, however, needs to be pursued in the widest possible context, including the insights of science but going beyond them, in the direction of the deepest and most comprehensive encounter with reality.

An Exchange between Steven Weinberg and John Polkinghorne

[*After their formal presentations, Steven Weinberg and John Polkinghorne engaged in a discussion of the issues of science and religion and design as well as responding to questions from the audience. Owen Gingerich moderated the discussion. The following is a transcript of their comments in which an effort has been made to retain their voices.*—EDITOR]

WEINBERG: First, let me respond to the parts of John's eloquent talk that dealt with a few scientific issues. I don't agree that the two metaphysical approaches to quantum mechanics are the probabilistic theory of Bohr and the deterministic theory of Bohm. With John, I don't think much of Bohm's reworking of quantum mechanics. What I would take as an alternative to the probabilistic view of quantum mechanics is a thoroughly deterministic view. After all, quantum mechanics in its basic equations is completely deterministic. The Schrödinger Equation tells us that if we know the wave function in any instant, we know precisely what it is at any future instant. There is no chaos, as in Newtonian mechanics, because the equations are perfectly linear. The problem always has been how to represent the observer in the deterministic evolution of the wave function. This is a problem that is increasingly being solved, although not yet completely. But that's the opposition I would make: between the modern deterministic view, which sees the observer as part of the reality described by the wave function, and Bohr's view, in which the observer was something separate.

POLKINGHORNE: That is another way of setting it up, but you would agree that it is a metaphysical argument about…

WEINBERG: Yes, that's right and…it's just I don't think you should take David Bohm, as the representative of…

POLKINGHORNE: …but, of course, you are right.

WEINBERG: Another scientific question is about fine-tuning. As I said in my talk, I am not terribly impressed by the examples of fine-tuning of constants of nature that have been presented. To be a little bit more precise: the energy levels of carbon are the most notorious example of fine-tuning that is cited: there is an energy level that is 7.65 MeV above the ground state of carbon. If it was 0.06 of an MeV higher, then carbon production would be greatly diminished and there would be much less chance of life forming. That looks like a 1% fine-tuning of the constants of nature. If the energy level were lower, then there would be even more carbon produced. But it is striking that it could not be more than a percent higher. However, as has been realized subsequently after this "fine-tuning" was pointed out, you should really measure the energy level not above the ground state of carbon, but above the state of the nucleus beryllium 8 (^{8}Be) plus a helium nucleus. And it is only 0.28 MeV above that. In other words, the fine-tuning is not 1% but it's something like 25%. So, it's not very impressive fine-tuning at all. I'm not saying that none of these examples of fine-tuning will survive or that we won't discover others. I'm also not saying that the many-universe picture has been established. These are open questions. Are the constants of nature remarkably well adjusted to allow for the presence of life? We don't really know. And can they be explained by having many sub-universes? We don't really know that either. And indeed at any moment we may get evidence of a supernatural supervisor of the universe. I mean suddenly in this auditorium a flaming sword may come and strike me for my impiety, and then we will know the answer.

POLKINGHORNE: Actually, we won't. But that's by the way.

WEINBERG: I don't agree that this is a metaphysical question. If there are these many terms in the wave function as Coleman[a] describes or many separate big bangs, as Linde[b] would have it, this is something that will appear in our scientific theories. It hasn't yet. We don't know this for sure. It has appeared as a possibility but it is not something that's going to remain a matter of metaphysical choice. Either our theories will show us that this is the case or they will not. And it will not be a matter of personal taste. At this moment the question is open.

To answer the question John raised—another question, one that he has raised before in his writings and that I thought I had answered in my talk. He asked "Why is it that we are so fortunate to be able to do quantum mechanics and do mathematics? Is it really necessary from evolution that we should be able to do this?" And I would answer; no, it is not, that we are able to have such abstract thoughts and to be able to have the leisure to sit around talking about it. It may be, as I explained, that in the great majority of planets where life arises and evolves, only that measure of intelligence evolves that is strictly necessary for breeding and eating. However, those animals are not discussing the issue and the fact that we are discussing the issue creates a bias and naturally the people who are discussing the issue have intelligence and the leisure to discuss it.

Now, the points I have covered so far are, I think, minor points, but John also makes the big point that science is not everything. There is metaphysics

[a]Coleman, Sidney, *Nuclear Physics* B, Vol. 310, p. 643 (1988).

[b]Linde, Andre, "The self-reproducing inflationary universe," in *The Magnificent Cosmos,* a *Scientific American* special issue (March 1998).

in addition to physics—I agree, but I do look differently at the examples he gave.

It is certainly true that scientists are helped in their work by non-rational processes including an aesthetic sense of beauty, which is very well developed among mathematicians. Mathematics is the science of order and they see order in an abstract, inner-directed way, which often turns out, quite spookily, to be relevant to the real world. This poses a problem. I would try to explain it as the sort of learning that goes on whenever one has long experience—one learns things that one cannot express in words. Our long experience is the many centuries of experience of scientific work in which we have learned what sort of thing is beautiful; we've learned what sort of mathematics is possible. Not all forms of order are possible. And we've learned what sort of ideas might be relevant to the real world. I think that's what we mean by beauty and, in fact, our sense of beauty has changed.

Today, beauty based on symmetry, on invariance under some group of transformations of change of point of view, is regarded as a highly beautiful part of a theory—something you would be proud to base a theory on. At the beginning of this century that wasn't true; in fact, Lorenz criticized Einstein for basing his special theory of relativity on a principle of symmetry, which is not something that he thought should be taken as a starting point of a physical theory. So, in this sense, I think, we are in the grip of a learning machine which is gradually beating into us a sense of beauty. which is a very important part of the work of science.

Finally, though, I must admit that science isn't everything. It certainly isn't. There are things that are outside the scope of science and that are still terribly important to human beings. There is metaphysics of a sort that goes beyond the kind of learning processes I mentioned. There is also aesthetics and morality. It seems to me that there's an unbridgeable gulf between statements with the word "is" and statements with the word "ought." There is no way science can ever tell you how you ought to behave. It may tell you, if you have some fundamental moral principles, how you can satisfy them, how you can bring about what you take as a desired goal. But it can never tell you what your goals ought to be.

There is a moral order. It is wrong to torture children. And the reason it is wrong to torture children is because I say so, and John says so, and probably most of us say so—but it is not a moral order out there. It is something we impose on others, and bully for us. In this respect I think religion is no better. Suppose you knew that the universe had been created by a designer who watched our progress and intervened and behaved very much like the god of the Old or the New Testament. That has no moral implications. That god may set down moral principles which are wrong. Wrong from what point of view? Our own point of view—what other point of view could there be? And indeed, that is what I would say is the case. I don't believe in torturing children but god apparently, according to many religious faiths, believes that children who were not baptized or who, when they got older did not accept god or come to him through Christ or through the Koran are subject to eternal damnation. I think that even those who believe in a god still have the responsibility to answer the question, "What is right?" And they have to answer it for themselves and, if they accept the morality provided by god, that is their choice, so that they, like the atheist scientist, have to make a free choice of moral behavior which is not dictated by a theory of the universe, religious or scientific.

POLKINGHORNE: I want to make a few comments on your remarks. This question of the nature of morality is a very important issue. I don't think we just make it up. I don't think that Steve and I make up one sort of morality and Hitler and Stalin make up another sort of morality. On what basis does Steve say, "Bully for us," and deny them. There has to be something that transcends human construction, otherwise these senses of value would not function the way they do.

Theology does make progress, very slowly. That's one of the reasons why I swapped out of physics and into theology. It is a more stately subject. But in the 19th century Christians first began to question and have continued to question and dispose of the question really, most of them anyway, that the god of love would condemn people to infinite torment for finite transgressions. That doesn't mean that Hell is gone, but that Hell is no longer thought of as a place of torment into which an angry God has cast people. Rather it is thought of as colored gray rather than red, as a place of boredom to which people have condemned themselves by their own choice of excluding the divine life. That's just a little point on Hell and I think that is actual progress in theology to have reached that.

But, Steve, I think the fundamental difference between you and me is this: We both want to take human persons seriously, but we take them seriously in radically different ways. You see human persons as constructing a world of meaning that is a sort of oasis in a vast desert of a hostile and meaningless universe. I see us not constructing meaning—although, of course, there is a constructive element in it—but as *discovering* meaning also, which for me is a clue to the nature of reality more generally. It is not an internal good of the human community, of sections of the human community; it's a perception of the real. And therefore I see that as a clue that we are not defiant inhabitants of an island of meaning in an ocean of meaninglessness, but that in fact the world has a meaning that extends beyond us. That is the basic difference between us, I think.

WEINBERG: Well, I don't disagree with that and I don't disagree with your characterization. If, in fact, there is out there, built into the structure of the universe, an objective meaning, an objective moral order, that would be really quite wonderful. And perhaps part of my passion about this arises from regret that it isn't true. But, if it isn't true, then surely it's better that we not kid ourselves into thinking that it is. It's better that we at least salvage what we can from the satisfaction of creating some meaning around us.

POLKINGHORNE: I would agree with you that, if it isn't true, then it is better that we know it. The central religious question is the question of truth. Religion can do all sorts of things for you, console you in life and at the approach of death, but it can't really do any of those things unless it's actually true— not in some knock-down sense— because, of course, the divine reality will always exceed our finite thoughts. But if there isn't a benevolent divine will or purpose behind the world, it is better for us to know that than to live in a sort of happy illusion. I don't see religion as a way of keeping our spirits up and keeping us through life in a happy illusion in that way.

WEINBERG: Yes, in this respect I think John and I represent what must be in today's world a minority—we are probably the wrong people to be debating each other. I often find that many people who claim to be religious actually have no beliefs to speak of. They have, to quote Susan Sontag, "piety without content." I spoke to some Buddhists a while ago and asked, "So, do you really believe in reincarnation?" And they said no, they didn't think they believed in reincarnation. And I couldn't imagine what it meant to be a Buddhist and not to believe in reincarnation: for them Buddhism was just a flag, rather than a belief. So, in this respect John and I are not the right people to argue.

GINGERICH: Steve Weinberg, what meaning, if any, do you see in the fact that many scientists see their own work, solutions to problems, as coming from outside themselves and even perhaps having a religious dimension?

WEINBERG: I don't think that's true of most scientists. In my experience, just talking to my fellow physicists at lunch, I find that most of them have not only no religious faith, but also no interest in the issue. I am a little unusual in being interested in the question. I think again there is a selection effect. Scientists who do talk about supernatural influences on their work are the ones who are likely to get published and win prizes endowed by Mr. Templeton. But, I think the public is getting a rather misleading view. I think most scientists are not atheists because they don't think about it enough to be atheists. There are scientists who are quite religious, my friend here and others, but I think they are thin on the ground.

POLKINGHORNE: Can I just briefly comment on that? This is anecdote swapping. My impression is somewhat different. I certainly agree that the majority of scientists are not religious believers in some traditional sense. The majority, in my view, and I'm thinking of my friends, are people who can neither take religion or throw it away. They are slightly wistful in relation to religion. They'd like to think there is a deeper meaning and purpose behind things. But they're wary of religion because they think religion involves accepting things on authority. I want to always say that religious belief isn't shutting your eyes, gritting your teeth, believing six impossible things before breakfast because the Bible tells you that's what you have to do. It is a search for motivated belief. The search is difficult, and different people will reach different conclusions about it. But you don't have to commit intellectual suicide to be a religious believer; otherwise I wouldn't be one.

GINGERICH: The questions which have come in are quite interesting but I think many of them are what you would consider simply debating points and not genuine questions. But here is one for both of you. What is the point of continuing to live in a universe that has no ultimate purpose? That's for you, Steve.

WEINBERG: Well, if you don't see the point then, too bad for you. I feel there is a point. How can I say it? There's nothing in science that says we should look at life as not worth living any more than there is something that tells what there is about life that is worth living. It is left as an open question for us to decide on any grounds we like. I enjoy life and there are things I value very much about being alive and that's the point it has. I remember in the preface to one of his plays, I think it was *Heartbreak House*, George Bernard Shaw said, "Darwin has knocked centuries of dusty theology out of the room and now we don't have any of that anymore but at the same time he has knocked out morality. And now because of Darwin's work there is no basis of any moral principle." I disagree. I don't think Shaw was right about that. I think Darwin—and science in general—perhaps took away the idea that there was a supernatural plan which imposes a moral order, but it did not say that we must behave immorally. We are left to make moral choices or not and we are free to make them. And, in fact, not only to make moral choices for ourselves but for others just as we would condemn someone else who tortured children. We're free to find a point and to make moral choices. We don't get them from an objective supernatural world order.

POLKINGHORNE: There was a German atheist philosopher, Max Horkheimer, who said there was a deep longing in the human heart that the murderer should not triumph over his innocent victim. And some of us entertain that hope that the murderer will not ultimately triumph. But those that can't entertain that hope and who live a life of austere nobility in the face of a hostile world, I believe their position to be quite admirable.

GINGERICH: Here is a question for you, John. Could you imagine an ultimate argument that God is not existent? If we cannot refute this mode of explanation, then it is just a matter of belief. But how can we be sure or convinced that it is not just wishful thinking?

POLKINGHORNE: That's a very interesting question. I think that certainty, in the sense of logical proof, is a pretty spare quantity. There isn't too much of it around. Kurt Gödel has told us that even mathematics has its *aporia*, as the theologians say, its uncertainties. And I think it is also the case that there is a sort of complementary relationship between things that are really interesting and things that can be proved. So, I think we shouldn't worry about proof and certainty.

That doesn't mean that anything goes. We should search for motivated beliefs. But I think this is true of both science and religion and everything that lies between them: we will attain beliefs that are motivated, but we will never be certain. One of the best books on the philosophy of science written in the 20th century is Michael Polanyi's *Personal Knowledge*. Polanyi was a distinguished physical chemist before he became a philosopher. That means he's not been very well accepted in the philosophical community, I am afraid to say. He wrote in this book—and he was talking in relation to his scientific beliefs—to explain how I can hold to what I believe to be true knowing that it might be false. And that is the human condition, whether it is concerns science, religion, or anything else. So, I think we shouldn't become fixated upon certainty. I don't think that you can prove God exists; I don't think you can prove that God does not exist.

GINGERICH: Steve, you stated that there is a mysterious realm that science will never explain and that a similar statement is true about religion. What do you see as the nature of this mysterious realm? How is it different from your conception of a religious realm?

WEINBERG: The realm I refer to is something that may be discovered in the next century. I am not certain about this, but I do believe that sometime in the next century or so we're going to find that all our physical theories converge to a fundamental theory, maybe something like the string theories that people are talking about today, or maybe something deeper, from which, in principle, all other scientific generalizations that don't simply rely on historical accidents can be inferred. The mystery will then be: why is that true? The theory will be something very specific, crystallized into a clear scientific statement. We will then wonder why it is true. As I said there may be a chance of answering the question why it's not slightly different, but we probably will not ever know why the truth is not totally different.

There is the possibility that we may be able to show that there is no other logically consistent theory that would allow a rich enough universe for people to be raising the issue. That doesn't completely satisfy me, but it may give some satisfaction. For instance, when you look for alternatives to quantum mechanics, you think of Newtonian mechanics, which don't allow for atoms, and I can hardly imagine how life could evolve in a purely Newtonian world.

But, on the other hand, the religious mystery is a mystery: we will never know whether any of it is true. Unless the flaming sword descends, and unless miracles start happening again in a reproducible way that they haven't… there'll never be any way of being certain about religion, and the truth, as the religious thinker finds it, will always be flexible, it will always be something that can vary indefinitely.

POLKINGHORNE: May I just say that, God forbid, if a flaming sword were to come and decapitate Steve before our very eyes that would pose a very big theological problem. Because that would be the capricious act of a magical, vengeful god and that's not the God of my belief. You see the problem with miracles…

WEINBERG: … it is the god, however, of your religious tradition.

POLKINGHORNE: I wouldn't say that the religious tradition is unsullied, but it's certainly not the sole strand within that belief. The problem with miracles is the problem of divine consistency. God is not capricious, but God is not condemned equally to dreary uniformity.

WEINBERG: Well, it would pose not only a theological problem but a janitorial problem as well…

GINGERICH: Steve, here's a final question for you. You completely reject any notion of a divine designer, but on what basis beyond faith can you justify the idea of multiple universes being more valid?

WEINBERG: I thought I had answered that, but I would be happy to say it again. I don't maintain that that idea is true—it's a possibility that has emerged and it remains a possibility. When I become convinced of its truth, it will be because the equations of physics that unify the various forces—the equations of quantum mechanics, relativity, all that—have that as a consequence. It won't be an act of faith. It will be a deduction from laws which we, unfortunately, at present don't know. Now you may say that it is an act of faith because we will not be able to observe these other big bangs, or these other terms in the wave function. But that's the dilemma that science has been in for a long time. We don't really observe quarks and we never will see the track of a quark. And yet we believe in quarks because the theories that have quarks in them work. And in the same way, if we come to that—and we have not yet come to that—we will believe in these other big bangs or these other terms in the wave function because the theories in which they appear work.

GINGERICH: Ladies and gentlemen, let me just remind you that this is the very room in which in April of 1920 the very famous Shapley/Curtis debate on the scale of the universe took place; a debate that has gone down in astronomical lore ever sense. You have been priveliged to have been present at this debate today, which may also assume mythic proportions.

Is the Universe Designed? Yes and No

DAVID RAY GRIFFIN

Center for Process Studies, Claremont School of Theology, Claremont, California 91711, USA

ABSTRACT: Addressing the title's question from the perspective of Whitehead's process theology, fourteen basic notions of which are explained, I argue that the universe is not designed in six common senses of that notion: It was not created out of nothing, all at once, through punctuated creationism, from a blueprint, solely for humans, or even with humans specifically in mind. But it is designed in two looser senses of the term: It reflects a divine aim at richness of experience, and it involved a divine establishment of this cosmic epoch's fundamental contingent principles—an idea that is consistent with process theism's view of divine power as purely persuasive.

KEYWORDS: cosmology; creation out of chaos; *creatio ex nihilo*; design; mind–body problem; omnipotence; panexperimentalism; problem of evil; process theology; science and religion; time; Whitehead, A.N.

I have been asked to address the question "Is the universe designed?" from the perspective of process theology, which is based on the philosophy of Alfred North Whitehead (1861–1947). The name "process theology" is derived from the title of Whitehead's major work, *Process and Reality*,[1] which he wrote in the late 1920s after coming to Harvard to teach philosophy. Whitehead had been educated at Cambridge University in England, where he wrote a dissertation in 1884 on Maxwell's *Treatise on Electricity and Magnetism* and then taught mathematics, including mathematical physics, until 1910. In some circles, Whitehead is best known for the major work of this period, *Principia Mathematica,* which he co-authored with his former pupil, Bertrand Russell. In the next period of his life, spent in London, Whitehead increasingly devoted himself to the philosophy of nature. After coming to Harvard in 1924, he turned to metaphysics, which, as he understood the term, differed from the philosophy of nature by including the human subject within the scope of that which is to be explained. The most important task of meta-

Address for correspondence: Dr. David Ray Griffin, Professor of Philosophy of Religion, Claremont School of Theology, 1325 N. College Avenue, Claremont, CA 91711-3154, Voice: 800-626-7821, ext. 1231.
davraygrif@aol.com

physical cosmology, he came to believe, is to reconcile our scientific intuitions with our religious, ethical, and aesthetic intuitions, which can only be done by developing a world view that is equally satisfactory for the scientific and the religious communities.[2]

Whitehead had been an atheist, or at least an agnostic, during most of his adult life. But shortly after beginning to develop his metaphysical cosmology, he came to the view that, if we are to give a fully rational account of the universe—meaning one that is both coherent and adequate to all the relevant facts, including the various dimensions of human experience—it is necessary to posit a non-local actuality. Although Whitehead used the term "God" for this actuality, the divine reality to which he referred, unlike the deity of traditional theism in the West, did not possess omnipotence as traditionally understood. Whitehead's turn to a form of theism did nothing to lessen his antipathy to the idea of a divine being who, while having the power to prevent evil, refuses to do so.

If faced with our question "Is the universe designed?," Whitehead's answer, like the answer to most Yes or No questions about complex issues, would have been Yes *and* No. This ambivalent answer reflects the fact that the notion of a designed universe has many connotations, not all of which imply all the others. I will deal with eight possible meanings of this notion, suggesting that, from the perspective of Whiteheadian process theology, the answer to six of them is No, but that there are two senses in which we can speak of the universe as designed.

SOME BASIC WHITEHEADIAN NOTIONS

To explain the Whiteheadian position on these issues, I will need to presuppose many of the fundamental ideas in the system. Unfortunately, to explain these ideas would take several hours, if not days. All I can do here is briefly sketch them, hoping that this sketch will be sufficient to make the ensuing discussion at least somewhat intelligible. Another problem for my assignment arises from the fact that Whitehead's position involves a rejection of many of the ideas of modern philosophy, some of which have been widespread in scientific circles, so that Whitehead's alternative notions may strike many of you as implausible, if not outrageous. Unfortunately, to defend these notions would take days, if not weeks. So, although I have argued elsewhere that all of these ideas can be defended as more plausible than their alternatives,[3] I can here only ask you suspend incredulity, granting these basic Whiteheadian notions as premises for the sake of discussion. I will briefly discuss fourteen of these notions.

(1) The most fundamental units of which the universe is composed are momentary spatiotemporal events, rather than enduring things. This notion, which Whitehead shares with Buddhism, fits with the dis-

covery that some of the so-called elementary particles exist on the order of a billionth of a second, so that they would more appropriately be called events.

(2) Each momentary event is an embodiment of creativity, from which the physicist's energy is an abstraction. By enlarging the notion of energy to include all that Whitehead meant by "creativity," we could say that the universe is made up of energetic events. This is true even of so-called empty space, which means that Whitehead's ideas here are consonant with recent thinking about the ("virtual" or "false") vacuum.

(3) One respect in which this more inclusive energy, this creativity, goes beyond energy as usually conceived in physics is that it includes an element of internal determination, or self-determination, so that no energetic event is wholly determined by the forces acting upon it from without. The epistemic indeterminacy of the world at the quantum level reflects an element of ontological self-determinacy.

(4) The energetic events are, of course, not simply embodiments of raw, unformed energy, but of in-formed energy, with the different types of things being different because they contain different forms, different in-formation.

(5) Each event prehends aspects of prior events, and thereby aspects of their informed energy, into itself. The term "prehend" is simply "apprehend" without the prefix, meant to indicate that this response to other things need not be a conscious process. The crucial point here is that each event is internally related to prior events. That is, rather than being a solid piece of stuff, or a Leibnizian monad devoid of windows, each event is internally constituted by its relations to prior events. Here Whitehead's view seems virtually identical with some Buddhist understandings of the "dependent origination" of all things.

(6) Enduring things, such as electrons, protons, and photons, exist because a particular form of energy is repeated by a long series of energetic events, perhaps dozens, hundreds, thousands, millions, or even billions of times per second. That is, although each event is influenced by all prior events to at least some slight degree, an event in an enduring individual is primarily constituted by its prehension and thereby internalization of the form of energy that was embodied by its predecessors in the enduring individual to which it belongs. The proton endures, in other words, because each of its protonic events essentially repeats the form of its predecessors, with this repetition going on, not quite endlessly, but for many billions of years. (I illustrated this point with protons, because they seem to have an especially high degree of tolerance for monotony.)

(7) Low-grade enduring individuals can, in certain combinations, give rise to higher-level enduring individuals, as quarks and gluons give rise to protons and neutrons, and these latter individuals combine with electrons to give rise to atoms and molecules, with still higher forms of enduring individuals perhaps being macromolecules, prokaryotic cells, organelles (which may be captured prokaryotic cells), eukaryotic cells, and the psyches of animals, from gnats to human beings. These higher-level enduring individuals are, by hypothesis, not simply complex arrangements of lower-level individuals. Rather, they involve higher-level energetic events, with their own unity of response to their environments. This emergence of higher-level units is possible because of internal relations. That is, because each event is internally constituted out of the things in its environment, a more complex environment can provide the basis for more complex events and thereby more complex enduring individuals.[4]

(8) The most complex enduring individuals on our planet, evidently, are the psyches of human beings. Although there is no ontological difference between the psyches of humans and those of other animals, as some dualists hold, or between animal psyches and lower-level enduring individuals, as other dualists hold, or even between living and non-living individuals, as vitalists hold, there are enormous differences of degree in terms of capacities for prehension and self-determination. Because our own existence is not entirely different from that of lower-grade enduring individuals, there are some features of our existence that can be generalized all the way down, to the simplest types of enduring individuals.

(9) The most general of these features is experience, this feature being presupposed by the two features already mentioned, namely, prehension and self-determination. I call this position, accordingly, panexperientialism. This notion is one of the features of this position that is often thought to make it self-evidently subject to one-word refutations, such as "implausible,"[5] because we all know that sticks and stones have no experience and exercise no self-determination. The "pan" in panexperientialism, however, does not mean all things whatsoever but only all true individuals—the things I have been referring to as energetic events and enduring individuals. Even then, the power of the modern worldview, which was adopted in the seventeenth century in opposition to views suggesting that matter involves sentience and spontaneity,[6] is such that most philosophers, scientists, and theologians refuse to entertain this idea seriously. One result of this refusal is that dualism and materialism, the two posi-

tions allowed by the modern worldview, have made little advance on the mind–body problem beyond the stand-off between Descartes and Hobbes three and a half centuries ago. I have recently shown that Whiteheadian panexperientialism can, at long last, resolve this problem, incorporating the strengths of dualism and materialism while avoiding their weaknesses.[7]

(10) Another of these generalizable features, both presupposed and implied by experience, is time, or temporal process. Because each event prehends into itself aspects of prior events, irreversible time obtains even for the most elementary individuals. Time as we know it—that is, as an asymmetrical, irreversible process—did not have to wait for the emergence of human experience, as some think, or for life, as others think, or even for aggregations of atoms subject to entropy, as still others think. Rather, time is already real for individual atoms, even for their constituent electrons, protons, and quarks.[8]

(11) Indeed, time is real even prior to the existence of enduring individuals. For Whitehead, the ancient idea that the origin of our universe involved the emergence of a particular form of order out of chaos—an idea that was suggested by Plato, the book of Genesis,[9] and many other ancient cosmologies—is essentially correct. For Whitehead, the chaos would have been a situation in which extremely trivial energetic events happen at random, meaning that none of them would have been organized into enduring individuals, not even individuals as simple as quarks. Since by a "thing" we usually mean an enduring thing, which retains its identity through time, the chaos prior to the creation of our world was a state of no-thing-ness. In this sense, we can say that our world was created out of nothingness. But, as Russian Orthodox philosophical theologian Nicholas Berdyaev put it, this was a state of relative nothingness, not absolute nothingness. In any case, in this chaotic situation, there would still have been time, or temporal process, because each random event would have prehended prior events and been prehended by succeeding events. (It should not be surprising, of course, that a position known as "process theology" would consider temporal process to be ultimately real.)

(12) In addition to all the local events constituting the universe, there is an enduring individual comprised of an everlasting series of nonlocal, all-inclusive events.[10] Rather than existing outside the universe, in the sense of existing independently of any realm of finite entities, this nonlocal individual is essentially the soul of the universe, providing the unity that makes it a universe. This everlasting individual is the home of all possibilities. By virtue of being prehended by all

local events, it is the primary source of both order and novelty in the universe. Being good, in the two-fold sense of having friendliness and compassion for all sentient creatures, as well as being ubiquitous, everlasting, and the source of the world's order, it can be considered divine.

(13) The influence of this divine individual, rather than ever involving supernatural interruptions of the world's normal causal processes, is a natural part of these processes. The fact that process theology regards the God–world relation as a fully natural relation is due in part to its panexperientialism. One of the reasons for the decline of theism since the seventeeenth century has been puzzlement as to how a cosmic mind could influence nature, understood in mechanistic or materialistic terms. The God–world problem was to some extent simply the mind–body problem writ large. Panexperientialism, by showing how our minds can influence our bodies, simultaneously shows how a Cosmic Mind could influence the physical world.

(14) Although this divine individual, being ubiquitous, exerts influence on all finite events, it cannot fully determine either the inner constitution or the external effects of any of them. Although creative power, which is the two-fold power to exercise self-determination and then to exert efficient causation on others, is embodied by this divine individual, this two-fold power is also embodied by all finite events. The power of the divine individual in the world, accordingly, is the power to evoke and to persuade, never the power to coerce, in the sense of the power unilaterally to determine.

As this brief summary indicates, Whiteheadian process theology is not simple. But, as Whitehead observed, all simple theologies "are shipwrecked upon the rock of the problem of evil."[11] This point is central to our topic, because the decline in the belief that our universe is in any sense designed has surely resulted from the problem of evil as much as from any scientific developments. In any case, given these fourteen notions of process theology, I turn now to the question of whether our universe is designed. I will begin with six senses in which, from a Whiteheadian perspective, the universe is not designed.

SIX SENSES IN WHICH THE UNIVERSE IS NOT DESIGNED

1. *Not created out of absolute nothingness:* Sometimes the idea the our universe is designed means that it was brought into existence *ex nihilo*, with the *nihil* in this phrase taken to mean absolute nothingness, so that even the mere

fact that there are finite actualities and temporal processes is due to divine design. Process theology rejects this view, holding instead that our universe, with its contingent laws of nature, is a particular instantiation of the universe, which exists eternally, embodying necessary, metaphysical principles. So, with reference to the papers by Jaroslav Pelikan and Anindita Balslev, we can say that the idea that the universe had a beginning, associated with Jerusalem, and the idea that the universe has always existed, associated with Athens and India, are both correct.

2. *Not created all at once:* Sometimes the idea that the universe is designed means that it, with all its present species of life, was created all at once, or at least virtually so. Process theology rejects this idea, agreeing instead with the consensus that the present form of our universe has come about through a long evolutionary process.

3. *Not progressively created out of nothing:* Sometimes, as in the thought of Alvin Plantinga and Phillip Johnson,[12] the idea that the universe is designed means that, although our present world came about over billions of years, each new species along the way was created *ex nihilo.* This view, known as "progressive creationism"—or, more fashionably, "punctuated creationism"—is rejected by process theology, which accepts the evolutionary view that all new species have arisen through descent with modification from prior species. With regard to the common questions as to why God created the world in such a simple state and then took so long to bring it to the present state, process theology's answer is that this is the only way that God could create a world.

4. *Not preprogrammed from the outset:* Some theists, such as Rudolf Otto early in the twentieth century, have held that, although God has never intervened in the world since its creation, every detail of the evolutionary process is designed, because every evolutionary sequence was preprogrammed.[13] Even the thought of Charles Darwin, with its deism and determinism, was not free from this implication, although this implication of Darwin's thinking existed in strong tension with his belief in the contingency of evolutionary developments.[14] Process theology rejects this notion of deistic design, holding more consistently than did Darwin to the contingency of every development in every evolutionary sequence, grounding this ubiquitous contingency in the doctrine that all individual events involve an element of self-determination. Evolutionary developments thereby involve chance in an ontological, not merely an epistemic, sense. Because the self-determination that exists at the quantum level is magnified, rather than being canceled out, in higher-level individuals, the contingencies increase in the later stages of evolution. The present world cannot be considered, even approximately, as simply the inevitable outworking of the Big Bang.[15]

5. *Not created solely for human beings:* Sometimes the idea that the universe is designed means the anthropocentric notion—held by William Paley, the utilitarian theologian studied by Darwin—that the universe was designed

solely or at least primarily for the sake of human beings. According to this notion of design, the value of other species is their utility for human beings. Process theology rejects this anthropocentrism, holding instead that every individual of every species has both intrinsic value, meaning value in and for itself, and ecological value, meaning value for the ecosystem. These two forms of value would have existed if human beings had never appeared and will continue to exist after we have departed.

6. *Human beings not inevitable:* Sometimes the idea that the universe is designed means that it was designed to bring forth our own species, just as it is. Process theology's rejection of this connotation is implied by its insistence on contingency, rooted in the self-determination that has pervaded the evolutionary process. In the evolutionary sequence that led to *Homo sapiens*, there were countless contingent developments. If a different possibility had been actualized in any of these cases, beings exactly like us would not exist. If a different possibility had been actualized in any of the more crucial cases, no beings even remotely similar to us would exist.

Now, having clarified several senses in which process theology does not think the universe is designed, I turn to two senses in which process theology thinks that it is.

TWO SENSES IN WHICH THE UNIVERSE IS DESIGNED

The idea of divine "design" is not one that process theologians naturally use, because it suggests that the creation of our universe came about in accordance with a detailed blueprint, prepared in advance. But if we understand the term "design" in a looser sense, to mean that the universe reflects some sort of purpose, then process theologians can speak of the universe as designed in two senses. First, the evolutionary process is viewed as reflecting a divine aim at increasing richness of experience, a directionality that is reflected in the rise of life and then the more complex forms of life. Second, the fact that our universe was able to bring forth life presupposed a basic cosmological order that can, with less qualification, be described as designed. I will discuss these two types of design in order.

Design in the Sense of a Divine Aim Towards Richness of Experience

Physicists, we are told, think of the universe as a physics experiment. Whitehead came to regard it as an aesthetic experiment, with the physics experiment being simply an aspect of this larger project. To explain: Experience is the only thing that is intrinsically valuable, meaning valuable in and for itself. Every individual, by hypothesis, has at least some slight degree of experience and thereby some slight degree of intrinsic value. But the intrinsic value of the simplest individuals, judged in terms of the aesthetic criteria of harmony, complexity, and intensity of experience, must be extremely trivial, compared with the intrinsic value of a human being, or even a bat. If, as Thomas Nagel has emphasized, we cannot imagine what it is like to be a bat,[16] far less can we imagine what it is like to be at atom, or even an amoeba. The divine aim, by hypothesis, has been to bring about conditions that allow for the emergence of individuals with more complex modes of experience and thereby the capacity for greater intrinsic value. This aim is reflected in the increasing complexity that, even allowing for all necessary qualifications, clearly characterizes the evolutionary process.[17]

The slowness of this process reflects the fact that the power behind this aim is not omnipotent in the traditional sense, not essentially the only center of power. Each event, having its own power of self-determination, can either adopt or resist divinely proffered novel possibilities through which the present situation could be transcended. And this present situation is supported by the power of the past, which weighs heavily on the present. Charles Peirce and William James had suggested that the so-called laws of nature are really its most long-standing habits,[18] which would mean that any type of enduring individual, such as a proton, a DNA molecule, or a living cell, would be a more or less long-standing habit. Peirce held that the longer a habit persists, the stronger it tends to become. This idea led him to the conclusion that the universe would become increasingly deterministic, as the habits of nature became stronger and stronger, thereby imposing themselves more and more heavily on the present. Whitehead, while endorsing the idea that the laws of nature are habits,[19] avoided the idea of increasing determinism partly by means of his doctrine of the divine reality as constantly presenting alternative possibilities. This divine influence, however, cannot unilaterally determine either what new possibility, if any, will be evoked or when this development will occur, because the divine evocative power is always competing with the power of the past embodied in the habits of enduring individuals. It may take, accordingly, hundreds, thousands, or even millions of years for an alternative possibility to be evoked into existence.

Although this hypothesis is consistent with both the tempo and the direction of the evolutionary process, it might be thought that it goes against a scientifically established randomness in the process. But the idea that all variations are random has more than one meaning. I have already endorsed one possible meaning, which is that variations involve chance in the ontological sense, an idea that Darwin himself and some neo-Darwinists have rejected. A meaning that many neo-Darwinists do insist upon is that variations are random in every other possible sense, which would exclude their being due even in the slightest to any sort of aim that would give a bias toward variations of a particular sort, such as variations that lead to greater structural complexity and thereby greater richness of experience. But the neo-Darwinian insistence that evolution is random in this sense is simply philosophical dogma, not grounded in any empirical discovery. The randomness that is central to neo-Darwinism as a scientific theory is randomness in a third sense,[20] according to which there is no tendency for variations to be adaptational, that is, advantageous for survival in the environment in which they occur. And the kind of tendency that process theology posits is not in conflict with randomness in this strict sense, because there is no necessary correlation between increased richness of experience and success in the struggle for survival. To give a human example: The emergence of the capacity to do higher mathematics, while it may have increased the satisfaction of some early human beings on the savannas of Africa, would not have increased the likelihood of their sowing their wild genes. The criterion of greater richness of experience is not in tension with neo-Darwinian randomness, except insofar as this randomness is used as a pseudo-scientific front for antitheistic bias.

The divine aim towards greater richness of experience means that there is, in spite of what I said earlier about the contingency of human beings, a sense in which we can regard ourselves as intended. That is, insofar as human experience involves dimensions that give it the capacity for greater intrinsic value than that enjoyed by our evolutionary predecessors, we can say that we reflect the divine aim. Although human beings as such were not intended, human-like beings were, insofar as they were possible. This would mean that, on some other planets in the universe with the conditions for life to emerge and to evolve for many billions of years, we should expect there to be creatures that, no matter how different in physical constitution and appearance, would share some of our capacities, such as those for mathematics, music, and morality, or, more generally, truth, beauty, and goodness.

These capacities, however, imply the capacities for lying, for ugliness, and for immorality, as exemplified, say, by genocide and ecological destruction. Must we not conclude, therefore, that the divine individual of process theology is as responsible for evil as was the deity of traditional theism? It is true that process theology's deity is responsible for evil in one sense, namely, that if human beings had not been evoked into existence, the world would have been free from all the evils caused and experienced by human beings. The question of theodicy, however, is whether the divine reality is responsible in such a way as to be indictable, that is, blameworthy.

With regard to this question, process theism differs from traditional theism in two crucial respects. In the first place, given the omnipotence attributed to the deity of traditional theism, that deity could have created beings who were identical to us in virtually all respects, having the capacity for realizing most of the values we enjoy, differing only by having much less, or even no, power to bring about evil. Because this traditional deity created our world *ex nihilo*, all the principles of our world were freely chosen. There were no metaphysical principles lying in the nature of things, beyond divine volition. In process theology, by contrast, such principles do exist, and one of these principles is that an increase in the capacity for richness of experience is impossible without a correlative increase in freedom and the power to affect other beings. This principle means that every increase in the capacity for good entails an equal increase in the capacity for evil. Any being with our capacities to experience and create good, therefore, would necessarily have our capacities to experience and cause evil. Insofar as we think of the divine individual as confronting a choice with regard to the existence of human-like beings, the choice was only between having beings approximately like us, with our capacities for evil as well as for good, or no human-like beings whatsoever. The deity of process theology can be indicted because of human evil, therefore, only by those who can honestly say that our planet would have been better without human-like beings altogether.[21]

A second crucial difference between the two types of theism is that, according to traditional theism, every instance of evil that has occurred could have been unilaterally prevented by God. One version of traditional theism, to be sure, says that God gave us genuine freedom, so that we can freely choose to do evil. It remains the case, however, that the deity of traditional theism could always intervene either to determine our decisions or to cancel out the natural effects thereof—hence the anger of virulent antitheists such as Stephen Weinberg. In process theism, by contrast, the divine power cannot do either of these. Although the human degree of freedom would not exist if the divine power had not led the evolutionary process to bring human beings into existence, now that we do exist, the divine power cannot cancel out our power to make our own decisions and to inflict them on others. The sense of meaning that comes from seeing the evolutionary process as divinely influenced is not, therefore, undermined or rendered horrible by the conclusion that the "divine" influence is actually demonic, or at least indifferent.[22]

Design in the Sense of the Establishment of the Most Fundamental Contingent Principles of Our Cosmic Epoch

I will conclude by briefly explaining the second sense in which process theology can regard our universe as designed. This second sense involves the much-discussed idea that our universe from the outset evidently embodied a number of "cosmic constants" that give the impression of being finely tuned in relation to each other, because if any of them were slightly different, life could never have evolved. And they do not seem to be simply "habits," as usually understood—that is, to be modes of behavior that have developed gradually and are only usually, rather than always, followed. Some traditional theists have used this fact as new evidence that our universe is the product of Omnipotent Intelligence. Such theists might argue that, even if process theism, with its non-omnipotent deity, can do justice to the world's evil, it cannot do justice to the best scientific account of how our universe originated. A divine being whose power can be resisted by the creatures could not, they might argue, have imposed all of these mathematical values with sufficient precision to pull off an initial creative event, such as a big bang, that would bring about all the conditions necessary for life to be possible in portions of the resulting universe. Although that conclusion might at first glance seem to follow from what I said earlier, I argue that it does not.

My argument is that, in a chaotic state prior to the beginning of our cosmic epoch, the two reasons why there is usually so much resistance to divine ideals would not apply. One of these reasons is that, as the evolutionary process increasingly brings forth more complex individuals, the world thereby has creatures with increasingly greater capacity for self-determination and thereby increasingly greater capacity to resist divine influence. In a chaotic state between cosmic epochs, however, the events would be extremely trivial, with a vanishingly small capacity to exercise self-determination.

The second reason why divine influence usually encounters so much resistance is that the divine intention to instill new ideals, meaning new possible modes of being and interacting, is usually in competition with the power of the past, the modes of being that constitute the essence of enduring individuals. However, in the postulated chaos between the running down of one cosmic epoch and the starting up of another, there would, by definition, be no enduring individuals, and therefore no entrenched modes of being to force themselves upon present events. The chaos would not be absolute, to be sure, because events would still exemplify the necessary, metaphysical principles, which by definition obtain in all possible worlds, including the relatively chaotic periods between cosmic epochs. But there would be no contingent cosmological principles constituting well-entrenched habits. In this situation, therefore, the divine influence, in seeking to get a set of contingent principles embodied in the universe, would have no competition from any other contingent principles.

In the first instant of the creation of a particular universe, accordingly, divine evocative power could produce quasi-coercive effects. A divine spirit, brooding over the chaos, would only have to think "Let there be X!"—with X standing for the finely-tuned set of contingent principles embodied in our world at the outset. To say this is not to suggest that this effect would necessarily have occurred immediately. It is also not to deny the possibility that our universe might have been preceded by a number of brief universes, which were not sufficiently fine-tuned to last very long. But it is to suggest an alternative to the three major ways of thinking of the laws of physics of our universe: that they are necessary, that they exist purely by chance, or that they are the product of an Omnipotent Designer. This alternative possibility is that a creator without coercive power could, in a chaotic situation, produce quasi-coercive effects. From then on, however, the divine persuasive activity would always face competition from the power embodied in the modes of being reflecting these contingent principles, so that divine power would never again, as long as the cosmic epoch exists, be able to produce quasi-coercive effects. In this way, process theism, while maintaining that God's agency in our universe is always persuasive, can nevertheless account for the remarkable contingent order on which our particular universe is based. This suggestion, I should add, will not be found in Whitehead's writings. But it does seem consistent with his position.

In sum: Although Whiteheadian process theology shares with late modern thought the rejection of many of the senses in which the universe had traditionally been thought to be designed, it can speak of our universe as designed in two significant senses. In doing so, furthermore, it can arguably do justice to the best scientific evidence about cosmic and biological evolution without being undermined by the horrendous evils that have resulted from the creation of life, especially human life.

NOTES AND REFERENCES

1. Alfred North Whitehead, *Process and Reality*, corrected edition, ed. David Ray Griffin and Donald W. Sherburne (New York: Free Press, 1978 [original ed. 1929]).
2. Ibid., p. 15; *Science and the Modern World* (New York: Free Press, 1967), p. vii.
3. For my most extensive recent defenses, see *Unsnarling the World-Knot: Consciousness, Freedom, and the Mind-Body Problem* (Berkeley & Los Angeles: University of California Press, 1998); *Religion and Scientific Naturalism: Overcoming the Conflicts* (Albany: State University of New York Press, 2000), and *Reenchantment without Supernaturalism: A Process Philosophy of Religion* (Ithaca, NY: Cornell University Press, 2001).
4. See the discussion of "compound individuals" in chapt. 9 of my *Unsnarling the World-Knot.*
5. I have discussed the allegation that panexperientialism (usually discussed under the term "panpsychism") is implausible in chapt. 7 of *Unsnarling the World-Knot.*
6. I have discussed the theological and sociological motives behind the modern notion of matter as inert and insentient in chapt. 5 of *Religion and Scientific Naturalism.*
7. *Unsnarling the World-Knot*, especially chapts. 6, 8, and 9.
8. On this issue, see my "Introduction: Physics and the Ultimate Significance of Time," David Ray Griffin, ed., *Physics and the Ultimate Significance of Time: Bohm, Prigogine, and Process Philosophy* (Albany: State University of New York Press, 1988), pp. 1–48, or my "Time in Process Theology," *KronoScope: Journal for the Study of Time*, Vol. 1/1-2 (2001), pp. 75–99.
9. For extensive discussions of the fact that the doctrine of creation *ex nihilo* is not a biblical doctrine, see Jon D. Levenson, *Creation and the Persistence of*

Evil (San Francisco: Harper & Row, 1988), and Gerhard May, *Creatio Ex Nihilo: The Doctrine of "Creation out of Nothing" in Early Christian Thought*, trans. A. S. Worrall (Edinburgh: T & T Clark, 1994).

10. For arguments for the existence of a divine reality as conceived by Whiteheadian process theology, see chapt. 5 of my *Reenchantment without Supernaturalism*. For the argument that this reality should be understood, with Charles Hartshorne, as an everlasting temporal society of events, rather than, with Whitehead himself, as a single everlasting actual entity, see chapt. 4 of that book.
11. A.N. Whitehead, *Religion in the Making* (1923; New York: Fordham University Press, 1996), p. 77. I have offered a Whiteheadian solution to the problem of evil in *God, Power, and Evil: A Process Theodicy* (Philadelphia: Westminster Press, 1976; 2nd edition with a new preface, Lanham, MD.: University Press of America, 1991) and *Evil Revisited: Responses and Reconsiderations* (Albany: State University of New York Press, 1991). I have explicitly discussed the connection of this theodicy with the rejection of *creatio ex nihilo* in "Creation out of Nothing, Creation out of Chaos, and the Problem of Evil," in *Encountering Evil: Live Options in Theodicy*, 2nd edition, ed. Stephen T. Davis (Philadelphia, PA: Westminster/John Knox, 2001), pp. 108–125.
12. I have discussed Johnson's position in "Christian Faith and Scientific Naturalism: An Appreciative Critique of Phillip Johnson's Proposal," *Christian Scholars Review,* Vol. 28/2: 308–328 (1998). A modified version of this critique, dealing also with Plantinga, is contained in chapt. 3 of *Religion and Scientific Naturalism.*
13. A discussion of Otto's position is included in chapt. 3 of *Religion and Scientific Naturalism.*
14. Darwin's deism and determinism are discussed in chapt. 8, "Creation and Evolution," of my *Religion and Scientific Naturalism.*
15. In *A Brief History of Time: From the Big Bang to Black Holes* (New York: Bantam, 1988), Stephen Hawking, while giving up the older ideal of a completely deterministic science, still says that the goal of science should be "the discovery of laws that will enable us to predict events up to the limits set by the uncertainty principle" (p. 173). He evidently holds, furthermore, that those limits are purely epistemic; as he suggests we can "still imagine that there is a set of laws that determines events completely for some supernatural being, who could observe the present state of the universe without disturbing it" (p. 55).
16. Thomas Nagel, "What Is It Like to Be a Bat?" in his *Mortal Questions* (London: Cambridge University Press, 1979).
17. For recent discussions of whether we can speak of progress in the evolutionary process, see Matthew H. Nitecki, *Evolutionary Progress* (Chicago & London: University of Chicago Press, 1988). I have discussed this issue in chapt. 8 of *Religion and Scientific Naturalism.*
18. See William James, *The Principles of Psychology* (1890; New York: Dover, 1950), I: 104–105; and Peter Ochs, "Charles Sanders Peirce," in David Ray Griffin *et al.*, *Founders of Constructive Postmodern Philosophy* (Albany: State University of New York Press, 1993), pp. 43–87, esp. pp. 67–68, 73–75.
19. A.N. Whitehead, *Modes of Thought* (1938; New York: Free Press, 1968), p. 154.
20. I have discussed these different meanings of randomness in chapt. 8 of *Religion and Scientific Naturalism.*

21. I have discussed this point in my three discussions of the problem of evil mentioned in note 11.
22. Antitheists sometimes charge that revisionary theists can overcome the problem of evil only by revising the conventional understanding of theism so drastically that the resulting position is no longer intelligibly called "theism," because the resulting referent of the word "God" has been redefined out of all recognition. That charge does indeed apply to many modern theologies. But, by showing that the deity of process theism embodies all the features of what can be called the "generic idea of God" in cultures primarily shaped by biblically based religions (*Evil Revisited*, pp.10–12), I have shown that this charge does not apply to process theism.

Cosmic Design from a Buddhist Perspective

TRINH XUAN THUAN

Department of Astronomy, University of Virginia, Charlottesville, Virginia 22903, USA

ABSTRACT: The Buddhist view of the origin of the universe is discussed. One of the basic tenets of Buddhism is the concept of interdependence which says that all things exist only in relationship to others, and that nothing can have an independent and autonomous existence. The world is a vast flow of events that are linked together and participate in one another. Thus there can be no First Cause, and no creation *ex nihilo* of the universe, as in the Big Bang theory. Since the universe has neither beginning nor end, the only universe compatible with Buddhism is a cyclic one. According to Buddhism, the exquisitely precise fine-tuning of the universe for the emergence of life and consciousness as expressed in the "anthropic principle" is not due to a Creative Principle, but to the interdependence of matter with flows of consciousness, the two having co-existed for all times.

KEYWORDS: cosmology; Buddhism; universe; anthropic principle; consciousness

SCIENCE AND BUDDHISM

When the developers of this volume asked me to provide a paper about science and Buddhism, they probably wanted the point of view of someone who has been trained in the scientific method of the West and is a practicing scientist, but who happens also to have some knowledge of the Buddhist tradition; in other words, someone who is familiar with two totally different ways of exploring the nature of the phenomenal world: one that relies on the rational method and uses physics and mathematics as tools and the other that relies on an analysis of phenomena through the contemplative method. Yet both share a common thread: both are based on experience and observation.

It is not my purpose in this paper to use science to justify Buddhism, nor Buddhism to give a mystical meaning to science. Both exist independently of one another and stand on their own: Buddhism is a science of the awakening,

Address for correspondence: Dr. Trinh Xuan Thuan, Department of Astronomy, University of Virginia, P.O. Box 3818, University Station, Charlottesville, VA 22903-0818. Voice: 804-924-7494; fax: 804-924-3104.

txt@virginia.edu

and whether the Universe is expanding or not cannot have any bearing on the philosophical underpinnings of Buddhism. On the other hand, science is perfectly self-sufficient and accomplishes well its stated aim—that of giving a coherent description of the physical processes operating in Nature—without the need of a philosophical support from Buddhism or any other religion. Yet both science and Buddhism aspire to describe reality, and if their approaches are both coherent and valid, their respective visions should not contradict and exclude each other, but rather should complement and re-enforce each other.

The description of the phenomenal world is not the main aim of Buddhism. Buddha is a physician of the soul and his main concern is to show the way to enlightenment. He has greater preoccupation with peace of mind, kindness, compassion and the joy and happiness in ourselves and others than in knowledge that does not contribute directly to a lessening of sorrow and suffering. Knowing that the Earth is round rather than flat, or that the universe had a beginning (or not) does not contribute directly to awakening. However, in order to analyze the causes of unhappiness, Buddha uses the methods of contemplative science that allow one not only to see clearly into the nature of the mind, but also to considerably refine our view of the phenomenal world. According to Buddhism, a correct analysis of the phenomenal world is necessary because an incorrect perception of reality may result in suffering and unhappiness. For example, if we are convinced that the material world has an intrinsic and permanent existence, then we may develop a strong misguided attachment to that material world which can cause frustration and suffering.

INTERDEPENDENCE IN BUDDHISM

In order to understand Buddhist cosmology, we have to comprehend one of the key concepts of Buddhism, that of "interdependence." One of the aspects of that interdependence is the relationship between humanity's consciousness and the reality we perceive around us. According to Buddhism, all the proprieties that we attribute to the phenomenal world are not necessarily intrinsic to the object itself, but are conceived by our mind and filtered through our perceptions. Thus the same reality may appear differently to different intelligences. Objects are thus devoid of intrinsic and autonomous properties and do not possess solidity and permanence. That is the profound meaning of "vacuity." It must be emphasized that vacuity in Buddhism is not nothingness as the word has sometimes been misunderstood—Buddhism has at times been accused totally wrongly of nihilism. Vacuity is the absence of independence and autonomy of things. Because of interdependence, there is the potential and capacity for phenomena to vary in an infinite number of ways, to develop in infinite directions. The only real nature of phenomena is thus their "interdependence." Vacuity is the ultimate nature of things because phenomena are devoid of an existence that is permanent and independent of the observer.

In Buddhism, there are thus two distinct levels of reality: that of conventional reality, which we are all familiar with in our daily lives, and that of ultimate reality, which has the quality of vacuity. Conventional reality concerns the transformation and change of things in the phenomenal world. These changes are governed by causal laws that are similar to the physical laws discovered by science in Nature. In that sense, the Buddhist view of conventional reality is very much like that of a scientist, with the difference being that, in addition to the physical laws, Buddhism introduces the laws of karma, which say that the consequences of our acts, be they positive or negative, will lead unavoidably to our future happiness or suffering. But conventional reality is mere appearance. On a deeper level, phenomena do not have an objective existence. Using poetic language, Buddha often compared reality to mirages, magic illusions, or dreams.

This interdependence between the nature of reality and the mind of the observer is not totally foreign to the scientist himself, although we usually think of science as being totally "objective." The information that nature sends us is inevitably altered by the instruments used for observation and analysis, be it a telescope, bubble chamber, or computer, and by the brains of the observers who interpret it. Reality is filtered through a nightmarish web of electronic circuits; it is manipulated, digitized, and reconstituted by powerful computers and complex mathematical treatments.

In 1609, when Galileo first pointed a telescope toward the sky, he had, at the beginning, a very hard time convincing his colleagues that the wonders visible through his telescope were not optical illusions. The problem of the veracity of images is a thousand times worse in modern astronomy. There have been so many steps between the raw signals and the final image that it is quite legitimate to wonder what "objective truth" remains in the image. Fortunately, there is a way to weed out erroneous observations in science. A result or observation is not accepted until it has been verified independently by other workers, using other techniques or other measuring instruments. It is highly unlikely that the same error would be repeated each time, or that the instruments or machines should fool us on every occasion.

Thus, in principle, technical difficulties are surmountable. If we could rely upon machines alone, reality could, in theory, be rendered as objective as possible. But what cannot be avoided is the human brain. Human beings cannot observe nature in an objective manner. There is a constant interaction between our inner world and the outer world. The inner world of the scientist is full of concepts, models, and theories acquired during his professional training. This inner world, when projected onto the outer world, prevents the scientist from seeing the "bare" objective facts, free from any interpretation. We only see what we want to see. On that subject, Charles Darwin, the father of the theory of evolution, told a charming story: He spent a whole day on a river bank and saw nothing but stones and water. Eleven years later, he returned to the same spot, searching for traces of earlier glaciation. This time, the evidence stuck out like a sore thumb. Not even an extinct volcano could have left more visible traces of its past activity than this old glacier. Darwin discovered what he was looking for as soon as he knew how to see.

Science goes even further: the very act of observing can modify reality. The science of quantum mechanics, which describes the behavior of subatomic particles, says so. The properties of a particle are unavoidably disturbed when it is observed because one has to shine light on it. Light and particles going through two holes behave like waves when the observer does not attempt to find out which hole the light or particles have gone through, but they both behave like particles as soon as one attempts to find out their precise path by placing detectors after the holes. This interdependence between observer and reality has been emphasized many times by the founders of the science of quantum mechanics.

Let's listen, for example, to Heisenberg, who remarks: "What we observe is not nature in itself but nature exposed to our method of questioning.," or to Bohr who says: "As our knowledge becomes wider, we must always be prepared, therefore, to expect alterations in the points of view best suited for the ordering of our experience. In this connection, we must remember, above all, that, as a matter of course, all new experience makes its appearance within the frame of our customary points of view and forms of perception."

Not only is there interdependence between the observer and the observed, but there is also interdependence between particles in the subatomic world. This is shown by a famous thought experiment, known as the EPR experiment, proposed in 1930 by Albert Einstein and his colleagues Boris Podolsky and Nathan Rosen. Imagine, they said, that a particle disintegrates spontaneously into two photons A and B. Nothing allows us to say *a priori* in which directions these two photons will propagate. There is one certainty, however: because of symmetry, they will leave in opposite directions. If A goes toward the west, B will go toward the east. Let us set up our instruments and check. Yes, A goes west and B goes east. It is as expected.

But this does not take into account the indeterminacy of the subatomic world. Quantum mechanics tells us that A has no precise direction before it is captured by the measuring instrument. It was wearing its guise as a wave and could take any direction. It is only after it has interacted with the detector that A turns into a particle and "learns" that it is going west. If A did not "know" what direction to take before being captured by the measuring instrument, how could B "guess" in advance the direction of A, and arrange its trajectory so that it would be captured at the same time in the opposite direction? This does not make sense. Einstein and his colleagues concluded that quantum mechanics had therefore gone wrong. But this not the case. Laboratory experiments have always confirmed quantum mechanics, and the theory does truly account for the behavior of atoms. How then are we to resolve the EPR paradox?

The paradox exists only because we assume that reality is "localized" on each of the two particles. The paradox is no more if we accept the idea that the two photons, even if they are separated by billions of light-years, are part of a single reality before they are recorded by the measuring instruments, and that they are in permanent contact with each other by some sort of mysterious interaction. Everything is interdependent. Reality is no longer local, but global.

BUDDHISM AND COSMOLOGY

What are the consequences of the concept of interdependence on cosmological ideas in Buddhism? The concept of interdependence implies that the elements of the conventional reality we are all familiar with do not possess an existence that is permanent and autonomous. This thing exists because something else exists; that happens because this has occurred. Nothing can exist by itself and be its own cause.

Everything depends on everything else. Suppose that there is an entity that exists independently of all the others. This implies that it is not produced by a cause, that is, either it has always existed or it does not exist at all. Such an entity will be unchanging since it cannot act on others and others cannot act on it. The world of phenomena could not function. Thus interdependence is essential for phenomena to manifest themselves.

Because the concept of interdependence implies that nothing can exist by itself and be its own cause, it goes against the idea of a creative principle, a First Cause or a God that is permanent, all-powerful, that has no other cause than itself, and which created the universe. In the same vein, Buddhism rejects the idea that the universe can be born out of nothing—a creation *ex nihilo*—because the universe has to depend on something else to emerge. If the universe was created, it is because there was a potentiality already present. The coming into being of the universe is merely the realization of that potentiality. One can thus interpret the Big Bang as the manifestation of the phenomenal world emerging from an infinite potentiality already in existence. In a poetic language, Buddhism speaks about "particles of space" which carry in them the potentiality of matter. This is strongly reminiscent of the vacuum filled with energy that is thought to have given birth to the material content of the universe in the modern Big Bang theory. Material phenomena and things are not "created" in the sense that they go from a state of non-existence to one of existence. Rather they go from an unrealized state to a realized state. Once it has come into existence, the universe goes through a series of cycles, each composed of four stages: birth, evolution, death and a state where the universe is pure potentiality but has not manifested yet itself. This cyclic universe has no beginning nor an end.

COSMIC EVOLUTION AND IMPERMANENCE

The cosmological theory that best describes the current observations of the universe is that of the Big Bang. It is now believed that the universe began its existence some 15 billion years ago with an enormous explosion from an initial state that was extremely tiny, hot, and dense and that spawned space and time. Since then, there has been a relentless ascent towards increasing complexity. Beginning with a vacuum filled with energy, through the primordial soup of elementary particles, the universe has woven an immense cosmic tapestry composed of hundreds of billions of galaxies, each made in turn of hundreds of billions stars. In one of these galaxies named the "Milky Way," on a planet near a star about two-thirds of the way from the Galactic center toward the edge, humanity appeared, capable of marveling at the beauty and harmony of the cosmos and of asking questions about it.

One of the most remarkable changes of paradigm that has occurred with the advent of the Big Bang theory is that the universe has acquired a historical dimension. We can now speak of the history of the universe, with a beginning and an end, with a past, present and future. That the universe has a history was not always accepted. Some 24 centuries ago, the Greek philosopher Aristotle thought that the Heavens, because they were perfect, had to be unchanging and eternal in contrast to the changeable and imperfect world of the Earth and the Moon. Newton's universe in the 17th century was static, unchanging, and devoid of history. As late as the 1950s, the steady-state theory, which says that the universe is on the average unchanging both in space and time, was considered a serious rival to the Big Bang theory. The Big Bang theory thus introduces the fundamental idea of cosmic evolution. Stars are born, live their lives, and die. In the course of their death throes, the massive stars eject gas that shall serve as seed for the birth of a new generation of stars. Those life and death cycles last from a few million years to several billion years. The universe itself may go through cycles of births and deaths, Big Bangs followed by Big Crunches, although it is not yet clear whether the universe contains enough dark matter for its gravity to halt its present motion of expansion and reverse it. The idea of ceaseless change, of constant evolution due to the never-ending chain of causes and effects, is also central to Buddhism. It is called "impermanence."

A UNIVERSE CONSCIOUS OF ITSELF

Modern cosmology has rediscovered the ancient covenant between humanity and the cosmos. Humans are the children of stars, the siblings of wild animals, and the cousins of plants and flowers; we are all star dust. Astrophysics teaches us that the emergence of life from the primordial soup depended on an extremely delicate adjustment of the laws of Nature and the initial conditions of the universe. A minute change in the intensity of the fundamental forces, and we would not be around to talk about it. The stars would not have formed and started their marvelous nuclear alchemy. None of the heavy elements that constitute the basis of life would have seen the light of day. The precision of the fine-tuning of the physical constants and of the initial conditions is astonishing. It is similar to the precision that a marksman has to exercise in order to put a bullet through a square target of 1 cm on a side located at the edge of the observable universe some 15 billion light-years away. This fine-tuning is at the basis of what is called the "anthropic principle," from the Greek "anthropos" which means "man."

The laws of physics are special from an even more subtle point of view. Not only did they permit humanity to step on the stage, but they also conferred on us the ability to be conscious and understand the world in which we live. The fact that humans do not simply and blindly endure the laws of Nature without understanding them is highly significant. Darwinian selection certainly played a role in fashioning our brain to help us cope with the many challenges of life, but the ability to ask questions about the universe and understand the mathematical laws governing it is not necessary and seems to have come as a bonus. Does this mean that humanity has reclaimed a central place in the universe? Hardly! The physical and chemical processes that unfolded on Earth and led to life and consciousness are probably not unique to our planet. An extraterrestrial intelligence endowed with scientific and mathematical knowledge would be just as suitable to give the universe a meaning.

CHANCE, NECESSITY, OR INTERDEPENDENCE?

How can we interpret such an astonishing fine-tuning of the physical constants and initial conditions of the universe that make it possible for life and consciousness to emerge? From a non-Buddhist point of view, there are two possible alternatives. One can evoke either chance or necessity.

If chance is the right answer, then the very precise tuning of the laws of physics and the initial conditions, so as to allow consciousness to arise, could be explained by the existence of a multitude of parallel universes. These parallel universes would contain all possible combinations of physical laws and initial conditions. Virtually all these universes would be barren and incapable of harboring life and consciousness ... all except ours, which, by pure coincidence, would have the winning combination, with us as the grand prize! Quantum mechanics allows the existence of such parallel universes; every time a choice or decision must be made, the universe could split into two: in one universe the Declaration of Independence would be written, in another America would remain a colony of England. In one universe the Berlin Wall would be torn down, in another it would remain. The observer himself would divide in two. There are also some Big Bang models that allow the idea of parallel universes: our universe would be only one small bubble among a multitude of other bubble-parallel universes within a meta-universe. On the other hand, if we choose the "necessity" option (i.e., reject the parallel universe hypothesis and adopt the one of a single universe, our own), then in order to account for the extremely precise fine-tuning, we must postulate a Great Architect who adjusted from the outset the laws of physics and initial conditions in order for the universe to become conscious of itself.

Both options are possible, and science cannot settle the issue. Like the 17th century French philosopher Blaise Pascal, we must make a wager: either humanity emerged by chance in an indifferent universe that is totally devoid of meaning, or our ascent was preprogrammed at the very beginning so we could give meaning to the universe by understanding it.

Buddhism offers a third alternative to account for such a precise fine-tuning for the emergence life and consciousness. As we have seen, it is not necessary to invoke a First Cause, a creative principle that has regulated everything from the start. There is no need for an "anthropic principle" or for a notion of design. According to Buddhism, consciousness has co-existed, co-exists and will co-exist with matter for all times. The same goes for the animate with the inanimate. Neither the universe nor consciousness had a beginning or end. Because they are interdependent, it is not surprising that the properties of the universe are compatible with the existence of consciousness. Two interdependent entities cannot exclude each other, but must be necessarily in harmony with each other.

This cosmic vision is in contrast to the usual picture of an ascent of the pyramid of complexity, where there is the formation of ever more complex forms of matter with the passage of time, which forms as they pass a complexity threshold become animate and endowed with consciousness. This does not mean that Buddhism rejects the Darwinian idea of evolution. Rather Buddhism would interpret the whole sequence of Darwinian evolution of ever more complex organisms as simply an increase in sophistication of the material support of a stream or a continuum of consciousness going from one form of material support to another.

❧

In summary, the cosmological view of Buddhism rests on the basic concept of interdependence. Because everything depends on something else, there can be no entity that exists independently of all the others. Thus, there is no First Cause and no creation *ex nihilo*. There is also no need to invoke an "anthropic principle" or any notion of design. The universe must be such as to harbor consciousness simply because the two are interdependent. This concept of interdependence is strongly reminiscent of the properties of interconnectedness and non-locality found in the science of quantum mechanics for subatomic particles.

What Did the Mystic Say to the Hot Dog Vendor?

Six Neo-Kabbalistic Metaphors for Cosmic Design

LAWRENCE KUSHNER

ABSTRACT: The universe is not designed by a deity outside, beyond, and other than the universe. The design of the universe is experienced in its unity, dynamic interdependency, and interconnectedness. The designer is to be found in the experience of such design, not beyond it. Within Judaism, the Kabbalah is one expression of this religious vision of the universe.

KEYWORDS: design of the universe; connectedness; unity; nothingness; Kabbalah

HIDDEN SIGNATURE[1]

I remember the first computer game we ever got. Since hardly anyone owned computers in those days, you had to use your television set for a monitor. By plugging a cable into the antenna jack, you turned the TV into a primitive video arcade. Then—are you ready?—you could play ping pong... in monochrome. You operated a little paddle that moved up and down along one side of the screen. The ball—actually a white square—moved horizontally. With enough coordination, you could get your paddle to intersect its trajectory, whereupon you heard the game's sole sound effect: "Bip." After you got good at it, you could crank up the speed: "Bip. Bip. Bip. Bip. Bip." People in our home, who shall remain nameless, played it for hours.

A few years later, Atari became all the rage. As I recall, Atari initially had four different games. My favorite was called "Adventure." It was your basic "dungeons and dragons" genre, with different castles and rooms, a key, hidden doorways, a bat that could steal the key, even whole areas where the obstacles were invisible.

Our whole family really got into it. The kids, of course, quickly surpassed their parents. They would come home from school with new tips and tricks.

Address for correspondence: Rabbi Lawrence Kushner, 309 Goodman's Hill Road, Sudbury, MA 01776. Voice: 978-443-5268.
kushner@acunet.net

Some of them were even "undocumented," which is computer-talk for saying that such maneuvers were not written down in any manual. One of the most amazing came home from junior high with my daughter: In a particular place inside the "black castle," the diligent searcher could find a small white dot that was too small to be noticeable as a normal game object. Indeed, if you did find it you would think it was only a defective pixel in your video monitor.

By taking this dot back to the starting screen however, you could enter a hidden, otherwise inaccessible room. Entering the room had absolutely nothing to do with playing the game. All you would find in the room was a rainbow and the name of the person who invented the game. For all I know, every computer programmer does something like this. *Somewhere, behind some hidden wall, available only to the initiated, there is another room.* And in that room is the name of the artist.

What I want to know is this: If the signature of the Creator is not just in some hidden room but in every created thing, why can't we see it?

I once heard of a man whose dental work made it possible for him to actually hear radio broadcasts. Somehow the combination of fillings in his teeth accidentally turned his mouth into a primitive receiver. But he found the sounds so distracting that he had the fillings replaced. The radio signals were still there, he just chose not to hear them any more.

VIRTUAL REALITY

A few years ago one of my sons brought home a new, and much more sophisticated computer game. "It is a new breed, Dad," he explained, "called 'virtual reality.' You play it by entering it. Your only chance at winning is by imagining that you are actually inside it. You have to ask yourself, 'What would I do if I really lived in this world?'"

At the beginning of this game (called "Myst"), you look at the screen and find yourself on an island. There's a dock, a forest, buildings, stairways. The graphics and sound effects are impressive and convincing. There is no manual, no instructions, no rules. You "go" places by aiming a little pointing finger and clicking. You can look up and down, turn around, climb stairs, wander all around the place. Everywhere your curiosity leads you, there are things to discover, learn, and remember. There are machines you can operate, and a library full of books you can actually open and read. After a while, the dedicated player will discover how to leave the island and go to other mysterious places. Devotees say the game is properly played over weeks and even months.

And the purpose of it all? Why, of course: To figure out what you're doing there. But to do that, you must first figure out how the place works.

What fascinates me here is not yet another sophisticated and clever way to waste time in front of the computer screen. (I can do that with File Manager.) It is the concept of a game whose purpose is for the player to discover the purpose. *Virtual* reality, *schmirtual* reality, this is no game. What's going on here? Why am I here? What are the rules?

Upon hearing about all this, Alan Feldman, a friend who is a professor of English, suggested that it seemed a lot like childhood. I'd go farther. It may be a lot like adulthood, too. We all find ourselves in "this world" and the "object" seems to be to figure out what we're doing here. Unfortunately, most of the ways one thing is connected to or dependent upon another thing are not immediately apparent.

After all, meaning is primarily a matter of relationship. If something is connected to absolutely nothing—symbolically, linguistically, physically, psychologically—it is literally meaning-less. And in the same way, if something is connected to everyone and everything, it would be supremely meaning-full. I suppose it would be God: The "One" through whom everything is connected to everything else, the Source of all meaning. Religious traditions are the collected "rules of the game." They tell us how the world works. And if you "play by them," you are rewarded (hopefully before it is time to leave) with an understanding of why you are here; with what is otherwise known as the meaning of life.

What if there were a virtual reality computer game that was programmed to approximate real life? If you could design such a program, what would be "the object"? The way I see it there are only five rules.

The first rule of the game of life is that you cannot decide when to begin playing. One day, out of the blue, you realize it: "Uh-oh, I'm playing the game!" Someone or something else determined when your game would begin. And it wasn't your parents. They may have known about the birds and the bees and even set out to conceive a child, but they didn't have a clue it was going to be you. And now that they've had a chance to meet you, while they most likely love you, they'd probably have picked somebody else. In religious language, this means that you are a creature. Someone else made you. And you are neither its partner nor its puppet: You are its manifestation, its agent, its child.

The second rule is that you cannot decide when to stop playing the game, either. One day, out of the blue: You're dead. For a slogan on the box of the game of "Life," we could use something I saw on a T-shirt: "Life: You're not going to make it out alive!" That means there's no way you can "win" the game by staying in it forever. No matter how many points, toys, honors, conquests, dollars you accumulate, sooner than anyone expects or wants, the game is abruptly over. You hear a little buzzer, the keyboard freezes, the screen goes blank. The game ends without warning. But there's good news: Dying does not mean you lose. It's what you do before you die that determines whether or not you win when you die.

The third rule—just to keep you on your toes—is that each player is issued apparently random, undeserved gifts and handicaps throughout the progress of the game. Figuring out why you got the combination package you did transforms all disabilities into gifts, just as refusing to figure out why you were issued what you received, transforms all gifts into disabilities. My father used to say that all men are not created equal. Some get dealt a full house; others, a pair of twos. The question therefore is not whether you deserve the hand you were dealt, but how you choose to play it.

The fourth rule is that points are awarded whenever you can discern the presence, or the signature, of the Creator, and then act so as to help others see it too. The signature is not just in objects, but in actions and thoughts and feelings; not just in sunshine and happiness, but in agony, struggle and death. Remember: Finding the signature and then acting in such a way as to help others find it too, is the only way to accumulate game points.

And the last rule is that everything is connected to everything else. And for that reason, life is supercharged, permeated, and over-brimming with purpose and meaning. Most of the time we are oblivious to it. We go about our lives as if every event were an accident. And then something happens and we see the connection, the signature of the Creator, a design. For a just moment it is unmistakable. We are astonished that we couldn't see it until now. All Creation is one great unity. There are no coincidences.

Throughout all Creation, just beneath the surface, joining each person to every other person and to every other thing in a luminous organism of sacred responsibility, we discover invisible lines of connection.

Now that's *my* idea of a game.

READING MUSIC

Last fall I turned fifty-five. To increase the likelihood that this milestone would only be a *joyous* event, my wonderful wife, the bride of my youth, after months of clandestine research, to my utter surprise and dumbfounded astonishment, presented me with a concert grade, Buffet, B-flat clarinet. Now this would just be the story of another extravagant and expensive gift were it not for one other fact. Prior to that moment last fall, I had never, in my entire life even touched a clarinet! I wasn't even conscious I wanted one.

"How did you know?" I sputtered.

"I just listened," she replied, "For twenty years you've been muttering that someday you'd like to learn how to play the clarinet. I checked with the kids, they heard it too. Apparently you're the only one who never knew."

Well, I've been taking lessons now for six months. (I come right after an eleven-year-old Asian kid.) I am a little disappointed that I still haven't had any feelers yet from major symphony orchestras, but it's only been six months. And besides, I still can't read music very well. Reading music, it's becoming increasingly clear, will definitely take a few more months.

There are so many nuances: All those little Italian abbreviations; Every Good Boy Does Fine; dotted eighth notes; keys with six sharps or six flats; funny little squiggles and marks everywhere! And then there's keeping time. To help you count the rhythm, at the end of every measure there is a little vertical bar line. I count: one, two, three, four, end of measure; one, two, three, four, end of measure.

My tutor says, "Why do you pause at the end of each measure?"

I say, "Because there's this little vertical line there."

She says, "You're not supposed to play *that*."

I say, "I'm not *playing* it."

She says, "Yes, you are. You're pausing at each one!"

I say, "But then how would anyone know it's the end of a measure?"

She sets down her clarinet and turns to face me. (This means she is exasperated and something important is about to come.) "The bar lines are not there. Yes, I know they're written in the sheet music. They're there to make it easier for you to count time but the divisions are only arbitrary super-impositions. They're not *in the music*."

This reminded me of something I learned from, Professor Daniel Matt, currently in the fifth year of a twenty-year-long project of translating the Zohar. We have a word, he once explained, for a leaf, a twig, a branch, a trunk, roots. The words make it easier for us to comprehend reality. But we must be careful not to allow our selves to fall into the habit of thinking that just because we have words for all the parts of a tree that therefore a tree really has all those parts. The leaf does not know when it stops being a leaf and becomes a twig. Nor does the branch know when it is no longer a branch and now the trunk. And the trunk is not aware that it has stopped being a trunk and is now the roots. Indeed, the roots do not know when they stop being roots and become soil. Nor the soil the moisture, nor the moisture the atmosphere, nor the atmosphere the sunlight. All our names are only arbitrarily superimposed on seamless reality.

The Kabbalists explained it this way: There are two worlds. The *Olam haPrayda*, the world of separation—the one we inhabit most of the time, this world, with its infinite array of discrete and autonomous parts, each with its own name (and, if its human, with its own agenda). And then there is the *Olam haYihud*, the world of unity, a radical monism, wherein there are neither parts nor names, where everything is one. Or perhaps more accurately, everything is The One.[2]

It will also come as no surprise that we will predictably understand our relationship with God depending on whichever of the two worlds we happen to inhabit at the moment. Most of the time the world is subdivided up into measures, each bordered by vertical bar lines, or parts, each distinguished by its own unique name and geographic coordinates. Our relationship with God here is likewise personal, one of two discrete, autonomous, and independent actors. And as in virtually all classical Western metaphors, God is *other* than the world—creating, designing, supervising, and hopefully running the place.

But the world of separation lies *within* the bosom of the world of unity.[3] Here we have another way to understand how we relate to God. In this world of unity there are no names, no parts, no separations and therefore no relationships, no bar lines, no measures. It is all one, The One, The One of All Being, you know, God. In the Yiddish, *"Alles ist Gott!"* And, just as music happens only when we are no longer aware of the discrete notes, measures, rhythms, in the same way, meaning comes when we comprehend the unity which is the substrate of all being, when the world of separation gives way to the world of unity. When, as in the imagination of the Kabbalists, we are finally able to pronounce all of scripture as one, long, interruptible Name of God.

But alas, to learn how to make music, you must first subdivide the whole score into smaller pieces, each one separated from the other by a little vertical line.

DROPS IN THE OCEAN

Look at it this way: Since any *thing* must necessarily be bounded by other things, then only something that is a *no-thing* could encompass every *thing*. Only the bound-less-ness of no-thing-ness can comprehend all being. In this way God is, as the Kabbalists said, the *Ayn Sof*, the Infinite One, the One without End, the Holy Nothingness. And the ultimate fulfillment of consciousness, the discovery of *the design*, occurs when you realize your presence *within* the divine, when you realize that it also includes you—the one hearing these words right now. But in order to do that, you must be willing to surrender your autonomy, your independence, your name and your boundaries. You must, in the words of the ancient metaphor, become as a drop of water fallen into the ocean. What did the mystic say to the hot dog vendor? Make me one with everything.

The contemporary theologian, Richard Rubenstein, describes this radical, monism: "God is the ocean" he says, "and we are the waves. In some sense each wave has its moment in which it is distinguishable as a somewhat separate entity. Nevertheless, no wave is entirely distinct from the ocean… The waves are surface manifestations of the ocean. [Even as] our knowledge of the ocean is largely dependent on the way it manifests itself in the waves."[4]

That reminds me of something I read in a sailing magazine: "Waves are not lumps of water moving along the sea but rather pure energy moving through and exciting the water."[5]

So, if you mean by the question, "Is the Universe designed?" "Is there a God somehow outside, beyond, or other than the universe?" then I think the answer is, no. If the twentieth century has taught us anything it is that God does not work like that. But, if you mean by the question, "Is the Universe designed?" "Is everyone and every thing joined to one another through invisible lines of connection into one great luminous organism and human consciousness is itself a dimension of that great unity?" then the answer is yes.

The Hasidic master, Rabbi Yehiel Michal of Zlotchov, offers a teaching he learned from his teacher, Dov Baer of Mezritch: "It's just the opposite of what everyone thinks. They assume that when they do not merge with their Creator but instead cleave to the things and matters of *this* world, that they amount to something.... They imagine that they are important. But how could anyone who might not wake up the next morning be important?...."

"In this way, if you think you are something, then alas, you are nothing. [Don't try this at home!] On the other hand, if, because of your fusion *with* God you think of yourself as nothing...then you are very great indeed... [You are] like a single drop of water fallen into the sea. It has returned to its source. It is one with the ocean. Now it's no longer identifiable as an independent thing in any way whatsoever."[6]

ANACORTES FERRY

Last summer I was invited to give some lectures out in Seattle so we took it as a sign and went out a few days early to explore the San Juan Islands of Puget Sound. The port of departure for most of the island ferries is the city of Anacortes, Washington. During weekends in the height of the tourist season travelers can expect wait times of a few hours.

And sure enough, when we dutifully arrived a hour and a half early, there were already a few cars ahead of us. They had just had the frustrating experience of watching the previous ferry slip away from the dock minutes earlier. One driver of a white Volvo wagon was still fuming about his near miss. Others commiserated with exasperating stories of their own. But each new tale only seemed to exacerbate his pain. And, as if the poor man didn't have enough problems already, fate—or was it something else?—had him parked next to an Avis rental car driven by a Jewish theologian and his wife from Massachusetts who were on a little vacation.

"Damn it!" He grumbled turning toward me, "If I had only left home another three minutes earlier, I'd have made it."

"No you wouldn't," I opined. "You would have gotten stuck in traffic or got a flat tire. You weren't *supposed* to make that boat. Please don't misunderstand me, but it doesn't look like that boat had your name on it, otherwise, you wouldn't be here now."

"Oh," he said, startled for a moment. "Thanks, I feel better."

(I got his card and sent him a bill.)

I do not mean here to advocate "going with the flow" or passive submission—as if we were free to do so—indeed what is set before us often requires stubborn, autonomous, courageous, solitary action. We just right now have this clear sense that we are part of some very, very big cosmic web of interconnections and meaning. There are no coincidences. We pay close attention to whatever seems to be coming down. We understand, as in the Heisenberg principle, that our eyes, our hands, our awareness are part of the equation.

In the words of Kalynomous Kalmish Shapira of Piesetzna, who perished in the Holocaust, "I may not be able to see it right now, but the Holy One fills all creation, being is made of God, you and I, everything is made of God—even the grains of sand beneath my feet, the whole world is included and therefore utterly nullified within God—while I, in my stubborn insistence on my own autonomy and independence, only succeed in banishing my self from any possibility of meaning whatsoever."[7]

BOSCHES DEL APACHE

We read in Exodus 25:20 of the ark of the Covenant. It is to be made of acacia wood overlaid with gold and carried by two long poles. And on top of this box containing the tables of the law are to be placed two gold cherubim whose "...wings are spread out above, shielding the cover [of the ark] with their wings." These winged angels come up again in Ezekiel 1:24 and 10:14.

The prophet has a psychedelic vision of God on a chariot that's borne by an eagle, a cherub, a lion, and a human being. Each creature had four wings. Ezekiel says: "...[it was] like the noise of great waters, I heard the noise of their wings, like the voice of the Almighty..."

I am standing in the middle of the New Mexico desert last winter. It is thirty degrees and a half-hour before dawn. We are in the *Bosches del Apache*, the forest of the Apache, a bird sanctuary.

I am not a big time naturalist, but my wife is an avid birder. And so, like spouses do, I tag along.

"For this one," teeth chattering, while sipping coffee from the thermos, I whisper, "I want extra points."

She smiles.

In the distance, the horizon is becoming visible now as dark ribbon of deep red in the winter chill. And dawn comes. The whole sky explodes into bright orange.

And then, within the next fifteen minutes, we watch in awe-struck silence, as 25,000 snow geese, cranes, great blue herons and God only knows who else, awake from their sleeping on the water and fly off for the next leg of their migration.

They are so close and there are so many of them, I can literally feel the flapping wing-flung wind on my face. The park rangers call it, "the flyaway." Last evening they warned: You're never the same after the flyaway. And I understand: Somehow, simply being present to experience this event changes your perception of what it means to be a creature.

And here's the thing chastens and humbles me: The birds do this all the time. Whether we're there to watch them or not, they land in the waters of the Bosches and come first light, they fly away into the dawn's early light, a sky-full of wings and beaks and feathers on their way to somewhere else. And they do it year after year after year. Just like the great whales do it through the waters of the sea and mitochondria do it through the fluid within our cells. Great flowing streams of life, currents of protoplasm—flying, swimming, running, moving, flowing, praying—doing what they know how to do—doing the only thing they know to do—doing what they were "meant" to be doing, doing what they're "supposed" to do. While I, in my ignorance—obsessed with completing some writing assignment—am doing what I'm supposed to be doing, what I am meant to do. All these creatures, moving on their ways, going about their business, like traffic on an expressway interchange at rush hour, one orchestrated flow of life. My God, I can still feel the wind of their wings on my face.

Probably the holiest ritual moment in ancient Judaism was on the Day of Atonement, Yom Kippur, when the high priest entered the holy of holies in the temple in Jerusalem. He only had one thing to do. He had rehearsed it for months. He had to pronounce one word: The ineffable Name of God. The Name made only of vowel letters. The Name made from the root letters of the Hebrew verb to be. A Name that probably initially meant something like, "The One who brings into being all that is."

And the room in which he would speak this Name was so sacred that if, God forbid, he should drop dead of a heart attack once inside, no one else would be able to go back in there to retrieve his corpse!

Rabbi Isaac explained in the Zohar[8] that they solved the problem by simply tying a rope abound his leg. Rabbi Judah further said that when the priest entered, even *he* closed his eyes—so as not to gaze where it was forbidden to gaze. But, as they sang their praises, he was able to hear the sound of the cherubim's wings…

NOTES AND REFERENCES

1. The first two sections of this essay appear in slightly different form in my book *Invisible Lines of Connection: Sacred Stories of the Ordinary* (Woodstock, VT: Jewish Lights Publishing, 1996).

2. To ask: “Why or how does The One become many?” is effectively to ask how (and why) did God create the world. Why would God, The One, create this world of separation?
3. They are still mutually exclusive, but to realize that one resides within the other is itself already a kind of monistic and mystical vision.
4. *Morality and Eros* (New York: McGraw Hill, 1970), pp. 86–87.
5. John Mellor, “Piloting Dangerous Waters,” *Cruising World*, August 1992, p. 18.
6. Yosher Divrei Emet, Jerusalem, 1974 #14.
7. B’nei Makhsahva Tova, Kalynomous Kalmish Shapira of Piesetzna, Jerusalem, 1989, 33 (Hebrew).
8. III Zohar 102a.

Are We Alone?

Lessons from the Evolution of Life on Earth

SARA VIA

Department of Biology, University of Maryland, College Park, Maryland 20742, USA

ABSTRACT: The understanding of life on Earth that we have obtained from the science of evolutionary biology offers clues to the qustion of what life might be like if found elsewhere. After presenting the basics of the evolutionary process, I discuss the factors that determine the outcome of evolution, the role of key innovations and extinction in evolution, and whether the evolution of human life is inevitable.

KEYWORDS: origin of life; evolutionary contingency; key innovations; extinction; extraterrestrial life

INTRODUCTION

Observations and experiments from the science of evolutionary biology have provided us with a solid base of knowledge about how life on Earth came to be. This is not to say that all the mysteries of the evolution of life are solved—far from it. However, the core of what we have learned about Earthly life from the study of evolutionary biology may provide important insights into the question of whether life exists elsewhere, and if it does, what form(s) it might take.

Diversity and adaptation are the defining features of life on Earth. The theory of evolution by natural selection provides an explanation for observations of diversity and adaptation in the living world. After presenting some basics about the evolutionary process, I will consider questions about a range of issues including: (1) What determines the outcome of the evolutionary process? (2) What is the role of evolutionary innovation on the course of evolution? (3) How do periodic extinctions alter the history of life? (4) Was the evolution of intelligent life inevitable?

Consideration of the course of the evolution of life on Earth puts the search for intelligent life in other universes into a context in which contingency is of

Address for correspondence: Dr. Sara Via, Department of Biology, University of Maryland, College Park, MD 20742. Voice: 301-405-8941.
sv47@umail.umd.edu

paramount importance. This raises issues about the possibility of intelligent extraterrestrial life that might otherwise be missed.

LIFE ON EARTH: DIVERSITY AND ADAPTATION

One of the hallmarks of life on Earth is its astonishing diversity. No one really knows for sure how many different species occupy the Earth, with current estimates falling anywhere between 10 and 100 million species.[1] Of those, only a small fraction are known to science even well enough to have a name (about 1,413,000 species have been named). For far fewer species is much of anything known about the details of the lives they lead on our planet.

From what we do know, however, it is clear that organisms make their living in an amazing array of different ways. Life on Earth may have one cell or many. Some organisms (plants) make their own food by capturing energy from the sun, others eat plants directly or eat the herbivores. Virtually every multicellular organism has other organisms that live in it or on it, and still others use dead biological material for food. The array of sizes ranges over or-

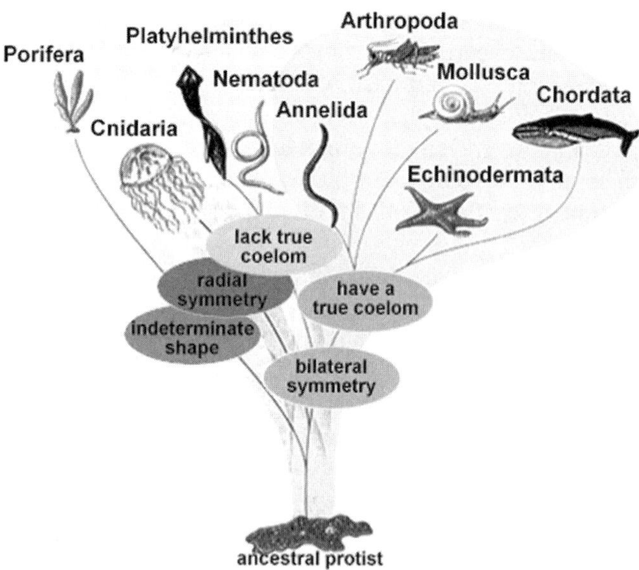

FIGURE 1. A simple classification scheme for the major phyla of animals. Some of the key innovations that occurred during animal evolution are indicated in the ovals. Note that the invertebrate lineages most closely related to Chordates are the starfish and sea urchins (Phylum Echinodermata). (From Audesirk and Audesirk.[2] Reproduced by permission.)

FIGURE 2. Praying mantis that closely resembles plants in its environment. (From Gould and Keeton.[3] Reproduced by permission.)

ders of magnitude, from microscopic bacteria to free-living animals that can weigh tons, to fungi in which a single organism may occupy hundreds of square miles of soil. Finally, organisms come in all shapes (FIG. 1), from those lacking any symmetry (such as sponges, Phylum *Porifera*) to the radially symmetrical jellyfish and corals (Phylum *Cnidaria*), to the bilaterally symmetrical higher invertebrates and vertebrates (Phylum *Chordata*). As if this weren't enough, the fossil record shows us that even more bizarre forms once lived on Earth. During the Cambrian period, and preserved in the Burgess Shale fossil deposits from British Columbia, we can see organisms with shapes not present in any living organism, that are so different they fit into no phylum that is currently alive. So, life on Earth has been even more diverse than it is today. Of these millions of different species, only one, *Homo sapiens*, is classified as "intelligent life." For various reasons discussed below, the probability that life in any form could occur in other universes is a much broader question than the probability that "intelligent" life could be found elsewhere.

The second defining feature of life on Earth is adaptation. An adaptation can be defined as a biological characteristic that suits an organism to the particular environment in which it lives. Adaptations may be seen in nearly all aspects of the size, shape, and coloration of organisms, as well as in the biochemical and physiological details of how organisms work. For example, the insect shown in FIGURE 2 (praying mantis) is shaped and colored just like the plants in the habitat it occupies. This resemblance to the background (crypticity) permits the mantis to effectively stalk its prey while concealing itself from its own predators. The converse of crypticity is seen in plants with

FIGURE 3. Pit viper; arrow points toward unique heat-sensitive organ used to locate mammalian prey at night. (From Gould and Keeton.[3] Reproduced by permission.)

brightly colored flowers that attract pollinators or in dangerous and/or distasteful creatures with orange or yellow and black warning coloration, such as bees and wasps. These warning colors signal "Don't mess with me or you'll be sorry." Both of these types of coloration (cryptic and warning) are adaptations that provide these organisms with enhanced survival or reproduction in their particular environment. Adaptations can also be seen in how organisms work. The mechanics of flight in insects and birds has lead to exquisite engineering modifications of structure with accompanying adaptations in muscle physiology, circulation and respiration. The ability of a male moth to detect the scent of a female from several miles away, using highly elaborated sensory organs on the antennae, is an adaptation, as is the organ in pit vipers that permits detection of warm-blooded prey objects in the dark (FIG. 3).

DIVERSITY AND ADAPTATION EXPLAINED

Diversity and adaptation are features of our world that demand explanation. The only explanation that is consistent with the range of scientific observations made over the past several hundred years in biology, geology and chemistry is the theory of evolution by natural selection. This scientific theory was originally formulated in the late 1800s by Charles Darwin and, simultaneously, by Alfred Russell Wallace (for a good description of the historical context within which the ideas about evolution by natural selection were developed, see chapter 2 in Futuyma[4]).

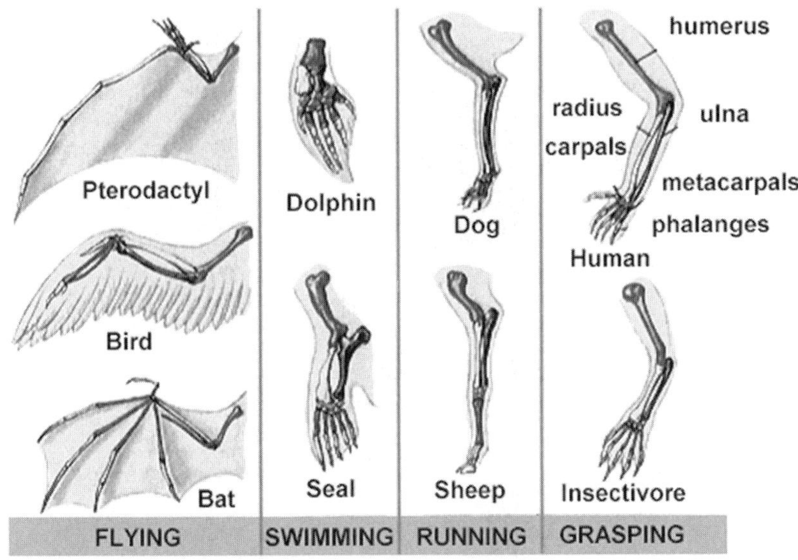

FIGURE 4. Homologous structures in the tetrapod limb, modified for various functions. Notice that the various bones, which are tinted different colors, have been modified in different ways to accomplish each type of locomotion (e.g., compare the modifications of the bones for flight). (From Audesirk and Audesirk.[2] Reproduced by permission.)

The process of evolution and its basis as an explanation for diversity and adaptation is two-fold. First, evolution is a change over time in either the kinds of species present on the Earth, or in their characteristics. Such changes can be clearly documented. For example, some types of organisms present in fossil layers deposited 400 million years ago have never been found in rocks that were formed more recently. In turn, many fossilized forms found in more recent sediments are never seen together with the kinds of fossils found in older sediments. For example, we know from fossil evidence that no humans were alive at the time of the dinosaurs—dinosaurs and humans are never found in rocks of the same age. This is evidence for evolutionary change in the composition of species over time.

Secondly, evolution is characterized by descent with modification. This means that organisms that share a common ancestor will be expected to also share many characteristics. During descent, however, variability in characters occurs, and features can (and have) become modified to take on different functions. For example, all life shares a common genetic code. This is powerful evidence that we are all descended from a single common ancestor. However, different organisms have very different genes and/or forms of

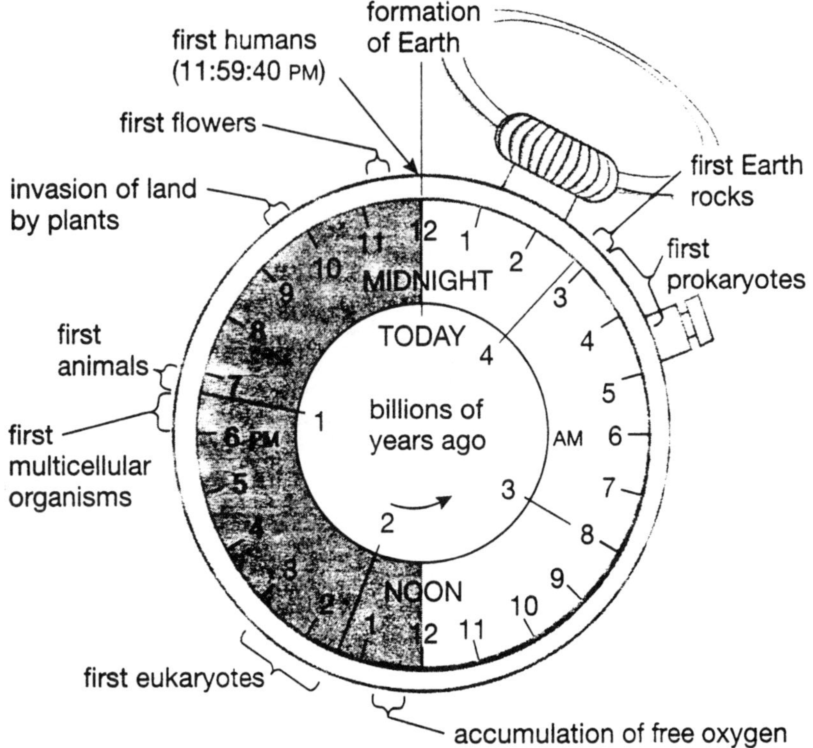

FIGURE 5. Earth's history depicted on a 24-hour clock. (From Audesirk and Audesirk.[2] Reproduced by permission.)

genes, and these differences in the "instruction manual" are responsible for many of their physical differences. This is descent with modification. A classic example of descent with modification is the observation that all four-legged vertebrates (tetrapods) share common bone types in their limbs. Such bones that are similar because of ancestry are called homologous. Looking at the limbs of tetrapods as different as bats, dolphins, sheep and humans reveals that the striking differences among their limbs have been evolved through modification of the same set of bones into wings, flippers, legs and arms (FIG. 4). This homology of structures with very different functions among organisms due to descent with modification is evidence for evolution. For further evidence of change over time and descent with modification, see the excellent treatment in chapter 3 in Ridley[5]).

Evolution is a fact. We can observe change over time and descent with modification. So why do we call the process that leads to diversity and adaptation the theory of evolution by natural selection? In science, a theory is "a statement of a natural law... QED." A scientific theory is not just a casual idea, or a set of untested hypotheses, but is instead a well-tested explanation of the natural laws that cause observed facts on Earth. To put the theory of evolution by natural selection into a context, another scientific theory that we all experience each day is the theory of gravitation. This theory explains why my keys fall to the floor when they are dropped. The explanation for the observation that the keys fall is based on well-tested and thoroughly substantiated knowledge of the causes (and effects) of gravity on the Earth. The explanation of the observations of change over time and descent with modification in the biological world is the similarly well-tested and substantiated theory of evolution by natural selection. For further discussion on evolution as fact and theory, see chapter 1 in Futuyma.[4]

EVOLUTION BY NATURAL SELECTION: HOW DOES IT WORK?

The evolutionary process has three basic components: variation, inheritance, and differential survival or reproduction. If these components are present in populations, then adaptive evolution will occur. First, nearly every characteristic of organisms varies slightly among individuals within populations. This variation is clear if we take any group of people, dogs, fish, plants, or any other group of organisms. Looking closely, we will see that aspects of size, shape, behavior, and other traits vary among individuals. Some of this variation is due to the environment that an individual has experienced. For example, getting a bit more food might make an organism slightly larger than others in its population. This kind of variation, caused by differences in the environment experienced by organisms, is not passed on to offspring and does not contribute to the evolutionary process. However, many differences among individuals are due to genetic differences, and this genetically based variation does contribute to evolution. For example, taller than average parents tend to have taller than average offspring because they pass on their genes for large size. Finally, when the differences among individuals affect their ability to survive and reproduce, then the individuals with the favorable characters will leave more offspring than others. This differential survival or reproduction of individuals with different characteristics is natural selection. If the favorable trait is inherited, then more of the next generation will have that trait, having inherited it from their successful parents.

This is the basis of evolution by natural selection: variation that is inherited (genetic variability) confers an advantage in survival or reproduction (natural selection) that leads to change over time and descent with modification. The traits that confer success are called adaptations, and these spread through populations because the individuals that bear them leave more offspring than do individuals with other characteristics. When inherited variation affects survival or reproduction, then evolution is inevitable.

WAS THE EVOLUTION OF LIFE AS WE KNOW IT INEVITABLE?

If variation is ubiquitous, and the variants that survive and reproduce best in a given environment come to dominate populations in an inevitable process, then wouldn't evolution proceed the same way again if it started all over? Did the process of evolution lead inexorably to life as we know it? I will argue that the answer is "no," that although the process of evolution always works in the same way, the outcome is highly contingent on a complex array of factors.

First, "life as we know it" is only a small slice of life on Earth. Most of us spend our days noticing only a few types of organisms—mammals (our families and pets), maybe other vertebrates (birds, fish), flowering plants (nearly everything in our yards except conifers), and perhaps fungi or "lower" plants like moss and algae, if we look hard. But fungi, plants, and animals are just the tips of the top branches on the tree of life, and even within these huge groups of organisms, only one species is usually considered as "intelligent life."

As mentioned before, most species on Earth have not yet been discovered or named by scientists—we may know most of the vertebrates, but in other groups, there are still many types of organisms that remain unknown to science. Of those organisms that have been named (about 1,413,000 species) most are insects (> 50%). About 20% are flowering plants and conifers, while mammals number only 4000 species.

The animal phylum in which vertebrates are found (Phylum *Chordata*) was the last animal phylum to diversify, and within this, the evolutionary path to mammals, and then on to primates (monkeys and chimps) and finally to humans has been a complex one. At many points, the outcome of evolution could have been different had even a small aspect of the complex environment of life been different. The evolutionary process is highly contingent. Natural selection modifies what is already present, given the genetic variation that is available, to promote traits that are favorable in the current environment. Evolution by natural selection is also opportunistic. Natural selection favors the best characteristics that are available. It does not lead to adaptation to future environments, nor preserve adaptations to past environments if such traits are no longer advantageous. Life could not have begun (and didn't) with complex organisms such as mammals, because life began as simple self-replicating molecules from which complexity evolved through an ongoing process of descent with modification.

The time course of the evolutionary path of life on Earth can be visualized as a 24-hour clock (FIG. 5). The evolution of life on Earth has been proceeding for 3.9 billion years, since the appearance of the first prokaryotes (single cells with no nucleus, such as bacteria). The immensity of time that passed between the evolution of these first life forms and the first multicellular organisms (2.8 billion years) is hard to fully grasp. For about the first 3.6 billion years of life, all living things occupied the oceans. The land was nothing but bare rock for all this time. Only 500–440 million years ago did organisms begin to colonize land. The clock metaphor makes the brevity of the existence of "intelligent" life on earth particularly obvious. If the Earth was formed in the first seconds of the "day," the first humans did not appear until 11:59:40 in the history of Earth—just a few seconds before the present (about 100,000 years ago).

Consider the impact of this evolutionary time scale on the likelihood that intelligent life would be found by an alien searching for life on our planet. Clearly, for most of Earth's history the only forms of life to have been found would be prokaryotes in the ocean (Domains Bacteria and Archaea). A time traveler that happened upon our world within the most recent "hours" of its existence (about the past 400 million years) might have been able to see simple land plants, amphibians, and reptiles. Mammals only began to proliferate after the extinction of the dinosaurs 65 million years ago. It would have to be a lucky time traveler indeed that happened upon our world during the "20 seconds" of its day that it has been occupied by humans. If "intelligent" life has evolved in other worlds, one could expect that the same immense amount of time may have elapsed there before its appearance. What is the chance that during the brief existence of intelligent life on two worlds, they would find one another? It seems much more likely that if we were to find life on other worlds, it would be some life form that were present during the bulk of the long time course of evolution, such as a prokaryote (see also paper by Nealson in this volume).

THE CONTINGENCY OF EVOLUTION: WHAT DETERMINES THE OUTCOME?

Earthly organisms must be made from *materials that are present on Earth*. Natural selection favors the individuals in a population that survive and reproduce the best *relative to others present*, in the *current environment*. The italics in the previous sentences highlight the three main contingencies of evolution: (1) Only certain materials were present in the early Earth to serve as raw materials for life; (2) natural selection is limited by the range of possible biological variation and by the variation present at a particular time; and (3) the environmental circumstances at a given time determine the characteristics favored by natural selection. Changes in the environment lead to changes in the type of characteristics that are favored. The outcome of the evolutionary process is determined by these three contingencies. Thus, it is worth examining them in more detail to determine whether we can reasonably expect the evolutionary outcome that has occurred so far on our Earth to be similar to the outcome of an evolutionary process in another world.

(1) The outcome of evolution depends on the materials available at the start of life. The elements that were common on Earth at the time that life was thought to have originated were carbon (C), hydrogen (H), oxygen (O), nitrogen (N), phosphorous (P) and sulfur (S). All biological molecules are combinations of these fundamental elements. Presumably, during the early years of the Earth's history, many different kinds of combinations of these fundamental elements occurred under various physical circumstances. Eventually, the molecules we know as nucleotides were formed, and then they combined into self-replicating RNA-based forms called ribozymes. If we take the ability to self-replicate as the defining feature of life, then these ribozymes were probably the first life on Earth. Further modifications led to the formation of the true nucleic acids (RNA and DNA). DNA is a giant polymer with a sugar-phosphate backbone of nucleotides and nitrogenous "rungs" made of different combinations of C, N, H and O. DNA forms the universal genetic code of information used by all life. RNA then plays a key role in the transfer of information from the DNA to the proteins that are the building blocks of every organism's body.

Once the trial and error process of early biochemistry produced self-replicating RNA, then it appears likely that the evolution of DNA as the genetic code may have occurred quite rapidly. At that point, other types of self-replicating molecules that could possibly have occurred were at a disadvantage, because a self-replicating system that worked was already beginning to dominate the use of resources. So, the first main contingency of life on Earth may be the genetic code itself on which all life is based. Though it is probably not the only way that information could be coded in a molecule, it is the way that became established on the early Earth, and nothing has come along to supplant it.

Another influence of the early materials present on Earth was the fact that water is the universal solvent. Water has very special chemical properties that we largely take for granted, and these properties have dramatically shaped the chemistry of biological life. It is important to recognize, however, that water is not the only possible solvent. Life based on a different set of initial elements, and a different solvent, might be entirely different from the life we know on Earth.

(2) Natural selection can only act on biological variation that is present at a given time. Most characteristics of organisms that matter to survival and re-

production can be described in a given population by an average value and a range of genetic variation around the current average. Evolution proceeds when natural selection favors certain variants that survive or reproduce better than others within their environment, and thus contribute more than their share of progeny to the next generation. This process of differential survival and reproduction of individuals with different characteristics leads to a change in the average value of characters so that the population comes to look more like the favored variants (change over time and descent with modification). The key here is that this variation involves slight differences from the current average. Thus, descent with modification means that as evolution proceeds, the average characteristics are modified into new forms. The modifications of the bones in the vertebrate limb for very different functions (FIG. 4) are an example of this type of evolutionary change. Even a huge evolutionary step like the colonization of land probably proceeded in this way, when a few lobe-finned fishes began to spend time scooting along muddy shores on their weak fins. Later lineages show modifications of this crude beginning of terrestriality, leading to the amphibians (only partially terrestrial), and later to the fully terrestrial reptiles. Change over time and descent with modification can be seen clearly in the evolution of terrestrial vertebrates.

Rarely, a variant form that is radically different appears in a population—one that does not fit within the usual distribution of variation. Such radically different forms may have been very important in the evolution of major innovations during the history of life. It is probable, however, that the first occurrence of a major new characteristic was only a crude version of its final evolutionary form that permitted organisms to take on a new function and that was then further modified by natural selection acting on small variations.

The key point here is that if a variant is not produced for some reason, that characteristic cannot and will not evolve in real populations, no matter how advantageous it might be. For example, it would be handy for large mammals to construct their skeletons from titanium instead of bone. This would allow a much lighter weight skeleton, and would probably permit the evolution of much larger body sizes, since the size of terrestrial vertebrates is largely limited by the strength of bone. However, titanium is not a material that is used by organisms (with the exception of a few prokaryotes, which appear to use almost every element on Earth [see Nealson's paper, which follows]). Thus, a variant within a mammal population that has a titanium skeleton is highly unlikely to ever appear. Instead, the best-working form of a skeleton that can be made with biological materials will be favored by natural selection.

(3) The environmental circumstances at a given time determine the form of natural selection and thus the outcome of evolution. Natural selection is the differential survival or reproduction of individuals within a particular population in a particular set of environmental circumstances. No single set of characteristics is universally favored in all organisms, because the characters that lead to high survival in one situation will not necessarily lead to good sur-

vival in a different environment. As the evolution of life on Earth has proceeded, there have been changes in the environment due to periodic physical changes on Earth. Radical climate changes have been caused by movement of the continents. Changes in orbital cycles and the earth's tilt have led to ice ages and dramatic changes in sea level. These enormous physical events have led, in some cases, to major extinctions, and they have certainly affected the characters that have been favored in organisms at particular times. Even more importantly, life has changed its own environment. As living things have diversified, they have become inextricably part of one another's environments. Consider the first life on land, colonizing bare rock. Soil was formed as the first bacterial colonists died and decomposed. This allowed plants to take root, whose decomposing bodies further added to the soil-building process. The evolution of flowering plants on land has provided not only food but also habitat for many different kinds of terrestrial organisms. The diversity of insects, for example, would not be at current levels were it not for the diversity of flowering plants that have evolved. In turn, plant diversity has increased through the effects of the insects themselves on speciation. In life on Earth, many species evolve together ("coevolve"). They acts as part of one another's environment, determining for each other the characteristics that lead to high levels of survival and reproduction.

KEY INNOVATIONS OPEN NEW REALMS

Occasionally during the history of life, the process of evolution by natural selection has produced characteristics that radically change the future course of evolution. These characteristics are called "key innovations" because they open up possibilities for life and the evolution of characters that were simply not possible before they evolved. One of the biggest breakthroughs in the history of life occurred 3 billion years ago. This was the evolution of pigments in bacteria that could accomplish photosynthesis, that is, the capture of light energy and the production of sugars from water and carbon dioxide.

Life's first cells subsisted by anaerobially (i.e., without oxygen) metabolizing inorganic compounds. This type of metabolism is not very efficient, and provides relatively little energy. Thus, the pace of early life was slow, and food supplies were limited to available (and non-renewable) inorganic molecules. The evolution of photosynthetic pigments within cyanobacteria that capture the energy of the sun and permit the construction of sugars, a renewable food source, was literally an Earth-changing event for several reasons.

First, oxygen is released as a waste product from the process of photosynthesis. Up until this time in the history of life, all life was anaerobic, and oxygen is poisonous to these cells. However, although oxygen was produced by the first photosynthetic bacteria, free oxygen was not liberated because iron in the Earth's crust took it all up, forming iron oxide. This process (the "rusting" of the world) lasted for 1.2 billion years, until the iron stocks were saturated. Then, free oxygen began accumulating in the atmosphere as O_2. This free oxygen relegated anaerobic bacteria forever to a minor role, as most of Earth's habitats became saturated with oxygen.

Secondly, for the first time on Earth, new food was made by the organisms themselves. Previously, all nutrition was obtained from existing inorganic molecules. The production of a renewable resource from the energy of the sun and commonly available molecules permitted a huge expansion in the extent and diversity of life on Earth.

The evolution of photosynthesis and the resulting change of the atmosphere from anaerobic to aerobic, along with the evolution of new food sources, was an irreversible change in the environment on Earth, one that would be part of the landscape for all of the major phyla of organisms (including all multicellular organisms) that evolved after that time. Photosynthesis initiated the inextricable interdependence of plants and animals. Plants are the only organisms that can harness the energy of the sun and use it to make food. The waste of plants (oxygen) is a requirement for animals, while the waste gas from animal respiration (carbon dioxide) is the raw material needed for photosynthesis. The required balance of plant and animal life on Earth was in place from that point onward. If photosynthesis or its equivalent had evolved using a different chemical mechanism, which seems possible under different conditions, the entire history of life would have been different. Is there any reason to believe that Earth-like photosynthesis would evolve on other worlds? It seems unlikely.

Following the evolution of photosynthesis, chance variation in biochemical pathways led to further key innovations. Some lineages of prokaryotes were able to utilize the waste oxygen for aerobic metabolism. Metabolism using oxygen is much more efficient than is anaerobic metabolism and liberates a great deal more energy, permitting the evolution of increasingly active life styles, including the evolution of the first predators, which were single-celled prokaryotes. Some of these prokaryotic predators ingested aerobically metabolizing bacteria. By chance, some of these prey were not digested, but came to live symbiotically within the "predator" cell. Eventually (about 2–1.1 billion years ago), these evolved into mitochondria, the powerhouses of the eukaryotic cell. A similar ingestion of a cyanobacterium in a lineage of protists leading to plants is thought to have led to the evolution of the chloroplast. With these two events, eukaryotic cells gained an enormous ability to make and process energy. The evolution of cellular organelles paved the way for the eventual evolution of plants and animals. This was a huge breakthrough in the history of life, without which the enormously active animal phyla could not have evolved.

So, long before organisms became multicellular, chance events had altered the trajectory of life—a particular genetic code was found and put into use by self-replicating nucleic acids, photosynthesis and aerobic metabolism evolved, and cells took on organelles (mitochondria and chloroplasts) that made new levels of activity possible. Would this same trajectory happen again on our Earth? Would the evolution of life on other worlds even be close?

After the evolution of multicellularity, about 1 billion years ago, natural selection within different lineages of single-celled Eukaryotes led to the evolution of plants, fungi and animals, the three main eukaryotic Kingdoms. Within animals, the evolution of the invertebrate phyla began, and during the "Cambrian explosion" between 544 and 505 million years ago, all of the animal phyla present on earth today (and evidently a great many more) appeared. Over time, various key innovations evolved from chance variation arising in the genetic code, which then spread through populations by the advantages that the variant characteristics provided to organisms that bore them. These included the evolution of tissues, bilateral symmetry (which permitted the evolution of a head end and the accompanying concentration of sensory organs there), the elaboration of specialized organ systems, the evolution of the exoskeleton (in insects, spiders and aquatic arthropods like crabs), and, eventually, the evolution of a vertebral column (see FIGURE 1, which shows the major animal phyla and the key innovations that spurred their evolution).

Even after the first evolutionary appearance of the vertebrates, there was still a considerable period of evolution required before the evolution and diversification of the mammals, the primates, and finally humans. Major innovations made the colonization of land and the evolution of mammals possible, including impermeable skin, the evolution of internal fertilization, the evolution of placental development of offspring, their nourishment through milk, and the evolution of homeostasis, which permits mammals to hold a constant body temperature. These characteristics produced lineages that provided great amounts of parental care to developing juveniles, permitting skills to be passed from parent to offspring. The tremendous biological diversity of the world into which mammals evolved may itself have favored the evolution of larger brain size through natural selection. No doubt other characteristics of the environment in which hominids evolved favored the evolution of large and active brains. Would these circumstances inevitably arise again during the course of a different bout of evolution?

Does the process of evolution itself necessarily lead to complexity, regardless of the details of exactly the forms of life that are evolving? It seems true that the very process of biological evolution is an elaborative one in which diversity of forms necessarily would be expected to increase. With that diversity comes new environments, new ecological opportunities, more biological possibilities. Would this always lead to complexity in the forms of life? In particular, would it always be expected to lead to intelligence?

The final kind of contingency that has played a significant role in the course of evolution of life on Earth has been periodic extinction. Though there have been six periods in the history of life severe enough to be termed "mass extinctions," many more minor periods of large-scale extinction have been noted. Mass extinctions have generally be caused by large-scale physical changes on the Earth, such as the fusing of the continents into a single land mass called Pangaea, or its subsequent breaking apart, though minor extinctions have been caused by a range of events, including activities of humans. Overhunting is thought to have caused the extinction of large mammals in North America about 10,000 years ago, and by all accounts, we are currently in the midst of a mass extinction caused by human activities.[1]

During each mass extinction, the majority of life forms present at the time have gone extinct. This wholesale elimination of types of organisms not only extinguishes those lineages, but also opens new opportunities for the lineages that manage to survive. In fact, each of the mass extinctions has heralded a new age in the history of life. For example, reptiles ruled the Earth for millions of years, but after the mass extinction 65 million years ago, the Age of Mammals dawned. During the time of the dinosaurs, small mammals had existed, but after the huge reptiles went extinct, mammals proliferated and diversified within a few million years. Ecological opportunities that had been dominated by reptiles became the purview of mammals. If the dinosaurs had not gone extinct, would humans have evolved? If we are currently involved in a mass extinction, what will be the next group to dominate the Earth?

The time at which a mass extinction occurs also plays a role in the subsequent trajectory of life's evolution. For example, when the dinosaurs went extinct, mammals were already present in low numbers and diversity. Thus, variants within mammal lineages were present and able to spread into the habitats dominated by dinosaurs through natural selection. If mammals had not yet evolved by 65 million years ago, those habitats might have been recolonized by a rediversification of reptiles or some other lineage. In an earlier extinction (about 440 million years ago), many marine invertebrates went extinct, opening the door for the diversification of fishes, leading to the Age of Fishes. Mammals could not have diversified and come to dominance then because the mammalian form had not yet evolved. The results of evolution after an extinction depend upon when it occurs.

DIFFERENT WORLD, DIFFERENT ORGANISMS?

This discussion suggests that much of the history of life on earth is due to a particular sequence of events that could have happened differently. It is tempting to ask what would have happened if key aspects of evolution on Earth had been different:

(1) What if water weren't the solvent, and the common elements were not C, H, N, O, P and S? Could life (a self-replicating entity of any form) evolve?

(2) What if life had evolved on land instead of in the ocean?
(3) What if a different genetic code had appeared in the first self-replicators?
(4) What if gravity were greater (or less)?
(5) What if different materials were available for use by organisms?
(6) What if there had been a different pattern or timing of mass extinctions?

Would life as we know it have evolved the same way? One would suspect that the trajectory of evolution would have been very different. Would Eukaryotes have even evolved?

It remains to be seen whether traces of life, present or past, will be found in other worlds. From this discussion, it seems likely that if signs of self-replication are found elsewhere, the forms of life may be very different than those seen on our planet. Our challenge may be not only to locate extraterrestrial life, but to recognize it.

FURTHER READING

Excellent general descriptions of the history of life on Earth and the process of evolution can be found in Audesirk and Audesirk,[2] Futuyma,[4] Gould and Keeton,[3] and Wilson.[1]

NOTE AND REFERENCES

1. E.O. Wilson, *The Diversity of Life* (New York: W.W. Norton, 1992).
2. T. Audesirk and G. Audesirk, *Biology. Life on Earth,* 5th edition (Upper Saddle River, NJ: Prentice Hall, 1999).
3. S.L. Gould and W.T. Keeton, *Biological Science,* 6th edition (New York: W.W. Norton, 1996).
4. D.J. Futuyma, *Evolutionary Biology,* 3rd edition (Sunderland, MA: Sinauer Press, 1998).
5. M. Ridley, *Evolution*, 2nd edition (New York: Blackwell Science, 1996).

Searching for Life in the Universe

Lessons from the Earth

KENNETH H. NEALSON

Jet Propulsion Laboratory, California Institute of Technology, Pasadena, California 91109, USA

ABSTRACT: Space programs will soon allow us to search for life *in situ* on Mars and to return samples for analysis. A major focal point is to search for evidence of present or past life in these samples, evidence that, if found, would have far-reaching consequences for both science and religion. A search strategy will consider the entire gamut of life on our own planet, using that information to frame a search that would recognize life even if it were fundamentally different from that we know on Earth. We discuss here how the lessons learned from the study of life on Earth can be used to allow us to develop a general strategy for the search for life in the Universe.

KEYWORDS: life; Mars; prokaryotes; eukaryotes; biosphere; protists; extremophiles; astrobiology; phylogeny; microbes; diversity

Within the next decade, NASA, in conjunction with colleagues from several nations, will embark on one of the most exciting missions yet undertaken in the exploration of space: the return of samples from Mars. The mission, which has been moved back several years, will almost certainly have an architecture similar to that originally planned for the Mars 03/05 mission (shown in FIGURE 1). That is, it will have two two separate launches, each of which will conduct experiments on the surface of Mars, retrieve and store samples, and put these collected samples into Mars orbit in two separate sample canisters. Subsequently, a Mars orbiter (sample return vehicle) will be employed to retrieve the orbiting sample canisters, and return them to Earth for analysis. Probably by 2011 we should have on Earth the samples from the surface of Mars on the order of 1–2 kg for scientific study. These samples will add an immense amount to our knowledge of the solar system, and of Earth itself, and will also be carefully scrutinized for the presence of indicators of present or past life on the red planet. Given the absence of any obvious features on Mars that suggest life, and the negative results (with regard to life) obtained in the Viking mission of the 1970, a skeptical observer might well ask, "Why send such a mission?"

Address for correspondence: Dr. Kenneth Nealson, Jet Propulsion Laboratory, California Institute of Technology, 4800 Oak Grove Drive, Pasadena, CA 91109-8099. Voice: 818-354-9219; fax: 818-393-4445.

knealson@jpl.nasa.gov

FIGURE 1. Architecture of a Mars Sample Return (MSR) Mission. The original architecture of the MSR as envisioned in 2000 shows the general plan and time scale for a joint mission concept between the U.S. and France. As originally planned, the mission consisted of two launches, one in 2003 with a lander and rover, and a second one in 2005, with a similar lander, rover, and orbiter (Earth ReturnVehicle or ERV). After landing on the Mars surface, the rover will collect samples (consisting of cores of both rocks and soil) and return these samples to a small cache in a rocket on the lander. These samples will be placed in a small sphere (orbiting cylinder or OS) and put into Mars Orbit, so that at the end of this phase, there will be two small "satellites" orbiting Mars. These will be retrieved by the ERV, and returned to Earth for analysis. The dates of this mission have been moved backwards, with a projected launch date of late in the decade.

In essence, the results of recent scientific inquiry in several areas provide the basis for increased optimism for finding life elsewhere in the universe, and if we are going to launch such a search: our nearby neighbor, Mars, is a reasonable first step. With regard to new knowledge: (1) we have learned things about the universe that have made the search for extraterrestrial life much more reasonable; and (2) our understanding of life on Earth has changed dramatically, altering our view of our own planet's biota, and therefore of the possibility that life might exist on places previously regarded as too hostile for life. While the astronomical discoveries (new suns, new planets, new information about the structure and history of Mars, the purported ocean on Europa, etc.) are important, it is the "lessons from the Earth" that I will concentrate on here.

As the director of the Center for Life Detection at the Jet Propulsion Laboratory, I was delighted to be asked to contribute to this symposium on science and religion. The connection with the purported conflict between these two areas is not at all obvious for a microbiologist interested in the early evolution of the Earth and its biota. From my perspective, both of these endeavors were possible only after humankind developed to a point where we had the honor and pleasure to practice science and to contemplate our own destiny, ethics, and morality. Thus, for most of the Earth's history our planet was dominated by life that had nothing to do with either science or religion—it really could have cared less! To this end, I believe it is appropriate to begin my presentation with our current view of life's history on our own planet, keeping in mind that any search for life elsewhere must be framed by our knowledge of the history of life on Earth.

In its early life-compatible stages, the Earth was still a fairly hostile spot for life as we know it now. It was warm, lacked oxygen in its atmosphere, and, because of the lack of oxygen, had little ozone to protect the planet from harmful ultraviolet light. Yet, it was in such an environment that life arose and left its earliest records. From recent studies of the Issua formation in Greenland, traces of metabolic activity (carbon metabolism) indicate that life existed on Earth as early as 3.8 Ga (billion years ago). This suggests that the invention of life took place rather rapidly, roughly within 200 million years of when the planet cooled and thereby became hospitable for carbon-based life.

These discoveries have triggered speculation about life in general (e.g., the problems associated with the invention of complex living systems), as well as about the possibility that similar living systems might have evolved on other planets. For example, it is generally agreed that in the early period of planetary development, and up until about 3.5 Ga, Mars and Earth may have shared similar planetary conditions. This has lead many to posit that life might have had adequate time, and the proper conditions, to develop on early Mars as well as here. Subsequently, however, Mars lost its magnetosphere, hydrosphere, and most of its atmosphere, making the surface of Mars, by Earthly standards, an extremely hostile environment. While the current conditions of high UV light, absence of liquid water, and sub-freezing temperatures suggest that extant surface life would be precluded, the possibility that it may have once existed cannot be excluded on the basis of on our knowledge of the history of the planet.

Most of the evidence of the earliest life on Earth is in the form of chemical tracers: there are few truly ancient fossils. This absence of a fossil record implies that simple, unicellular life dominated the early Earth, a fact consistent with what we know of unicellular life on the planet today, and of conditions necessary for fossil preservation. In fact, until about 2 Ga (billion years ago), there was little oxygen on the planet, and the development of complex eukaryotic cells, which live via oxygenic respiration, was probably not possible.

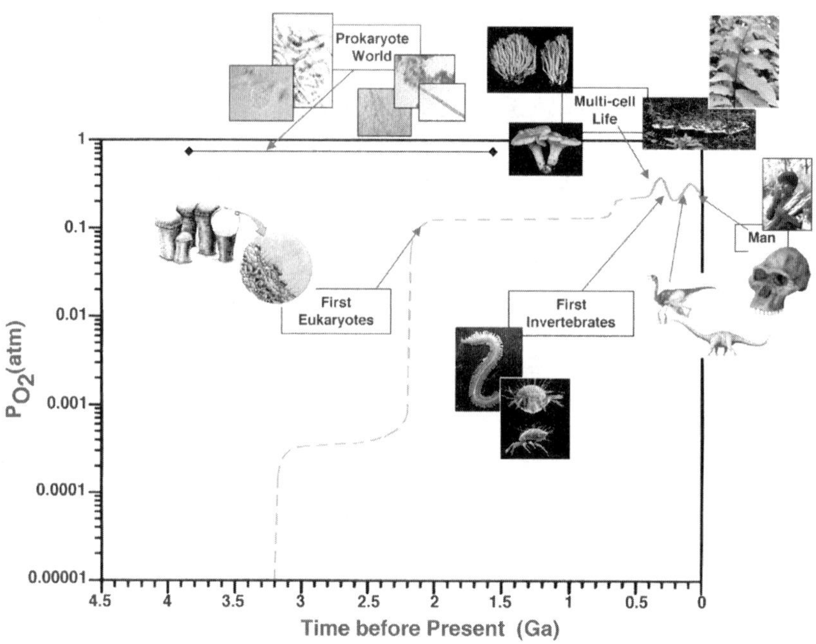

FIGURE 2. Evolution of life on Earth as related to the appearance of oxygen in the atmosphere. The plot of oxygen versus time is modeled after the data of Rye and Holland (1999), who have proposed this pattern as the most likely based on studies of ancient soils (paleosols). The pictures of organisms are meant to emphasize that the early Earth was colonized by simple organisms, probably prokaryotic in nature, and that complex organisms (multicellular large creatures) did not appear until the oxygen levels were near to what they are now, approximately 500 million years ago.

On the basis of the study of ancient soils, it is believed that oxygen first appeared and rose rapidly in the atmosphere approximately 2 Ga (FIG. 2) and that it was only upon this rise that the development of eukaryotic cells was possible. The Cambrian explosion of species and complex multicellular eukaryotes did not occur until approximately 500 million years ago, when oxygen reached current levels. From that point onward, the Earth began to take on what we would find a familiar appearance: occupied by plants, animals, and fungi.

However, even before the rise of oxygen, Earth was teeming with microbial life—this is the perspective that must be kept in mind when searching for life on other planets of unknown evolutionary age. Indeed, other planets could be in any of these stages, and the search for life cannot simply assume that a given stage of life or planetary evolution will have been reached. One should also keep in mind that planets evolve and life evolves, and that the interplay between these alters both: the evolution of Earth (on which the presence of life has made a strong impact) has had a drastic impact on the subsequent evolution of life. The oxygen we breathe is a product of the early evolution of photosynthesis, which supplies it. Without this innovation, the planet might well be alive, but its life would look much different from what we see today.

But what is life on Earth really like today—how well do we know the life on our own planet? The answer to this question is probably, "Very poorly," especially if we consider the full complement of life on Earth. To do so, we must take into account not only the "higher" organisms, but also the single-celled, anucleate organisms collectively referred to as prokaryotes. These organisms need to be factored into any discussion of the nature of life on Earth, both past and present. To illustrate the way in which prokaryotes are viewed now in comparison to just a few years ago, I offer the following quotation from Dr. E. O. Wilson, the noted expert on animal evolution, from his book, *The Naturalist*:

> If I could do it all over again and relive my vision of the 21st century, I would be a microbial ecologist. Ten billion bacteria live in a gram of ordinary soil, a mere pinch held between the thumb and the forefinger. They represent thousands of species, almost none of which are known to science. Into that world I would go with the aid of modern microscopy and molecular analysis.

Dr. Wilson has made immense contributions to our understanding of macroscopic life on Earth, and in this quote he expresses the opinion that it is now time to move such thinking to the microbial level. To expand on these thoughts, I would point out that while microbial biomass is thought to account for 50% or more of the Earth's biomass, we have been able to culture and characterize only a small percentage of these prokaryotes, and thus know almost nothing about nearly half of our Earthly biota. This is a sad state of affairs for a society that claims to be ready to embark on a search for life in the Universe!

In addition to our new insights about the early appearance of Earthly life, a number of other biological findings have changed our perceptions of life on Earth. These new developments, which must be factored into any search for extraterrestrial life, involve the nature and diversity of life on Earth (e.g., the very definition of Earthly life), as well as new insights into the toughness and tenacity of life as we know it.

With regard to our view of the nature of life on Earth, major changes have occurred in the past two decades. We have moved from a peculiarly eukaryocentric view of life to one that openly admits that the small, single-celled creatures that were once ignored play a vitally important role in the metabolism of our planet. The view of life that most of us learned from our biology teachers is that commonly referred to as the five kingdoms (see below). It was derived through the work of Linnaeus and others in the mid-1700s.

This classification scheme relied upon observation of the visible features of organisms to give each a name (e.g., *Homo sapiens* for humans), and to

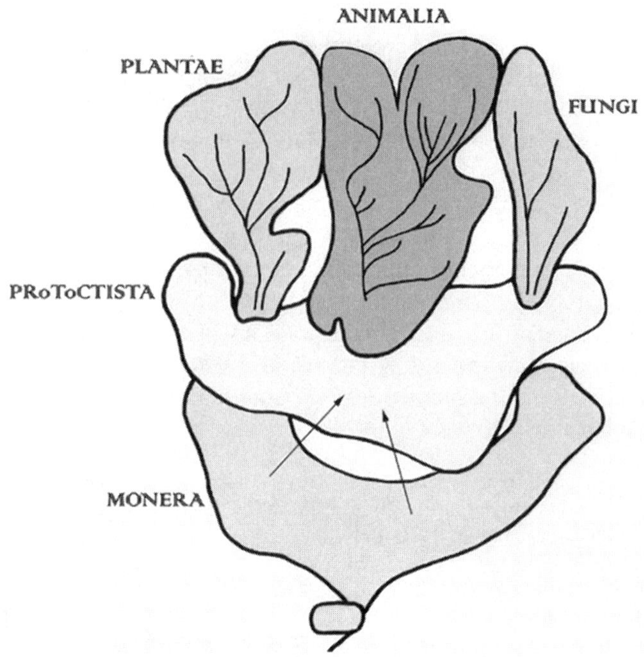

FIGURE 3. The five kingdoms of life. This view represents Earthly life as composed of five kingdoms, four of which are eukaryotic, and the fifth, the monera (what we now call prokaryotes), at the base of the tree. This view of life is based on structural and functional analyses of organisms, and a proposed evolutionary line in which life moved from simple to more complex, and smaller to larger.

group organisms of similar appearance together. The diagram in FIGURE 3 is called a phylogenetic tree; these trees are used to illustrate the evolutionary progression that may have occurred to result in the extant organisms (e.g., to answer the question of which organisms preceded which in time).

Largely because of the nature of the tools available (human eye, hand lens, and later, simple microscopes), it is not surprising that the such trees were dominated by the macroscopic, many-celled eukaryotes such as the fungi, plants, and animals. The tiny eukaryotic protists (amoebae, paramecia, giardia, etc.), being visible but not understood, were relegated to the next-to-the-bottom rung of the ladder, while the prokaryotes (bacteria) were handily put at the bottom, where they could be acknowledged, but not seriously so. This entire approach was reasonable at the time, in the sense that structural diversity was driving classification, and the single-celled, anucleate prokaryotes, as they are called, have little that is comparable with the structurally and behaviorally diverse larger organisms, collectively referred to as the eukaryotes.

This view of the biosphere has dramatically changed in the last decade with the advent of molecular taxonomy and phylogeny. The basic idea behind this approach is that there are some molecules common to all Earthly life (16S ribosomal RNA, for example), and that, if one could sequence such molecules and compare the sequences, it might be possible to use this chemical information to compare all life, even that which can be seen only with a microscope. While the germ of this idea is actually decades old, its demonstration was realized only recently with new development of techniques in sequencing of nucleic acids, and the use of this information for organismic comparisons.

The work pioneered by Dr. Karl Woese of the University of Illinois has changed the way we look at life on Earth.[7] From the point of view of the prokaryotes, which lack features that can be used to compare them to each other or to the eukaryotes, this molecular methodology allowed one, for the first time, to have a sense of the phylogeny (a natural history which had been previously lacking) of the various groups.[8] Not only could the prokaryotes be compared to each other, but also because the eukaryotes also contained these same molecules, the comparisons could encompass all of the five kingdoms. The results of this approach were quite dramatic: the four eukaryotic kingdoms were found to be quite homogeneous, while the prokaryotes were found to be very diverse, and thus were expanded to two separate kingdoms, referred to as Bacteria and Archaea (FIG. 4).

A quick glance at the molecular tree reveals that the major genetic variation among the eukaryotes is seen in the single-celled protists, while the three dominant kingdoms (plants, animals, and fungi) are actually clustered at the end of the eukaryotic assemblage, and display only a modicum of genetic diversity (see below). Apparently, it is possible to achieve structural and behavioral diversity (traits that have appeared only in the last 500 million years) while remaining genetically rather homogeneous. Given that multicellular eukaryotes evolved only recently, and that for nearly 3 billion years the prokaryotes dominated the surface of the Earth, one should not be surprised that the bulk of the apparent genetic diversity on the planet resides in this group.

Another notable feature of life on Earth is that of its toughness and tenacity. In order to consider this issue, we should briefly return to the discussion of prokaryotes and eukaryotes. Some of the key properties used to distinguish the prokaryotes from their more complex eukaryotic cohorts are shown in TABLE 1. The eukaryotes are defined by the presence of a nucleus and nuclear membrane (*eu* = true; *karyon* = nucleus), and in general are characterized by complex structures, complex behavioral features, and simple metabolism. Their metabolism is primarily via oxygen-based respiration of organic carbon, and the sizeable energy yields from these processes are used to support their complex structural and behavioral investments. Basically, plants make organic carbon via photosynthesis, and animals eat the plants (and other ani-

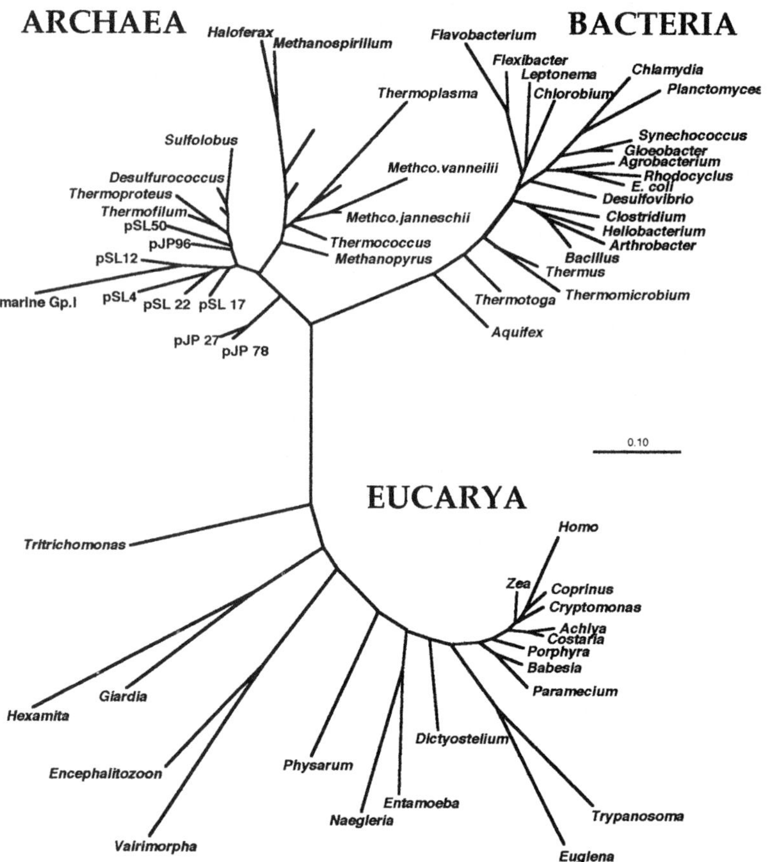

FIGURE 4. The three kingdoms of life. On the basis of sequence analyses of the 16S ribosomal RNA gene, which is contained in all Earthly organisms, it is possible to construct a molecular phylogeny that can quantitatively compare all organisms, even those that can not be cultivated in the laboratory. Such an approach has yielded quite a different view of life, as shown here, in which the major kingdoms of life (animals, plants, fungi, and protists) shown in FIGURE 3, are grouped into one kingdom, and the prokaryotes are expanded into two kingdoms, called the Archaea and the Bacteria.

mals), leading to the kind of complex communities we easily recognize under the general heading of trophic levels or predator–prey cycles. The very existence of complex structures (both intracellular organelles, and multicellular tissues and organs) renders the eukaryotes sensitive to environmental extremes often easily tolerated by their structurally simple prokaryotic relatives (e.g., above 50 degrees C, it is unusual to find functional eukaryotes).

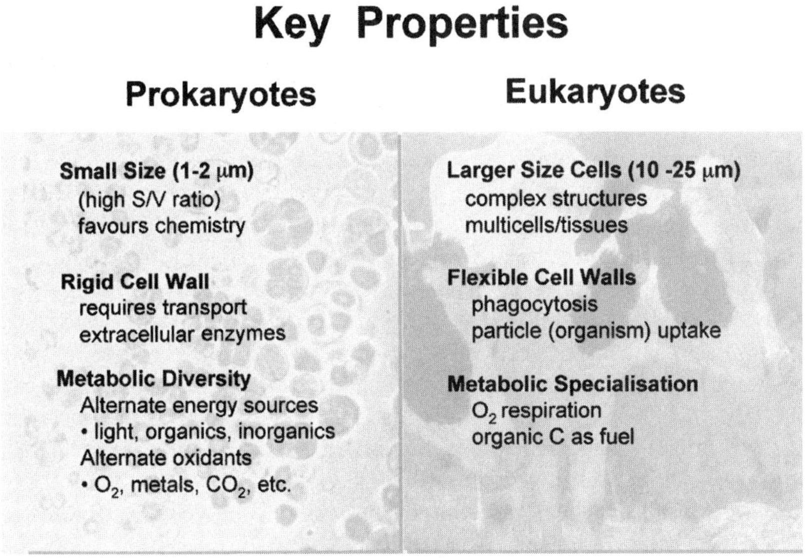

Key Properties

Prokaryotes	Eukaryotes
Small Size (1-2 μm) (high S/V ratio) favours chemistry	**Larger Size Cells (10 -25 μm)** complex structures multicells/tissues
Rigid Cell Wall requires transport extracellular enzymes	**Flexible Cell Walls** phagocytosis particle (organism) uptake
Metabolic Diversity Alternate energy sources • light, organics, inorganics Alternate oxidants • O_2, metals, CO_2, etc.	**Metabolic Specialisation** O_2 respiration organic C as fuel

FIGURE 5. Properties of the prokaryotes and eukaryotes. The small anucleate organisms known as prokaryotes share some properties that allow us to group them into functional domains that are quite different from their eukaryotic counterparts.

On the other hand, the prokaryotes are the environmental "tough guys," tolerant to many environmental extremes of pH, temperature, salinity, radiation, and dryness. I refer to these organisms as the sundials of the living world—tough, simple, effective, and nearly indestructible. Some of the fundamental properties that distinguish them from the eukaryotes are shown in FIGURE 5. First, they are small—they have optimized their surface to volume ratio so as to maximize chemistry. On the average, for the same amount of biomass, a prokaryote may have 10–100 times more surface area. Thus, for a human whose body mass may include a few percent (by mass) bacteria (as gut symbionts), the bacteria make up somewhere between 24 and 76% of the effective surface area! In environments like lakes and oceans, where bacterial biomass is thought to be approximately 50% of the total, the bacteria compose 91 to 99% of the active surface area, and in anoxic environments, where the biomass is primarily prokaryotic, the active surface areas are virtually entirely prokaryotic. In essence, if you want to know about environmental chemistry, you must look to the prokaryotes!

Prokaryotes have rigid cell walls, which preclude life as predators. They are restricted to life as chemists, and do their metabolism via transport and chemistry. This is in marked contrast to the eukaryotes, which are capable of engulfing (by a process called phagocytosis) other cells, and thus engaging in biology. In essence, the prokaryotes spurn life as biologists in order to optimize their skills as chemists. The full effect of such evolution is now easily visible through the genomic analyses of prokaryotes, where, in general, high percentages of the structural genes are involved with membrane and transport processes.

In many cases, up to 25% or more of the total genome deals with the interface between the cell surface and the environment and is involved with uptake, transport, or metabolism of environmental chemicals. In eukaryotes on the other hand, much of the DNA is devoted to the more biological concerns such as development, regulation, and differentiation. Finally, the prokaryotes are metabolically very diverse, while the eukaryotes are quite restricted in their abilities. The prokaryotes have developed a metabolic repertoire that allows them to utilize almost any energetically useful chemical available that is abundant on the Earth.[9] Being opportunists, these ingenious chemists have simply harvested every worthwhile corner of the chemical market, learning to utilize organic and inorganic energy sources of nearly all kinds. Let us look, for example, at the major sources of energy available on Earth today, as shown in FIGURE 6.

On the left one sees the potential energy sources, ranked from the most energy-rich at the top to the least energy-rich on the bottom. On the right are the oxidants that can be used to "burn" these fuels, with the best oxidant (oxygen) at the bottom, and the worst one (carbon dioxide) towards the top. Since a fuel needs to be "burned" to yield energy, we can estimate the amount of energy available simply by connecting a given fuel with an oxidant. If the arrow connecting any given so-called redox pair slopes downwards, it indicates that energy is available from this combination, and there is almost certain to be one or more microorganisms capable of using this combination. In marked contrast, the eukaryotes utilize only a few organic carbon compounds, and only molecular oxygen as the oxidant—they sacrifice diversity for high-energy yield, while the prokaryotes occupy the diverse, lower-energy habitats.

But what about the toughness of prokaryotes? The word "extremophile" has crept into our vocabulary in the past decade: invented to describe organisms that are resistant to, and even thrive in, extreme conditions. As shown in the FIGURE 7, these extremophiles can be resistant to chemical (pH, salinity), physical (temperature, dryness), or nutritional extremes, and it is seldom in nature that an organism encounters just one extreme. For example, under high temperatures, it is common to find anoxic conditions, as the solubility of oxygen is very low in hot water. Furthermore, due to high evaporation rates, warm systems are often associated with high salinity. Thus, desert ponds are often of high pH and salinity, as evaporating water and the minerals trapped there interact to produce extreme conditions.

The most notorious extremophiles are perhaps those associated with high-temperature environments—hyperthermophilic bacteria capable of growth at

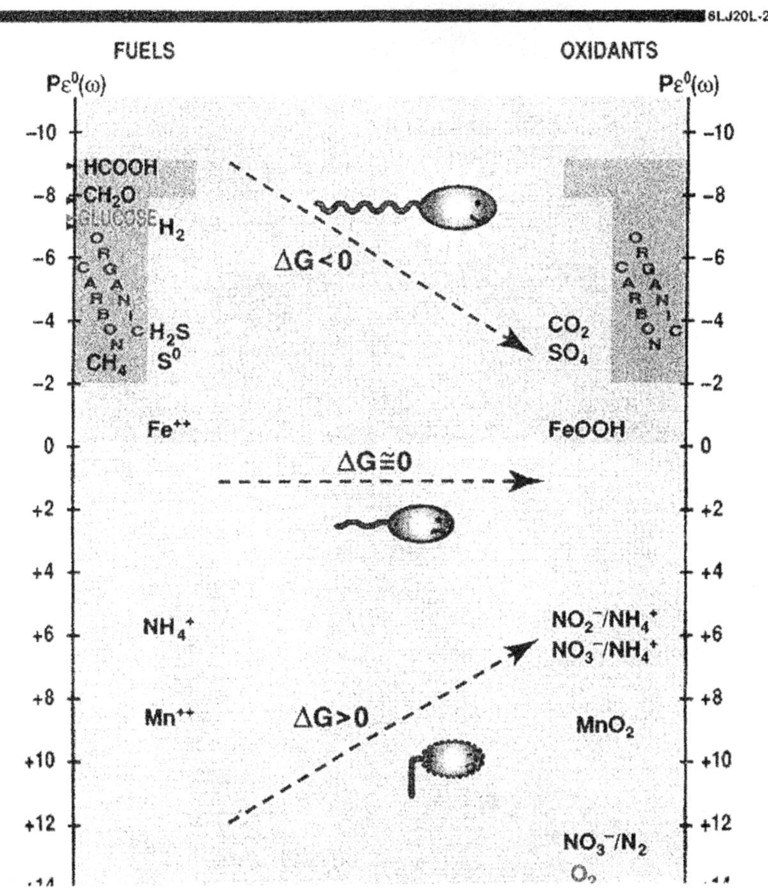

FIGURE 6. Sources of energy and oxidants on Earth. Some of the energy sources available to organisms on Earth are shown on the *left*, with the most energy-rich at the top and the least energy-rich at the bottom. On the *right* are shown the available oxidants for the burning of these biological fuels. The fuels and oxidants commonly used by the eukaryotes are glucose and oxygen, while those available to the prokaryotes are HCOOH, CH_2O, H_2, H_2S, CH_4, S^0, Fe^{++}, NH_4^+, Mn^{++}, and CO_2, SO_4, FeOOH, NO_2^-/NH_4^+, NO_3^-/NH_4^+, MnO_2, NO_3^-/N_2 respectively.

100 degrees C and above with a maximum temperature of about 115 degrees C, well above the boiling point of water. These organisms can only be grown under pressure where the water is stable, and will freeze to death at temperatures as high as 80 degrees C, temperatures that would result in severe burns

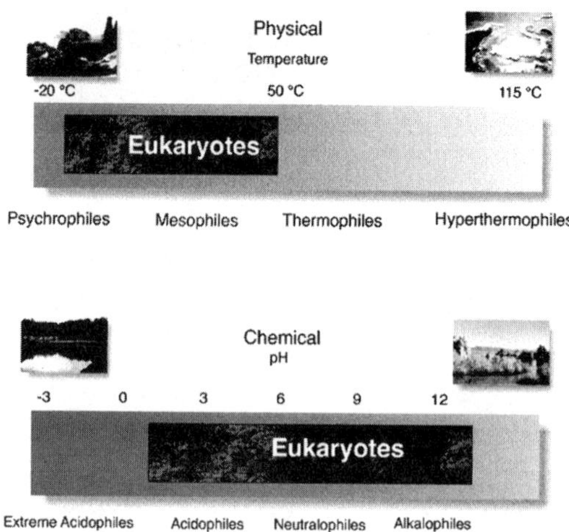

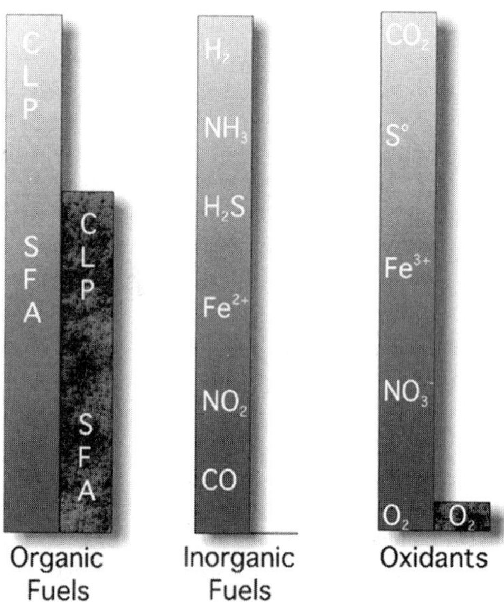

FIGURE 7. Extremophiles. This figure illustrates the ranges of environments encountered by various types of extremophiles, and to make the point that extreme conditions can be physical, chemical, or metabolic. It is seldom that one encounters just one of these conditions at a time.

FIGURE 8. Living in the rocks. This figure shows a layered community of organisms found in the tufa mounds (carbonate deposits) of Mono Lake, California. This environment is characterized by high pH (about 10) and high salinity, but in the rock environment, above the water level and shielded from the sun by the carbonate rocks, is a healthy and ubiquitous population of photosynthetic cyanobacteria, living in harmony with many other prokaryotic species.

to humans. Other bacteria are known that live in saturated salt brines, and at pH values as low as −1 and as high as +11.

One of the strategies of life that often emerges when things get tough is that of an endolithic lifestyle—the ability to associate with rocks, either on or just under the surface. In California's alkaline Mono Lake, for example, we can see that the tufa mounds that dominate the alkaline lake, and which appear to be dead, are actually teeming with life (FIG. 8).[10] A few millimeters under the rock surface are populations of cyanobacteria that hide from the intense sunlight, positioning themselves for optimum growth in the now-filtered light. A similar situation occurs in many hot and cold desert rocks and soils, where the photosynthetic microbes are found under the surfaces of rock layers.[11]

In addition to the physical and chemical extremes noted above, I would like to point out another property of prokaryotes, referred to here as nutritional extremophily. Given that eukaryotes are almost entirely limited to growth on organic carbon with oxygen as the oxidant, any set of conditions in which organic carbon or oxygen is absent is a potential life-threatening situation. For the prokaryotes, however, such environments are simply opportunities to continue living but with a different nutritional system.

While it cannot be said with certainty *when* such metabolic diversity arose on Earth, its very existence forces one who is hunting for life to include such "extreme" habitats in the search, and to broaden the definition of life to include metabolic abilities that a few years ago might have been summarily dismissed. The ability to grow on energy sources such as carbon monoxide, ferrous iron, hydrogen sulfide, or hydrogen gas implies that bacteria could inhabit worlds not previously considered as candidates by most scientists seek-

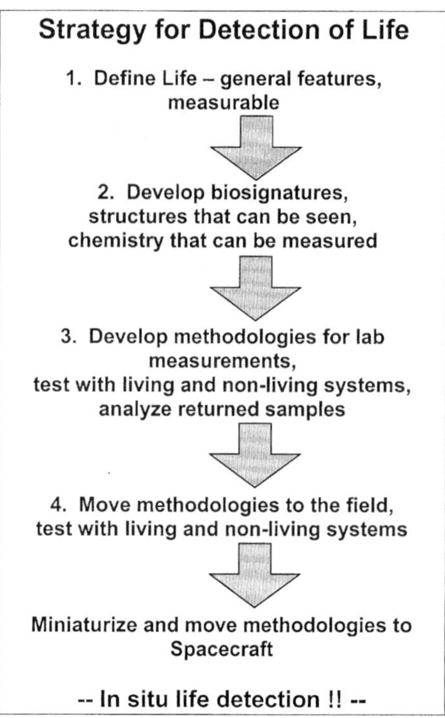

FIGURE 9. Searching for life in the universe. Searching for life when we do not know what it looks like may represent one of the truly great challenges facing human scientists. One must move to fundamental definitions of life, and develop biosignatures to aid in the search—biosignatures that are not dependent upon known earthly molecules, but which would never miss earthly life if it were encountered.

ing extraterrestrial life, and must be included in any emerging search strategies.

A final point regarding the prokaryotes relates to their tenacity and ability to survive for long periods of time. There are many examples of bacteria being revived after long-term storage, but perhaps none any more dramatic than those from the Antarctic permafrost, where soils that have been permanently frozen for 3 million years or more have yielded copious numbers of living bacteria. Dr. David Gillichinsky and his colleagues from Puschino, Russia, have been drilling in such sites for many years now, and a number of organisms have been "revived" from their carefully collected samples. It is not unusual to find 10^6 to 10^7 (1 to 10 million) viable bacteria from each gram of permafrost.[12] These are not cold-loving (psychrophilic) bacteria that have adapted to these freezing conditions, but simply mesophilic organisms that have been trapped within this icy storage facility for millions of years.

So, as we are poised to proceed to other celestial bodies in search of life, we find that our definition of habitability is quite different from what we adhered to just a few years ago. In response to this, we must: (*a*) consider that the physical and chemical conditions tolerant to life are broader than we once thought; (*b*) examine the potential energy sources available and look carefully for life forms utilizing any such energy; and, (*c*) be prepared for subtle, single-celled life that may not be obvious at first glance, even looking in places where life might have been preserved as dormant forms.

Given these "lessons from the Earth," is it possible to design a strategy that will allow us to search for life with some degree of confidence? The success of any search will depend upon the ability to define the general features of life, to develop methods for measuring such general features, and to employ these methods for remote sensing, for *in situ* studies, and for the analysis of returned samples.

Studies of Earthly life suggest that metabolic evolution, one of the keys to life's becoming a global phenomenon, was already in full swing more than 500 million years ago. Most of the Earth's geology, and many of its atmospheric properties, which we still see today, were in place by that time. If we want to search for life elsewhere, we must keep in mind that there is no guarantee that a particular planet will have evolved to the same advanced stages we have on Earth—a historical perspective is thus key to developing a strategy for life detection. To put this another way, we must know the early history of a planet in order to frame the search for life properly.

Since Earth is the only place where we are certain that life exists, it will serve as our laboratory for the development of the search strategy. The overall strategy is still in its early stage of definition, but a general idea of it involves three phases: (*a*) the development of non-Earth–centric biosignatures for life detection; (*b*) the testing of these biosignatures on Earthly samples to see just how good they are; and, (*c*) the eventual use of these biosignatures and tests for the analyses of extraterrestrial samples.

For most biologists this entire process is a new endeavor, asking new questions. It is rare that a biologist is handed a rock and asked: "Is it alive?" or, "From this sample, can you prove whether there was ever life on Earth?" Yet that is what we will be faced with in a few years when samples are returned from Mars. In fact, if another planet was teeming with life, as is Earth, this would not be a difficult task, even with very old rocks. It would be relatively easy to tell that Earth was (and is) alive from almost any distance, and especially so if samples were available for detailed physical and chemical analysis. However, if the signs of life are subtle or unfamiliar, then the task becomes much more difficult.

This difficulty is demonstrated by the present controversy surrounding the now famous Mars meteorite, ALH 84001. Four years ago, this 4.5-billion-year-old rock was reported to contain evidence for life on Mars. But even now, after extensive research, the jury is still out as to whether the evidence is convincing. The problems stem from many fronts, including the age of the sample, the difficulties in separating indigenous signals from those due to Earth contamination, and the very definition of life and how one proves that it is (or was) present. What this meteorite really has taught us is that we have a lot to learn about how to distinguish life from non-life.

Biologists may not even be the right group of scientists to answer the question, "Is there life in this rock?" As a biology student, I was never asked such a question. Rather I was given a frog and asked, "How does it work?" or "What is it made of?" These days, the questions have changed to "What genes are there, and how do they function?" but the general problem remains—biologists study life they already know how to detect, they do not seek to detect life they do not know about. This is a question inherently interdisciplinary in nature, and perhaps best suited to those who are willing to define life in general terms that would include all life on Earth, but not exclude life made of different types of molecules.

As a group, we biologists should be extremely well suited to detect life as we know it primarily because we understand its chemistry so well. There are molecules that can be detected at very high sensitivity, allowing us to find a single bacterium in a liter of water. However, if such key indicator molecules are not there, the search becomes much more difficult, and the likelihood that life elsewhere would contain the same key molecules certainly cannot be depended upon. The problem then takes on a different aspect: Because we rely on Earth-centric indicators of life, we biologists may unwittingly be the least well suited to detect life that might differ in its chemical make-up.

To this end, our astrobiology group is focusing on what we feel are two fundamental properties of all life, structure and chemical composition, both of which can be detected and measured. Historically, structures are the paleontologists' keys to recognition of past life on Earth: It is structures that characterize life as we know it and that should be expected to characterize any new forms of life we encounter. We do not know in advance the nature of the structures or the size scales over which to search, but we do expect there to be structural elements associated with any life forms. When one is hunting in a new spot, dependency on known structures has a number of potential traps, including the fact that one might discard structures simply because they are unfamiliar. It will be important to remain open-minded about the types and sizes of structures found in samples from new sites.

While we believe that life will be linked to some structural elements, structure alone will not prove the existence of life. However, coupling structural analysis with the determination of chemical content may well provide a tool for strongly inferring the presence of life. On Earth, life is carbon-based, with a peculiar and remarkably constant elemental composition (hydrogen, nitrogen, phosphorous, oxygen, carbon, etc.), which is remarkably out of equilibrium with the crustal elemental abundance of our planet. In other words, there are more or less of some elements than would be present if there were no life on Earth. Life is, almost by definition, a source of negative entropy: a structure composed of groups of chemical monomers and polymers that are not predicted to be present on thermodynamic grounds, given the abundance of chemicals in the atmosphere and crust of the Earth. The exact nature of these chemicals is not so important as the fact that they are grossly out of equilibrium with their surrounding geological environment. So, if methods were available for analysis of the chemistry of structures at the proper size scale(s), then the possibility of detecting extant (or even extinct) life would be greatly increased. While there are other properties of life that may be measurable (such as replication, evolution, and energetic exchange with the environment), and which may leave traces in the geological record, we believe that if life does or did exist, then it will be detectable by the existence of structures and their distinctive chemistries.

Ultimately, we would like to have samples from many places in our solar system and beyond, but realistically, we will probably need to make measurements remotely and *in situ*, and be satisfied with these as our indicators of life. As our ability to measure structures and chemistry improves, the possibility of answering the question of whether life does or does not exist beyond Earth will improve as well. A strategy for exploration, sample collection and return, and finally, sample analysis will be needed. Given the number of other solar systems already known to exist, and the emerging numbers of planets around faraway stars, it seems unlikely that life will not be found elsewhere. Development of the proper strategy, and definition of those conditions that do and do not support life, will be key to the ultimate discovery of extraterrestrial life. With the proper strategy and approach, the question seems to be not one of whether, but when.

NOTES AND REFERENCES

1. Klein, H. P., "The Viking mission and the search for life on Mars," in *Rev. Geophys.,* Vol. 17 (1979), pp. 1655–1662.
2. McKay, C.P., E.I. Friedmann, R.A. Wharton, and W.L. Davies, "History of water on Mars: a biological perspective," *Adv. Space. Res.*, Vol. 12 (1992), pp. 231–238; Golombek, M.P., "A message from warmer times," *Science*, Vol. 283(1999), pp. 1470–1471.
3. Mojzsis S.J., G. Arrhenius, K.D. McKeegan, T.M. Harrison, A.P. Nutman, and C.R.L. Friend, "Evidence for life on Earth before 3,800 million years ago," *Nature,* Vol. 384 (1996), pp. 55–59; Schidlowski M., P.W.U. Appel, R. Eichmann, and C.E. Junge, *Geochim. Cosmochim. Acta,* Vol. 43 (1979), pp. 189–199; Schopf, J.W., and C. Klein, *The Proterozoic Biosphere, (*Cambridge, UK: Cambridge University Press, 1992).

4. Knoll, A.H., "The early evolution of eukaryotes: A global perspective," *Science*, 256 (1992), pp. 622–627.
5. Rye, R., and H.D. Holland, "Paleosols and the evolution of atmospheric oxygen: a critical review," *Am. J. Sci., Vol.* 298 (1998), pp. 621–672.
6. Margulis, L., *Symbiosis in Cell Evolution* (San Francisco: W.H. Freeman & Co., 1981).
7. Woese C.R., "Bacterial evolution," *Microbiol. Rev.,* Vol. 51 (1987), pp. 221–271; Woese, C.R., "There must be a prokaryote somewhere: microbiology's search for itself," *Microbiol. Rev.,* Vol. 58 (1994), pp. 1–9; Pace, N.R., "New perspective on the natural microbial world: molecular microbial ecology," *Amer. Soc. Microbiol. News,* Vol. 62 (1996), pp. 463–470.
8. Olsen, G.J., C.R. Woese, and R. Overbeek, "The winds of evolutionary change: breathing new life into microbiology," *J. Bacteriol.,* Vol. 176 (1994), pp. 1–6; Stahl, D.A., "The natural history of microorganisms," *Am. Soc. Microbiol. News,* B., Vol. 59 (1993), pp. 609–613.
9. Nealson, K.H., "Sediment bacteria: who's there, what are they doing, and what's *new?,*" *Annu. Rev. Earth Planet. Sci.,* Vol. 25 (1997), pp. 403–434; Nealson, K.H., "The limits of life on Earth and searching for life on Mars," *J. Geophys. Res.,* Vol. 102 (1997), pp. 23,675–23,686; Nealson, K. H., "Post-Viking microbiology: new approaches, new data, new insights," *Origins of Life and Evol. Biosph.,* Vol. 29 (1999), pp. 73–93.
10. Sun, H., K.H. Nealson, and K. Venkateswaren, "Alkaliphilic and alkalotolerant actinomycetes isolated from Mono Lake, tufa," Abstracts from the 99th Annual Meeting of the American Society of Microbiologists ASM Press, (Washington, D.C.: ASM Press, 1999); Sun, H., and K.H. Nealson, K.H., "Endolithic cyanobacteria in Mono Lake," Abstracts from the 99th Annual Meeting of the American Society of Microbiologists (Washington D.C.: ASM Press, 1999).
11. Friedmann, E., Antarctic Microbiology, 1st ed. (New York: Wiley lnterscience, 1993); Friedmann, E. I., "Endolithic microorganisms in the Antarctic cold desert," Science, Vol. 215 (1982), pp. 1045–1053.
12. Shi, T., R.H. Reeves, D.A. Gilichinsky, and E.I. Friedmann, "Characterization of viable bacteria from Siberian permafrost by 16S RDNA sequencing," Microbial Ecol., Vol. 33 (1997), pp. 169–179.

Homes for Extraterrestrial Life

Extrasolar Planets

DAVID W. LATHAM

Harvard–Smithsonian Center for Astrophysics,
Cambridge, Massachusetts 02138, USA

ABSTRACT: Astronomers are now discovering giant planets orbiting other stars like the sun by the dozens. But none of these appears to be a small rocky planet like the earth, and thus these planets are unlikely to be capable of supporting life as we know it. The recent discovery of a system of three planets is especially significant because it supports the speculation that planetary systems, as opposed to single orbiting planets, may be common. Our ability to detect extrasolar planets will continue to improve, and space missions now in development should be able to detect earth-like planets.

KEYWORDS: extrasolar planets; the solar system; Doppler spectroscopy; space astrometry

There is good news and bad news in the quest for extrasolar planets as possible sites for the rise of intelligent life. The good news is that astronomers are beginning to find convincing evidence for planets in orbit around stars like the sun. The bad news is that none of the planets discovered so far is anything like the earth. Instead, they are all giants, at least 100 times more massive than the earth, and unlikely to be suitable for life as we know it.

The first good evidence for a *system* of planets orbiting a solar-type star was announced on April 15, 1999 by a team of astronomers from four research institutions.[1] The three planets in this system are also giants. But, if they are like the giant planets in our own solar system, they will have rocky moons, and one of those moons just might prove to be inhabited (but I would not bet on it). More importantly, now that we have more than just one example of a planetary *system*, it is easier to imagine that somewhere there may be other systems with earth-like planets orbiting in a habitable zone.

Address for correspondence: Dr. David Latham, Harvard–Smithsonian Center for Astrophysics, 60 Garden Street, Cambridge, Massachusetts 02138. Voice: 617-495-7215.
dlatham@cfa.harvard.edu

FIGURE 1. Earth.

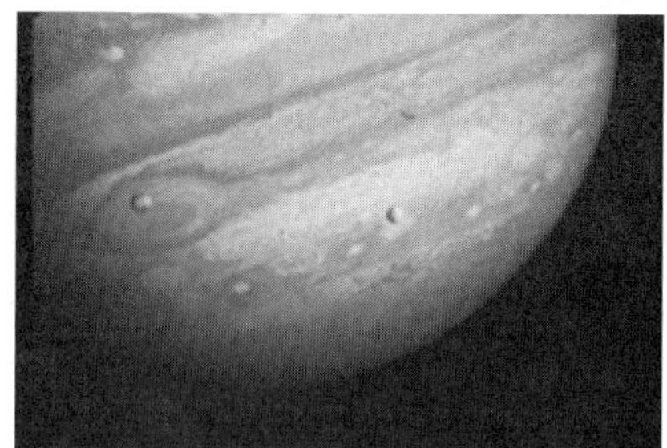

FIGURE 2. Jupiter.

FIGURE 1 shows what most people think we should be looking for: a blue-white rocky planet with a surface and atmosphere, and a stable orbit in a zone where water is liquid and life can be comfortable. Captain Kirk, find us some more of these! But, FIGURE 2 shows the kind of planet we think astronomers have been finding so far: Jupiter, 300 times more massive than the earth, with no surface and shrouded in a dense atmosphere composed of noxious gases such as methane and ammonia.

Moreover, the extrasolar planets found so far are all in orbits much closer to their parent stars than Jupiter is to the sun. This was completely unexpected ten years ago, when the first extrasolar planet candidate, the unseen companion of HD 114762, was discovered in an orbit similar to Mercury's.[2] The thinking at that time was dominated by the one example of a solar system that we knew about, our own, where the largest planets, Jupiter and Saturn, orbit majestically in wide, long-period orbits out in the icy nether regions where it was cold enough for them to form by the accumulation of gases and volatile compounds onto rocky cores 10 to 20 times the mass of the earth. Although Jupiter is the largest planet in the solar system, it is puny compared to the sun, and pulls with only one-thousandth of the sun's mass.

The inner terrestrial planets in our solar system are in turn puny compared to Jupiter. The earth is ten times smaller in diameter and is composed mostly of dense refractory materials, such as minerals and metals that were able to stand the heat close to the sun when the earth formed. Nobody expected to find giant planets so close to their parent stars, because nobody thought that giant planets could form in a region where all the ices would have melted into volatile gases.

There is a lot of empty space in our solar system. In a scale model where the sun is represented by a weather balloon 1 meter in diameter sitting in the Baird Auditorium of the Smithsonian's Museum of Natural History, Jupiter would be a grapefruit orbiting at a distance of 500 meters, as far away as the Federal Triangle Metro stop. The earth would be the size of a beer nut, orbiting well outside the museum, say at the distance of Constitution Avenue (see TABLE 1 for more details).

TABLE 1. Our solar system

Object	Period	Mass	Scale Diameter	Model Orbital Size
Sun		1000	1 m	
Mercury	88 d	0.0002	3 mm	40 m
Venus	224 d	0.0026	8 mm	70 m
Earth	365 d	0.0032	8 mm	100 m
Mars	687 d	0.0003	4 mm	150 m
Jupiter	12 yr	1	10 cm	500 m
Saturn	29 yr	0.30	9 cm	1 km
Uranus	84 yr	0.046	3 cm	2 km
Neptune	165 yr	0.054	3 cm	3 km
Pluto	249 yr	0.0005	2 mm	4 km

In contrast to our own solar system, the first extrasolar planets were found in tight, short-period orbits. For example, 51 Pegasi was found to have a companion similar in mass to Jupiter,[3] but orbiting with a period of only 4 days. In our scale model, that planet would be represented by a grapefruit with an orbit about the same size as the stage of the Baird Auditorium. In retrospect it should not have been a surprise that massive planets in short-period orbits were the first to be discovered, because they are the easiest to detect with the Doppler technique being used. This technique is indirect. We do not see the light reflected (or heat emitted) by the planet itself, but instead we detect the reflex motion that the gravitational pull of the planet induces in its parent star. Just as the planet sweeps around in its orbit, so the parent star must respond with a counterbalancing motion. Of course, the amplitude of the star's motion is much smaller and harder to detect by the ratio of the masses. Jupiter orbits at 12 kilometers per second, but the sun's reaction is 12 meters per second, not much faster than a sprinter can run. But, if you move Jupiter in 100 times closer, the orbital velocity must go up by a factor of 10. If you make the planet 10 times more massive than Jupiter, the Doppler signature goes up by a factor of 10.

Therefore, the first planetary companions discovered by the Doppler technique were in very tight orbits and/or were considerably more massive than Jupiter. This is illustrated in FIGURE 3, where I have plotted the 20 planets discovered so far using the Doppler technique, together with Jupiter and Saturn. The vertical axis is the semi-major axis of the planet's orbit (a measure of the size of the orbit) in astronomical units (AU, the distance of the earth from the sun). At 5.2 and 9.5 AU, Jupiter and Saturn (plotted as filled circles near the top of the figure) have considerably larger orbits than any of the extrasolar planets. The planets at the bottom of the diagram have such small orbits that

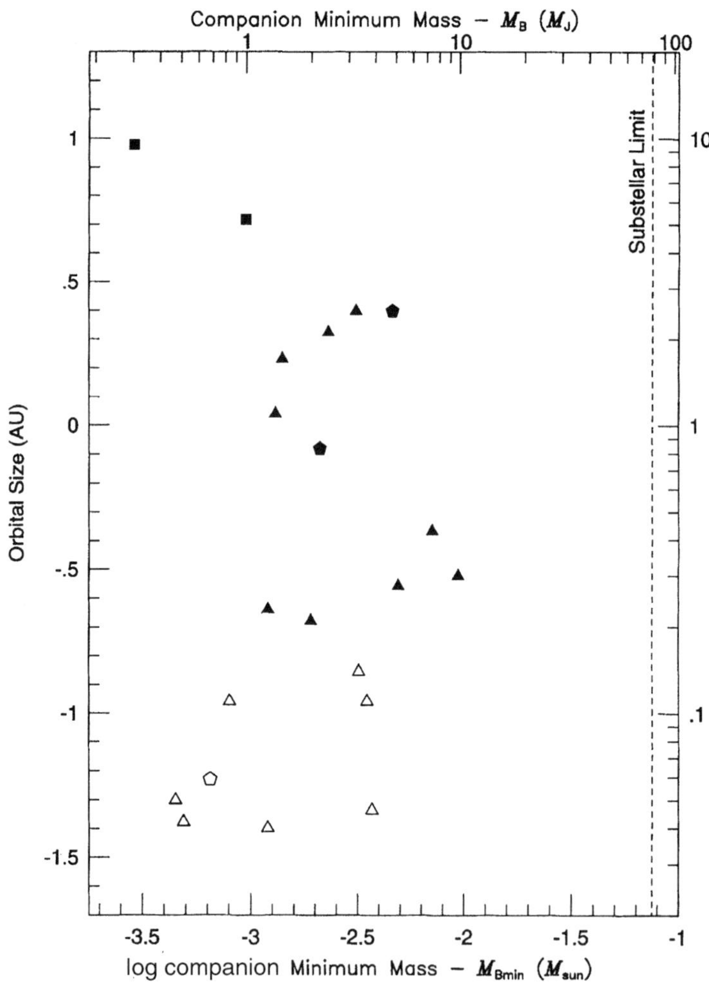

FIGURE 3. Orbital size relative to log companion in 20 planets discovered using Doppler technique.

the periods are as short as 3 or 4 days, and the shapes of the orbits have been circularized by tidal forces.

The horizontal axis in FIGURE 3 is the mass the planetary companion would have if the orbit happens to be oriented so that we view it edge-on. The actual inclination of the orbit to the line of sight, *i*, cannot be determined from Doppler measurements alone, and thus is usually unknown. If the orbit is actually tilted up to the line of sight, then our estimate of the mass of the planet is too small by the factor 1/sin(*i*).

In FIGURE 3 I have only plotted the planet candidates with minimum masses less than 10 Jupiter masses (M_J). There are also a few companions with minimum masses in the range between 10 M_J and the substellar limit at about 75 M_J, but they seem to be relatively rare. It is almost as if the process that makes planets does not produce companions much bigger than 10 or 20 M_J, while the process that makes stars does not produce companions much smaller than 75 M_J.[4]

So far, we have *not* detected any *true* Jupiters, planets the size of Jupiter orbiting with periods of 10 or 20 years. This does *not* mean that such systems do not exit. It does *not* mean that the configuration of our solar system is unique, with the terrestrial planets huddled in close to the parent star, and the giant planets patrolling the outskirts, warding off incoming intruders such as asteroids or comets that might be instruments of mass extinction. No, such a conclusion is premature, because we have so far barely achieved the Doppler precision that is needed to detect a true Jupiter, and with few exceptions the observations do not yet cover long enough time spans. Both these shortcomings are being addressed. New and better instruments and techniques are being developed, and if the telescope time-allocation committees continue to be wise, we will gradually accumulate enough additional data with sufficient time coverage to detect true Jupiters.

The fraction of solar-type stars with discovered planets is still rather low, just a few percent at most. But, this could prove to be just the tip of the iceberg. The Doppler technique is picking up the easiest planets first, the ones

FIGURE 4. Keck I facility in Hawaii.

that are the most massive and/or have the shortest orbital periods. As the Doppler precision improves and the time coverage grows longer, I am confident that we will find more planets, ones with smaller masses and wider orbits.

One aspect of this research that I find very interesting is that we may be discovering enough planets so that we can begin to decipher some of the general characteristics of the population. This means that we can begin to address important questions like how do planets form?, and how do their orbits evolve?, and what are the implications for finding rocky planets like our own in comfortable, stable orbits? But, the number of discovered systems is still woefully inadequate for this kind of interpretation to be reliable, and the selection effects are severe and dangerous. Thus, one of the main activities under way now is to carry out an initial reconnaissance of large samples of stars, to identify additional systems in large numbers, and to explore how the frequency and characteristics of planets depend on parameters such as the mass of the primary star, or its metallicity. Probably the best facility in the world right now for this kind of work is located in Hawaii, halfway to space (as the Hawaiians like to say), on the summit of Mauna Kea. Keck I, the first of the two 10-meter telescopes (FIG. 4) and the marvelous High Resolution Echelle Spectrograph built by Steve Vogt, are getting heavy use for planet searches and follow-up studies, supported primarily by NASA. Several large Doppler surveys are under way with this facility and with others around the world, and thousands of stars are being monitored. We can expect to see announcements of dozens of new planets over the next several years.

In my opinion it is not enough just to discover new planets. It is critically important to do follow-up studies. One of the most obvious types of follow-up study is to continue to monitor the systems where a first planet has been discovered, to see whether any additional modulations in the Doppler velocities emerge as you move to higher precision and longer time spans. Additional periodic modulations would then be interpreted as second or even third planets, a planetary system. This work is important because we want to know more about the formation and evolution of planetary *systems*, not just single planets. The problem is that there is little hope that the Doppler technique can ever attain the precision needed to detect planets like the earth, primarly because the tiny orbital velocity induced by an earth would be swamped by velocity jitter due to other astrophysical phenomena, such as star spots coupled with rotation, or macroscopic motions in the atmosphere of the star. But, if we can learn more about how *systems* form, then maybe we can invoke some theoretical arguments about where and when earth-like planets are likely to form, even though we can not detect them observationally.

As noted above, the first reliable detection of a system of three planets orbiting a solar-type star, upsilon Andromedae, was announced April 15, 1999. The innermost planet, with a period of only 4.6 days, was the first to be discovered in this system.[5] The residuals of the observed Doppler velocities from the orbital solution for this planet were not as good as expected, and the disagreement grew more serious as time passed and more observations were accumulated. Eventually it became clear that there were at least two additional periodic modulations, corresponding to a second and a third planet in wider orbits, with periods of 240 and 1267 days. There are two interesting patterns in the characteristics of these planets. The orbits grow progressively more elliptical with increasing period, and the minimum masses grow progressively larger. At first glance these patterns appear to support theories of planet migration, where giant planets form initially in the cool outer regions of a circumstellar disk, in a region where there is plenty of material for planet building, and then gravitational interactions move one or more of the planets into the inner regions of the system while others move outward, in some cases escaping from the system altogether. In this scenario the three planets orbiting upsilon Andromedae would all be gas giants similar to Jupiter, and unsuitable for life as we know it. However, if the analogy between these planets and Jupiter holds true, they will be orbited by rocky moons, just as Jupiter is, and maybe there is a chance of finding a habitable world among the moons.

The satellites of Jupiter and Saturn offer a rich variety of environments. The innermost of the four Galilean satellites of Jupiter is Io (FIG. 5). It is so close to Jupiter that it is tortured by massive land tides, which heat the moon and keep it active. There are spectacular volcanos on Io that spew sulfur and sulfur dioxide. The surface of Io is truly a place of fire and brimstone, like the traditional hell that I learned about in Sunday school, not the modern hell of dreary monotony, with no MTV or remote controller. The next Galilean satellite is Europa (FIG. 6), a small rocky planet covered by frozen oceans and laced with networks of cracks and ridges in the ice. Some planetary scientists speculate that there may be liquid water under the surface ice, and that this environment might support simple life forms. Next is Ganymede (FIG. 7), the largest of the Galilean satellites, but still only twice the mass of the moon, one fortieth the mass of the earth. Ganymede is a rocky moon, and it has tectonic activity that has created a series of mountain ridges on its surface. Callisto (FIG. 8) is a quieter world, made of more icy materials. It is less affected by the tidal forces from Jupiter, and some scientists consider it to be the most promising remaining site in the solar system for the evolution of simple life. Titan (FIG. 9), the largest satellite of Saturn, is also very interesting, because it has managed to retain a substantial atmosphere.

The wide variety of environments that we find on the satellites of the giant planets in our own solar system suggests that the satellites of extrasolar giant planets may be the best hope for finding habitable abodes elsewhere. In the case of upsilon Andromedae I am not very optimistic that habitable satellites will be available, because the orbits of the outer two planets are too eccentric and therefore are not able to provide the thermal stability that we think is a prerequisite for the rise of life.

FIGURE 5. Io.

FIGURE 6. Europa.

FIGURE 7. Ganymede.

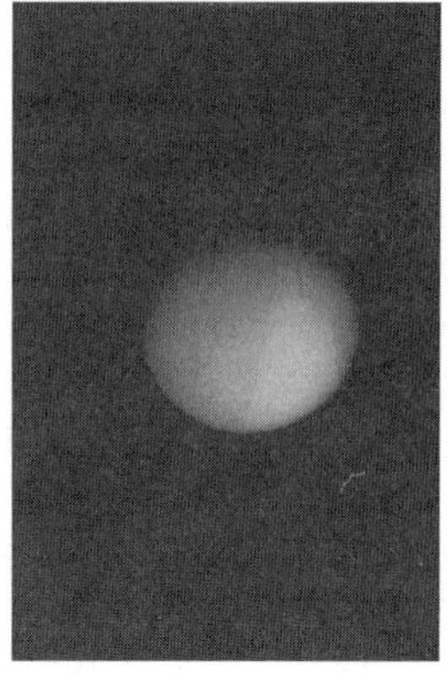

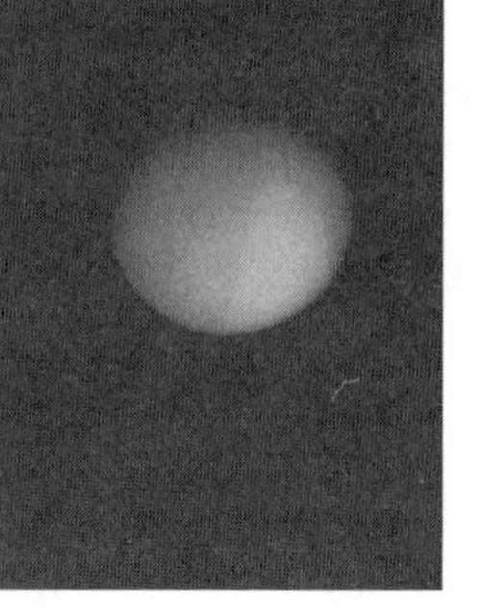

FIGURE 9. Titan.

FIGURE 8. Callisto.

Now that we have a second example of a planetary system orbiting a solar-type star (actually a third system, if you include the planets orbiting the pulsar PSR1257+12),[6] it is much easier to imagine that planets like to form in systems. But we need to show that this is actually the case by finding multiple planets in many more systems.

The Doppler technique is by no means the final word on how to find and study extrasolar planets. The human imagination is fertile, and people have come up with a wide variety of other ways to approach the problem, some new and some old. I will not mention photometric searches for transits by planets, or photometric surveys for microlensing events by planets, or the ambitious talk about using a nulling interferometer in space to cancel out the glare of the parent star in order to make a one-pixel image of the orbiting planet, so that its light can be fed into a spectrograph to see what kind of an atmosphere it has. However, I do want to acknowledge the several plans for space astrometry missions, because they show great promise for planet searches starting just a few years from now.[7]

The inspiration for several of these missions came from the remarkable success of Hipparcos, a European mission to survey more than 100,000 stars with an astrometric accuracy of 1 or 2 milli-arcseconds. Although this performance is impressive, it is barely adequate to detect the presence of brown-dwarf companions, failed stars just below the substellar limit. Two follow-up missions have been proposed for the near future, DIVA in Germany and FAME in the United States. These missions would use modern CCD detectors to extend the Hipparcos survey to much fainter stars, in much larger numbers, and with much better astrometric precision. These missions would be able to explore the transition region between brown dwarfs and giant planets.

GAIA is a much more ambitious European successor to Hipparcos, and aims to reach the astonishing astrometric accuracy of 4 micro-arcseconds, even for very faint stars. That angle is about the size subtended by a dime seen at the distance of our moon. GAIA should be able to explore extensively the realm of giant planets orbiting a wide range of stars. GAIA is a candidate for the European Space Agency's 2009 Cornerstone Mission.

In the United States, NASA is well along in its planning for the Space Interferometry Mission (SIM), which adopts a rather different strategy of pointing at specific targets rather than scanning the entire sky. This should allow SIM to achieve better astrometric accuracy and to reach fainter stars than any of the other proposed missions, but at the cost of a much smaller target list. SIM may even be able to achieve the exquisite accuracy needed to detect the reflex motion of the nearest stars in reponse to the pull of terrestrial-sized planets. Thus there is a reasonable prospect that in my lifetime we will discover the first extrasolar planet like the earth.

It is an exciting time for extrasolar planet research, both observationally and theoretically, and inevitably this will have an impact on our thinking about the prospects that intelligent life may have arisen elsewhere in the universe.

REFERENCES

1. Butler, R.P., G.W Marcy, D.A. Fischer, T.M. Brown, A.R. Contos, S.G. Korzennik, P. Nisenson, and R.W. Noyes, 1999, *ApJ*, Vol. 526 (1999), p. 916
2. Latham, D.W., R.P. Stefanik, T. Mazeh, M. Mayor, and G. Burki, *Nature*, Vol. 339 (1989), p. 38
3. Mayor, M. and D. Queloz, *Nature*, Vol. 378 (1995), p. 355
4. Mazeh, T., D. Goldberg, and D.W. Latham, *ApJL*, Vol. 51 (1998), p. L199
5. Butler, R. P., G.W. Marcy, E. Williams, H. Hauser, and P. Shirts, *ApJL*, Vol. 474 (1997), p. L115
6. Wolszczan, A., and D.A. Frail, *Nature*, Vol. 355 (1992), p. 145
7. Latham, D.W., in *Ultraviolet-Optical Space Astronomy Beyond HST*, eds. J.A. Morse, J. M. Shull, and A. L. Kinney, ASP Conf. Ser. Vol. 164 (1999), p. 134.

What is SETI?[a]

JILL TARTER

The SETI Institute, Mountain View, California 94043, USA

ABSTRACT: This paper describes the efforts of SETI that are going on around the world. It presents a brief rationale for the choice of methodology and the limits that can be set on the basis of such results to date.

KEYWORDS: search for intelligent life; astrobiology; exobiology

What is SETI—the Search for Extraterrestrial Intelligence? Perhaps it is best to start by saying what SETI is not. It is also not an investigation of UFOs or alien abductions. It is not a religion, or worse, a cult. It is not politically correct. And, perhaps unexpectedly, it is not actually a way of **directly** detecting intelligent life elsewhere in the universe. More about that later.

What is SETI? SETI is a suite of scientific explorations that attempt to answer the question, "Are we alone?" in the universe by doing experiments. It is the product of our anthropocentric experience and our limited understanding of the universe that we inhabit. It is an endeavor practiced by very pragmatic people. It is potentially a multi-generational exploration. It is not possible to predict when or if SETI will be successful. But, most of all, SETI is very important.

As an experimental exploration, what are the appropriate experiments? One possibility is active experiments, physically going to places where life might be found and then looking. We have visited the Moon and will likely visit Mars in this century. Such physical exploration is fine for the solar system, but the stars are too far away. Stars require more passive experiments, remote sensing of distant environments. This is what SETI is really all about—doing experiments to try to detect evidence of distant technologies and, indirectly, intelligent technologists.

At the same time SETI investigators try to keep an eye out for unexpected things. There could be little nanoprobes in our local environment or other extraterrestrial artifacts. More likely there could be the sorts of anomalies in our

[a]This article is based on a transcipt edited by Dr. James B. Miller.

Address for correspondence: Dr. Jill Tarter, Project Phoenix, The SETI Institute, 2035 Landings Drive, Mountain View, CA 94043. Voice: 650-960-4555; fax: 650-968-5830.

astronomical data that could turn out to be as fruitful as those little bits of scruff that Jocelyn Bell found when she discovered the first pulsars.

Nobel Laureate Christian de Duve has argued that life is a cosmic imperative—at least from a chemist's point of view. Can such a claim be substantiated? NASA's "Origins" and "Astrobiology" Programs and programs of the European Space Agency (ESA) are actively seeking to return to Mars and, further, to explore Jupiter's moons (Europa, Ganymede, Calisto) looking for life. Finding a second genesis of life in our solar system would greatly strengthen the arguments for the ubiquity of life in the universe.

Life is a planetary phenomena. A starting point for the search for life beyond our solar system is to find extra-solar planets. There are various observational techniques currently in use with terrestrial telescopes that allow this search to take place from the ground. At the time of the Cosmic Questions Conference 19 extra-solar planets had been discovered, some in multi-planet systems. Today, the number is 74 and still counting. However, none of these planets appear to be Earth-like. "Earths" are hard to find.

NASA and the ESA are developing ambitious terrestrial planet-finding (TPF and Darwin) programs. These programs are seeking to develop space-based observational capacity to be able to discern Earth-like planets in systems as much as 50 light-years away. Spectroscopy using the TPF systems will allow atmospheric chemists and biologists to use the relative amounts of carbon dioxide, water vapor, and ozone in a planet's atmosphere to find whether it someday could, did, or even now does support life.

In relation to the search for extraterrestrial life, these programs are very exciting. But what about extraterrestrial *intelligent* life? The chemical assay may be successful. Biologically relevant gases may be found, but that observation will not tell whether they came from algae or alumni.

Unfortunately, intelligence cannot be detected directly over interstellar distances. But if intelligence cannot be detected, why does SETI exist? In the first place SETI is something of a misnomer. Rather than trying to directly detect intelligence, SETI is actually seeking to detect manifestations of a distant technology. If such technology is detected, then the existence of intelligent technologists could be inferred.

If this is the intent of SETI's investigations, for what kind of technology can we look? If we imagine an advanced technological civilization, it would probably generate energy, power for different kinds of transportation systems and, unfortunately, perhaps energy for conflict and war. But the technological manifestation that has drawn the most attention is information transfer or communication. Given current terrestrial technology, this is the most detectable form of evidence for a celestially distant technology.

Thus, SETI has a very pragmatic definition of intelligence, that is, the ability to build a transmitter. This is the practical definition because this is what can be detected.

So, the search today is for electromagnetic signals. To engage in such a search it is necessary to answer the basic question: What sort of electromagnetic radiation does technology produce? In order to answer this question it is also necessary to determine what sort of electromagnetic radiation nature appears never to produce.

One possible signature of technology is extreme compression in the frequency and/or the time domains that allow high signal-to-noise ratios. In the frequency domain, nature does not seem to be able to become a coherent emitter, and astrophysics does not manifest itself on nanosecond timescales.

We have looked at natural radiation and the narrowest feature in the spectrum that we have been able to find at radio wavelengths is associated with an hydroxyl (OH) maser and occupies 300 hertz of the spectrum. There is no electromagnetic radiation found in nature that is more narrow. This is already non-thermal radiation. Maser emission is a non-linear amplification that narrows the frequency down to less than thermal. Yet, with contemporary technologies, we produce signals cheaply that are a fraction of a hertz wide all the time. These are very good detectable signals.

This, then, is where in the spectrum the search for SETI is taking place. Any electromagnetic radiation that is more narrow than 300 hertz is fair game for us to begin thinking about technology.

Now, it may be that what in fact we find is a natural phenomenon and leads to a whole new branch of coherent astrophysics that we did not think was possible. That, of course, in itself would be a good thing.

In terms of the time duration, we have not been able to find a comparable niche. Nature exhibits pulsed emission on a large number of time scales. We cannot find anything that would naturally say to us, "Technology." So, we are attempting to detect various pulse repetition rates and durations. We have to build our equipment to be flexible enough to look for lots of different temporal signal signatures.

There are two types of searches on the telescopes today. There are sky surveys in which you look every place that you can. However, you are not able to spend very much time at any one frequency at any point in the sky if you are going to look at the whole sky.

Suppose that the luminosity function for transmitters is like the luminosity function for stars. When you go out at night and look up at the sky, the bright stars that you see are not the closest stars. The close stars are the little faint ones. The bright stars are intrinsically much, much brighter than the nearby stars, and although they are much, much farther away, they appear brighter in our nightime sky.

It might be the same for extraterrestrial transmitters; the most detectable transmitter might be coming from very far away in a direction that you would not otherwise think to point your telescope. If you do a limited sensitivity survey of the entire sky you may stumble on it where you would miss it pointing your telescope in directions that you already knew about.

There are a number of such sky survey programs on the telescope today. The most systematic is the Serendip IV (Search for Extraterrestrial Radio Emissions from Nearby Developed Intelligent Populations) search. It is being carried out at the Arecibo Observatory, the world's largest radiotelescope. Serendip IV searches a hundred megahertz of the spectrum near the hydrogen line. In the new SETI@Home project, two megahertz of the raw Serendip data are put onto tape and sent to Dan Wertheimer and other researchers at the Space Science Lab of UC Berkeley, where it goes on a server. From there it will be distributed to ordinary folks running screensavers who want to participate in real-time sensitive signal detection. Serendip IV itself can only do a very first high-level cut at the data reduction. But on this small fraction of the bandwidth, by using distributed processing on home PCs, it is possible get enough compute cycles to do a very thorough frequency/time analysis of the data looking for signals. Serendip IV is funded by the SETI Institute and by the Planetary Society.

Another project is BETA (Billion-channel Extraterrestrial Assay), which uses a 26-meter telescope at Harvard College's Agassiz Station (Oak Ridge Observatory). The program is run by Paul Horowitz, professor of physics at Harvard, and it is also funded by the Planetary Society.

META (Megachannel Extraterrestrial Assays) was an earlier version of BETA and a clone of it is currently being operated in Argentina. There is also a Southern Serendip in Australia using the Parkes radiotelescope. The Southern Serendip is surveying the southern sky by piggybacking on a 13-beam focal plane array at Parkes Observatory.

Lastly, there is a new project being developed called project Argus. The SETI League will attempt to organize 5,000 volunteers around the world with backyard satellite dishes. These will form a network of very low sensitivity but continuous observation of the sky, looking for very strong but perhaps transient events. The project organization is based on the model of the Association of Variable Star Observers, a group of amateur astronomers who have been extremely useful in optical astronomy over the years.

The second major way of looking is to target, *a priori* to pick out the directions where you think there is a high probability of having a signal. This is what Project Phoenix, the project that I direct, is about.

Project Phoenix is a systematic microwave search from the frequencies between 1.2 and 3 gigahertz, excluding only those parts of the spectrum that are saturated with satellite and radar transmission. We target a thousand nearby stars; we look at the nearest 100 stars independent of their spectral type; and we look at all known exoplanetary systems. Then we select solar-type stars, preferentially those that are a few billion years old and that do not have any companions that might disrupt planetary orbits.

We are also following up on 11 events that the META search published a few years ago on the possibility that these might turn out to be extraterrestrial transmitters which, on average, lie below the detection threshold. Interstellar scintillation may have amplified them briefly so that they could be seen by the META system, but then they faded away and were never seen again. So we are going back to look with much better sensitivity at these targets and to look several times.

The name Phoenix alludes to the fact of rising from the ashes of congressional termination. Unlike the movie *Contact*, we do not wear headphones. But we do actually plan for success and there is a real bottle of champagne in the observer's icebox at Arecibo.

The signal processing that we do is based on full-custom digital signal processing gear. We cover 20 megahertz of dual polarization bandwidth. With 1-hertz resolution we have 56 million channels that we analyze every second. The full-custom electronics is difficult to replicate and maintain over a long period and we will never do it that way again.

We can package the equipment in a trailer so that we can take it to large telescopes around the world. It is the silicon intelligence of the system that does most of the work, although at first it takes a lot of people to get the system up and running. Then it takes fewer people as the system learns to do the job.

Eventually, you have a system that runs itself and runs the telescope. Of course, now that we have it all working, we are about to build a new system that has five times the bandwidth, has higher resolution, and should be coming on line at the telescope by the Fall of 2002.

There are some things that are unique about Project Phoenix. As a NASA program, when we first deployed to Arecibo Observatory in 1992, we learned a very important lesson: one telescope is not enough. In an RFI (radio frequency interference) environment, which is temporally varying, you cannot do an efficient search program with one telescope.

So, we now routinely use two telescopes, widely separated, in what we call a "two-element pseudo-interferometer." It is a pseudo-interferometer because we never actually bring the signals together and delay the phase of one and multiply them together. But it does work like an interferometer and it works very well.

We were in Australia in 1995 and spent a year and a half at Green Bank, West Virginia. We are now at Arecibo, our second telescope being in the United Kingdom, the Lowell telescope at Jodrell Bank.

At Arecibo we observe within plus or minus 6 hours of local midnight. We do this not because the sun is a strong radio source, but because of the solar wind. The charged particles leaving the sun can actually destroy the coherence of a narrow signal. So we are observing at night and every session we detect an extraterrestrial signal.

We detect the transmitter on board the Pioneer 10 Spacecraft, which has now left the solar system. It moves on the sky like a star. It has a transmitter of a few watts of power 6 billion miles away and we detect it very well.

What have we learned to date? For about 400 stars at frequencies between 1.2 and 3 gigahertz, we have not found any transmitters with at least 10^{12} watts of effective isotropic radiated power (that is equivalent to a strong terrestrial radar). Our observing is very efficient and we are continuing with the next 600 stars.

What are our future plans? We want our own telescope. We are tired of trying to share the telescope with others and getting only bits and pieces of time. We have signed a memorandum of agreement with the University of California at Berkeley and we are trying to design and build our own telescope. We call it the 1hT[b] (one hectare) telescope. It will have 10^4 square meters of collecting area. It will not be your average radiotelescope: It will be able simultaneously to conduct both SETI and regular radioastronomy observations because of its unique construction.

The 1hT will be the equivalent of a standard 100-meter telescope. With a set of small dishes it is possible to have a large field of view. Within that large field of view there are many SETI targets at any one time. If there is enough computing power, it is possible with such a telescope to form multiple beams on the sky.

The plan is to start with three beams. With the ability to cover a frequency range of 1 to 3 gigahertz (and looking at each target three times), it will take 6.3 years to look at about 100,000 targets. If we were more ambitious and could get more computing power quickly, we could go up to 12 beams and extend the frequency range up to 10 gigahertz. At this frequency the search would take about 8 years to compete the 100,000 targets.

In addition to serving the observational needs of SETI, it is our intention that the 1hT be a prototype for the SKA (square kilometer array) with a million square meters of collecting area. This is an observational instrument that the international astronomy community wants to build. By serving as a prototype, the 1hT can stimulate the work necessary make the SKA a reality. Building the SKA is also in SETI's interest because such an instrument will have a sensitivity a factor of 100 better than the 1hT. With the SKA, signals can be detected that are a hundred times fainter or are the same strength but 10 times farther away. With such an instrument, I can increase my target list.

What would we be able to observe with the sensitivity of a square-kilometer array? We could begin to be able to detect carriers emanating from the equivalent of a terrestrial television station in the nearby stars. But, with our current signal-processing equipment we would still only be able to observe

[b]This array is now being prototyped and is called the Allen Telescope Array (ATA) in recognition of the generous support from the Paul Allen Foundation.

100,000 stars over 9 years—unless I have more computer power. I would really like to observe a million stars.

If I could process 10 gigahertz of the spectrum all at once for each thousand-second observation, then I could get a million stars observed in a decade. This is really betting on Moore’s law that computing power will continue to grow exponentially. I think it is a good bet.

We have other plans. In addition to the radio searches that use traditional technology, we are trying to figure out how to build an omnidirectional sky survey device that can look for weak transient signals. Actually, we know how to build it. What we cannot do is afford the computing for it. It would take about as much computing as we currently have on the whole planet right now. Access to such computing power will come. This is another bet on Moore's law.

Extraterrestrial Intelligence? Not Likely

IRVEN DEVORE

Ruth Moore Research Professor of Biological Anthropology, Harvard University, Cambridge, Massachusetts, USA

ABSTRACT: The possibility that there exist extraterrestrial creatures with advanced intelligence is considered by examining major events in mammalian, primate, and human evolution on earth. The overwhelming evidence is that the evolution of intelligence in creatures elsewhere who have the capability to communicate with us is vanishingly small. The history of the evolution of advanced forms of life on this planet is so beset by adventitious, unpredictable events and multiple contingencies that the evolution of human-level intelligence is highly unlikely on any planet, including earth.

KEYWORDS: extraterrestrial intelligence; evolution of intelligence; evolutionary history; human uniqueness

An increasingly popular belief, spurred on by the recent discovery of other planetary bodies, is that there are incalculable numbers of sentient creatures throughout the universe. I agree that if there is the slightest chance that such creatures exist, we must use every means to discover the evidence. Actual contact with extraterrestrial beings would be a momentous event in human history; all other matters that now occupy human attention would shrink to insignificance. But based on the history of the evolution of intelligent life on this planet, I believe it is very unlikely that there are creatures out there either listening for us or signaling their own presence. For decades we have been sending episodes of "Dragnet" and "I Love Lucy" (not to mention "Third Rock from the Sun") across the galaxy and beyond. So far, there is no evidence that extraterrestrial fans are about to influence Nielson ratings.

As an aside, in the spirit of this volume, it gives me no pleasure at all to propose that we are unique and alone in the universe. The hubris of our species certainly needs no augmentation; we seem intent on destroying our own planet for ephemeral creature comforts, while at the same time murdering each other on an increasingly relentless scale. The history of the twentieth

Address for correspondence: Dr. Irven DeVore, Department of Anthropology, Harvard University, Peabody Museum, 11 Divinity Avenue, Cambridge, MA 02138. Voice: 617-868-4784; fax: 617-497-7227.

devore@fas.harvard.edu

century, when wars expanded from contests between male warriors to include attacks on civilian populations, has resulted in a slaughter on a scale that exceeds any other in the blood-spattered history of our species.

Centuries of progress in moral philosophy seem irrelevant in the face of attacking armies. If, as believed by some, *Homo sapiens* is the ultimate goal and crown jewel of evolution, one might wistfully ask why the process of evolution could not have produced a kinder, gentler, less rapacious, less murderous species. Unfortunately, the answer to such a question is deeply rooted in the same processes of natural selection—processes driven by amoral, ruthless competition—that produced all species, including ourselves.[1] Very few today believe that the "meek will inherit the earth." If there truly are experiments in intelligence in progress on other planets, one hopes that at least some of these experiments will produce more benign and less rapacious creatures than ourselves. Sadly, even this thin reed of hope seems quite unlikely; the overwhelming evidence from the history of this planet is that the blind, uncaring, amoral processes of natural selection pit organisms against each other in a struggle for survival. Although this process has been documented in every species that has been carefully studied, the process could not be more manifest than in the history and prehistory of our own lineage.[2] Furthermore, no scientist has proposed a credible model for evolution other than by natural selection.

IS THERE OTHER LIFE IN THE UNIVERSE?

In recent years astronomers and biologists have discovered that many of the molecules essential for life—purines, amino acids, etc.—seem to be widely distributed in the universe, and hence might gain a foothold on other suitable planets. Some maintain that these seeds of life, "transpermia," wafting through the universe are the most likely reason for the origin of life on earth. But whether life on this planet began by a unique, fortuitous combination of elements in Darwin's "warm little pond" or was jump-started by transpermia from afar, bear in mind that there is no evidence of any form of life during earth's first billion years. Only at about 3.8 billion years ago have we found evidence for the first life forms, prokaryotes—simple cells without an organized nucleus. The phrase "from amoeba to man" is commonly invoked to suggest the great span of evolutionary history, but biologists know that the development of organized, self-replicating organisms from unorganized lifelike beginnings was an even more momentous step in evolution. What emerged from the original "blue-green algae stage"? Bacteria and their kin, the eukaryotes, and they have thrived!

Stephen Jay Gould has famously remarked that, if we were honest, we would call this epoch not "the age of mammals," but "the age of bacteria." It is sobering to realize that each of our bodies contains more bacteria than the total number of all humans and proto-human hominids throughout all history. Today we devote enormous energy to cultivating the bacteria we like, and developing antibiotics against those we don't like. Because of selection the latter strategy is clearly beginning to fail. One could argue that the bacteria have expertly manipulated us toward *their* goals, not *ours*, and that an objective observer would not only declare bacteria "winners in the evolutionary race," but might also add that they are also the kind of organisms most likely to survive after a human-induced Armaggedon. From one point of view we are minor actors in a scheme of which we are largely unaware. From another point of view bacteria illustrate how very simple organisms can become spectacularly successful—aided immensely, of course, by invading complex organisms, and bending these hosts to their own purposes.

COULD E.T. CALL HOME?

In this essay, I am not addressing the question of whether there might be colonies of simple organisms elsewhere in the universe, but rather whether there are creatures capable of a two-way "conversation" with us. Two giants of modern evolutionary biology, Ernst Mayr and George Gaylord Simpson, arrived independently at the conclusion that the prospect of such an interchange is extremely unlikely. Ernst Mayr has written that no fewer than six of the eight conditions to be met for success in the search for extraterrestial intelligence are highly improbable.[3] When these improbabilities are combined, the prospects for successful contact seem even more remote. G. G. Simpson's essay "The Non-Prevalence of Humanoids" [4] makes many similar arguments. For example:

> Even in planetary histories different from ours might not some quite different and yet comparably intelligent beings—humanoids in a broader sense—have evolved? Obviously these are questions that cannot be answered categorically. ...(But) The factors that have determined the appearance of man have been so extremely special, so very long continued, so incredibly intricate that I have been able hardly to hint at them here, Indeed they are far from all being known, and everything we learn seems to make them even more appallingly unique. If human origins were indeed inevitable under the precise conditions of our actual history, that makes the more nearly impossible such an occurrence anywhere else. I therefore think it extremely unlikely that anything enough like us for real communication of thought exists anywhere in our accessible universe.

What makes evolutionary biologists so much less credulous than others in this debate? Probably the most important reason is that a great many educated persons, including many scientists not trained in biology, harbor a deep misconception about the meaning of evolution. Most believe that "progress and improvement through time" is an inevitable outcome of the origin of life and its evolution. Each succeeding generation, having been successful in the "survival of the fittest" is now measurably advanced over its ancestors. But such "progress" implies that there is a plan unfolding through time, leading to more and more perfected forms, and not the action of selection on different genetic combinations that we actually observe.[5] Darwin himself was concerned about the conflation of "progress" and "evolution" in his day. Partly due to the writings of Darwin's grandfather, Erasmus, early ideas about evolution already implied improved development over time, a meaning in common usage today, for example, in phrases such as "the evolution of Formula 1 racing cars from the horseless carriage." Darwin avoided the connotation-loaded "evolution," coining instead more specific, scientific terms like "natural selection" and "descent with modification."

Darwin was very aware that his bleak, mechanistic interpretation of evolution might cause an uproar in the Christian church. Many historians of science have inferred that Darwin delayed publication of his ideas for many years—rushing into print only after he received a letter from A. R. Wallace that briefly outlined Darwin's central argument. And, just as Darwin feared, materialistic and theistic explanations of life were at sharp odds from the beginning.

Just as important as "progress" in this controversy is the "argument from design," eloquently described by the eighteenth century theologian, William Paley.[6] His argument, in brief, was that the astonishingly intricate wonders of the world could not possibly have been the result of mere chance; instead one had to assume a cosmic "watchmaker." This argument continues to be invoked to the present day, supported by "proofs" that complex organs such as the eye could not have come about by chance alone. Richard Dawkins, in *The Blind Watchmaker,*[7] offers a sensitive but devastating rebuttal against "design" in evolution. In this position he joins a distinguished group of biologists like Mayr, Simpson, and countless others who critiqued the works of Lecomte de Nouy and Teilhard de Chardin, who were convinced that there was a plan and a final goal to evolution and whose conviction attracted many followers.

On the basis of the geological record, evolutionary biologists consider that the history of life on earth is exactly that: *history*—not a series of predictable outcomes. Natural selection, the engine driving evolution, is a blind, uncaring, unpredictable process. G. G. Simpson famously characterized evolution as following a "zigzag opportunistic course." As continental masses break up and recombine, as cosmic objects smash into earth, as continental glaciers appear, and as deserts become tropical forests, many adaptive niches disappear while novel ones are created. Besides these geological and climatological events, there is also the ever-changing web of potential death and competition from parasites, predators, and other species contending for one's ecological niche. Which populations will succeed in these races is utterly unpredictable. Indeed, it is often one of the least typical organisms in a transformed environment that adventitiously has the wherewithal to take advantage of the opportunities that are created by new circumstances—hence the "zigs and zags" in the fossil record. Far from being predictable events, these changed circumstances often lead to quirky results, and with no plan to follow, sometimes to extraordinarily "inefficient" or "fragile" new forms; if evolutionary processes emerged to help species survive over time, they have made an extraordinary botch of the matter. Paleontologists report that *more than 99% of earth's species have gone extinct*. Ernst Mayr estimates that there have been more than a billion species in earth's history, perhaps many more. Out of that huge number only one, *Homo sapiens*, developed the ability to create civilizations. And among those 25 or so civilizations, only one attained the technology to take advantage of the electromagnetic spectrum. The ability to transmit and receive signals from afar seems to have sprung from a very unlikely series of events.

CHANCE, COMPETITION, AND CATASTROPHE

The arguments above and the examples offered below can be heuristically thought of as chance, competition and catastrophe. All of these terms emphasize the historical contingencies that lie at the heart of the evolution of species. A change in a species is the result of an unpredictable series of antecedent events, and not "predictable outcomes from laws of nature." Change in living species is constrained by all preceding events; changes are limited by the genetic materials at hand, bequeathed to the organism by past generations. Differential reproductive success and mutation are the raw materials of morphological change, but the former is notoriously dependent on local, changing conditions and the overwhelming majority of mutations are deleterious. If a novel adaptation is sufficiently advantageous (for example, bipedalism in the hominid lineage), the resulting changes are by no means necessarily the "best solutions." Natural selection can operate only on the status quo. As a result, most changes in species are jury-rigged solutions, cobbled together by a tinkerer—certainly not what one would expect from a cosmic watchmaker. If we could be objective about ourselves, we would admit that there are much better body plans, and that any sophomore at Cal Tech could design a human body plan with far better engineering solutions to the pains and failures brought on by bipedalism (from hernias to backaches to dysfunctional feet and the difficulties of childbirth), not to mention equally inept "solutions" in the visual and vascular systems.

Evidence has surfaced in recent years that widespread catastrophes were more important in earth's history than previously thought. Sometimes a planet-wide event scours the earth of nearly all life. The great granddaddy of these mass extinctions occurred in the Permian (245 million years ago), and caused the extinction of 96% of all creatures on earth. The later Cambrian decimation, which eliminated 80–90% of the panoply of wonderful new creatures that had evolved, whose remains are richly preserved in the Burgess Shale, has been eloquently described in Stephen Jay Gould's book *Wonderful Life*.[8] In that volume Gould says that one purpose of his book is to make a statement about the nature of history and the awesome improbability of human evolution.

A mass extinction that has deservedly caught the popular imagination occurred at the end of the Cretaceous, when an interplanetary object smashed into the earth, and incidentally did the mammals a great favor; it eliminated all the dinosaurs. For a hundred million years dinosaurs dominated the planet: the land, the sea, and the air. (By contrast our hominid lineage has been around for only about 7% of that time.) And, despite their opportunities over this hundred million years, and the astonishing variety of forms that appeared, there is no evidence that dinosaurs showed any trend towards higher intelligence.

Had it not been for cosmic intervention, the small shrew- or vole-like mammals scurrying around between the dinosaurs' legs—picking up crumbs, so to speak, from the dinosaurs' table—would probably be about the same insignificant mammals today. It would seem that the unprepossessing early mammals could not successfully compete with the dinosaurs. But when a fortuitous event eliminated the dinosaurs, the myriad of new opportunities in the environment resulted in an explosive radiation of the mammalian species.

Beyond the kinds of catastrophes mentioned above, there have been many other major changes in earth's ecology. We are only beginning to understand the consequences of alternating tilts in the earth's axis, or the consequences when the magnetic poles reverse. We are also in a very early stage of understanding long-range weather patterns, but the geological record has already revealed that some major ecological changes can take place in less than a century, and some can be measured in decades. Note that a species that is most able to survive and take advantage of the opportunities following a catastrophe is often able to do so because it adventitiously had traits that, while not especially remarkable during the long period leading up to the catastrophe, were suddenly very advantageous.

THE RISE AND NEAR EXTINCTION OF EARLY PRIMATES

The history of evolution, like all history, is full of quirks, near escapes, sudden successes, and other unpredictable events. The earliest of our primate ancestors were among the earliest and most numerous of all mammals, dating from the beginning of the Tertiary, the "age of mammals," some sixty million years ago. One prominent candidate for "earliest primate ancestor" is a peculiar proto-primate that looks more like a rodent than a primate; it has large incisor teeth for gnawing, and claws rather than fingernails. If you passed one in a city park, you might well guess that you are looking at a kind of squirrel that you hadn't noticed before.

Although the earliest primates had a very promising start in the early Tertiary, they soon began to shrink dramatically in numbers. Not coincidentally, rodents, who appeared later in the fossil record, entered earth's history with an awesome radiation of their own—exploding into terrestrial environments throughout the world, becoming, along with bats, the most numerous mammal species. The rodents were a serious threat to the survival of the early primates. The early primates, living in forests (as do most primates today), were at a tremendous disadvantage competing with rodents for forage in the trees. Primates have only one set of adult teeth, and as gnawers they were seriously challenged by the rodent adaptation of growing incisor teeth throughout their lives.

The survivors from these early primates were atypical—they developed grasping hands and feet. With this adaptation even large primates, by using four grasping appendages, could distribute their body weight among the branches, and forage among the smaller terminal branches of a tree, harvesting fruits, nuts, and fresh leaves more successfully than much smaller squirrels could manage. So in many parts of forest ecosystems, grasping appendages could successfully trump continuously growing teeth. Uniquely among mammals, all members of the Order Primates have grasping appendages (usually all four) with fingernails (whose histology is anatomically quite distinct from that of typical mammalian claws). The grasping hand is a singular primate adaptation, and a symbol of the whole Primate Order, found in prosimians, monkeys, apes, and humans.

To illustrate the difficulty of predicting adaptations and "advances" from the fossil record, consider the ceolacanth. The ceolacanth is a living representative of an ancient group of lobe-finned fishes, the crossopterygians, who were believed to have been extinct for 65–70 million years. They are of special interest to paleontologists because they appear to be the best candidate for a creature that was a transition stage from fish to amphibian. The thought was, to caricature it, that ceolacanths and their kin were lurching along on stubby fins from puddle to pond, in the hope that their offspring might someday be amphibious. But when a live ceolacanth was unexpectedly caught in 1938, it was not mucking around in brackish ponds, intent on being a proto-amphibian. Instead, its habitat is now known to be 200 feet down on a coral reef; no one has a good idea why its stubby, powerful fins are an advantage down there.

"Romer's rule" summarizes arguments such as the above succinctly. Alfred S. Romer was a famous evolutionary biologist who wrote the "bible" on the evolution of the vertebrates. His specialty was the fossil history of the amphibians. Romer famously said that an amphibian is not a deliberate transition between fish and reptile at all; an amphibian *is not trying to conquer the land*, but is so desperately trying to remain a fish that it is willing to slog from pond to puddle in order to lay its eggs in water and allow its babies to grow up like proper fish. It follows that the initial survival value of a favorable innovation is conservative, in that it attempts to maintain a traditional way of life under altered circumstances. In this sense biological innovation that results in the emergence of new forms does so in spite of every effort by the species to sustain itself in its present niche.

To summarize, I believe that the enormous array of contingencies leading to ourselves is such that, if an ancient lobe-finned fish had swum up the south rather than the north fork of an estuary in ancient Gondwana, we would not be here.

THE IMPROBABLE PATH TO ADVANCED INTELLIGENCE

Astronomers interested in extraterrestrial intelligence often argue "Improbable though the evolution of advanced intelligence may be, you must admit that there is only one sample, one planetary history and that intelligence evolved on earth." But this is naïve. Biogeography and paleontology show that earth provides not one sample, but many millions of samples. Sketches of a handful of these samples follow.

Placental mammals never made it to Australia. Instead, the marsupials radiated there into a very wide variety of niches that remarkably paralleled their mammalian lookalikes elsewhere: marsupial dogs, cats, moles, flying squirrels, and kangaroo mice. Of course some of the niches in Australia were filled by unique adaptations. The kangaroo, for example, fills the grazing/browsing niches occupied by ungulates elsewhere. Not only did intelligence, or even placentation, fail to evolve among these marsupials (who after all covered a continent the size of the United States), but a form even remotely like our primitive primate ancestors also failed to appear. From the earliest days the homeland of the entire primate lineage has been in the trees, and it was not as if the "tree niche" in Australia had not been exploited by marsupials. The koala's adaptation to tree life makes it a marsupial's nomination for a monkey. If you've ever held a koala in your arms, you know that its brain and behavior could only be generously described as "somnolent." The other tree-exploiting candidate for primatehood in Australia is the tree kangaroo. If ever there were a hopeful monster, this must be it. Its secondary adaptation to tree life was cobbled together from its hopping adaptation to the ground. So long as there are no serious predators, the tree kangaroo can make a comfortable living in the branches. But when humans appear it becomes just another easy lunch. The Australian fauna illustrate a remarkable independent radiation of forms parallel to the mammals. But nowhere do these marsupials display any inclination toward advanced intelligence, or even important preconditions for it to develop.

Consider another example, the island of Madagascar. Some fifty percent of all primate varieties live on Madagascar, but they are all prosimians. Madagascar broke away from Africa before monkeys colonized it, and no monkey or ape has ever emerged on Madagascar.

This does not mean that prosimians were not successful. Far from it, they radiated into an enormous variety of niches. There are prosimians adapted to swamps, deserts, montaine forests; they are mostly found in trees, but some adapted to the ground and all habitats in between. Like other prosimians in Africa and Asia the prosimian's body and brain reflect the fact that they live in a world in which tactile and olfactory reception is as important as vision. Most have vibrissae, scent glands, and a basic mammalian body plan. Some have multiple young and some forage at night. Yet they also have the primate hallmark: grasping hands. Prosimians include some of the smallest mammals on earth, and in historic times there were lemurs as large as a donkey. Despite this rather astonishing radiation of forms and adaptations, prosimians never evolved what could even be generously called a "monkey-level of intelligence."

The tarsier, a prosimian of Asia, developed binocular color vision rather like the system that evolved in humans. Why should this be when other prosimians lack binocular vision. The tarsier is about the size of a rat; it bounds around in trees in semi-darkness in the Philippines and parts of southeast Asia. A unique trait in this group is the greatly elongated tarsal bone, which gives the tarsiers a great deal of extra leverage when they decide to hop. If humans had such an adaptation, we could all leap tall buildings at a single bound. During a long leap a tarsier can grab a moth in mid-air and still make a safe landing. Such acrobatics require the precision provided by stereoscopic color vision. It is ironic (but no great surprise in evolutionary history) that a characteristic like binocular color vision, that turns out to be so critically important in the development of our own intelligence and technology, also evolved millions of years ago in an utterly different context, and in support of a very different evolutionary path.

The evolution of monkeys provides two parallel examples. After South America split away from Africa, the independent evolution of New World monkeys reveals many close parallels with Old World monkeys, but significant differences as well. South American monkeys have exploited a wide variety of niches. There are many species of squirrel-sized, pair-bonding (in fact, often polyandrous) marmosets and tamarins. There are species with male-pattern baldness and no tails. Provocatively, the "ape niche" has been filled in South America by three genera of large monkeys, but in a manner utterly different from arm-swinging apes. Instead these monkeys developed a fifth and unique appendage, complete with fingerprints: the grasping tail. Their prehensile tails allow them to distribute their weight and harvest fruits and other goodies out on the terminal limbs of trees where the foraging is richest. So, the ape niche was filled in the South American rainforests, but not by an ape. Indeed, there is no reason to suppose that a creature the size and intelligence of the great apes would ever have appeared in South America.

In numbers and variety the Old World monkeys, found throughout Africa, tropical Asia and southeast Asia, are among the most successful of all mammalian species. The baboons of Africa and their close cousins, the macaques of Asia (e.g., the rhesus monkey) are well known to most people. I began studying baboons in 1959[9] because, quite unlike most monkeys and apes, baboons foraged on the ground in large multi-male groups, were reasonably good hunters, adroit at escaping predators, and had a social organization that could support a long period of infant dependency. In other words these large monkeys show many of the traits we know became important in hominid evolution. (But such comparisons can be easily stretched too far; baboon solutions to their terrestrial niche are still "monkey solutions," and quite unlike the way of life, strategies, etc. we know from human foragers).[10] Baboons seem to have many of the prerequisites for advanced intelligence: large brains, intense curiosity, and they exercise considerable insight and guile in social interactions. From their long fossil record we also know that there have been many species of baboons, some quite small, and some gorilla-sized. Despite their admirable successes—their ingenuity, social complexity, and long lives—as far back in prehistory as we can trace baboons, they seem to be built to the same basic plan. From the point of view of developing an ape or human level of intelligence, one can only say that baboons are mired in a baboon niche.

THE ORIGINS OF HUMAN INTELLIGENCE

What leads to intelligence of an advanced, human kind? First one needs a certain kind of brain; ours is built to the same basic plan as the Old World monkeys and apes, and all species in this group have developed visual acuity, with reduced dependence on tactile hairs, olfaction, and auditory signals. To produce human intelligence it would certainly seem to help if the sensory inputs to the brain can be combined with the motor skills available when one has an intricate organ like the hand (or, in the case of elephants, the trunk) in front of the eyes.

For selection to produce an ape brain the size of ours would appear to be trivial; simply one more division of brain cells in a developing chimp would produce a human-sized brain. Considering the enormous value of large brains to humans, why haven't the chimps opted for that one extra division of cells?

Such a brain would not be organized like the human brain, of course, but in any case the chimps' "metabolic budget" simply could not sustain such a brain. The brain is an enormously energy-devouring organ and the chimpanzee lifestyle can't afford it. Even asleep a human infant can require up to 80 percent of its metabolic energy just to maintain its brain. Developing and supporting such an expensive organ requires many conditions. For example, while marsupial reproduction has its own advantages, it does not have placentation to supply the rich resources to support a greedy infant's brain in the womb.

Once one begins to consider the background to human intelligence, it is readily apparent that this involves very special physiological adaptations, adventitious traits inherited from ancestors, the consequences of the dietary adaptation, right on through to a very special, supportive, social organization. Human adaptation had to allow the mother to not only be "bipedally effective," but also to give birth to a large-brained infant, whose brain then grows even larger after birth. During the period of late pregnancy and lactation both mother and infant are dependent and vulnerable. To sustain them requires a social group that is willing to provision them.

Humans and chimps are very closely related. We shared a common ancestor only 6–7 million years ago. Chimps and humans share 99 percent of their genes; by this measure we are more closely related to chimps than are horses to zebras or foxes to dogs or sheep to goats. If humans manage to wipe themselves out with some rampaging disease (leaving near relatives largely unharmed), could one or more of the great apes fill the human niche? Sadly, I think the "Planet of the Apes" scenario will remain just an entertaining fantasy.

Many of these issues were brought home to me vividly when I briefly babysat a young orphan chimp, Koby, at Jane Goodall's Gombe research station. Despite his small size and tender age, Koby's upper body strength was embarrassingly close to my own. This brought home to me the fact that the lifestyle of the great apes means that their young must consume energy in early muscular development, and there is not enough energy left over to also support a greedy brain.

THE HOMINID LINEAGE

Not so long ago it was possible to convincingly arrange all fossil hominids on an "ascending line" from the small early Australopithecines to *Homo sapiens* (a treatment that is still frequent in popular writing). Not only were our early dating methods crude, but also no one could possibly envision the explosion of hominid forms unearthed in the past two decades. In the past five or six million years, hominids rejoiced in their own impressive radiation—

showing many different adaptations, most of them resulting in dead ends. During these periods many of these species were contemporaneous, and there is little agreement on how they might have competed with each other. Over the millions of years various hominid ancestors zigged and zagged, finally coughing up *H. sapiens*. No one agrees which of the earlier ancestors should be anointed as "leading to *H. sapiens*," and it will require much more evidence before the histories of these competing species can be sorted out.

To properly explain and document the arguments outlined above would require a very large volume. A summary of the argument I have sketched is that evolution is understood by evolutionary biologists to be due to blind, uncaring, natural selection. Evolution is viewed as historical, including all the contingencies that such an interpretation suggests. Nothing in what scientists know about evolution provides a shred of evidence that a cosmic process has guided evolution to creatures like ourselves. On the contrary, because of the many extraordinary historical contingencies that led from molecules to prokaryotes to eukaryotes to chordates to mammals to primates and finally to *Homo sapiens*, the evidence strongly suggests that the events leading up to human intelligence were fortuitous and unique. I therefore believe that our chances for meaningful communication with other creatures in the universe are infinitesmally small.

ACKNOWLEDGMENT

Without the tireless energy, goading, and forgiving nature of Meg Lynch, this manuscript would never have been completed.

NOTES AND REFERENCES

1. G. G. Simpson, *The Meaning of Evolution*, (New Haven, CT: Yale University Press,1949); Ernst Mayr, *What Evolution Is* (New York: Basic Books, 2001); George C. Williams, *Adaptation and Natural Selection* (Princeton, NJ: Princeton University Press; Robert L. Trivers, *Social Evolution* (Benjamin/Cummings, 1985); Richard Dawkins, *The Extended Phenotype* (New York: W. H. Freeman, 1982).
2. Richard Wrangham and Dale Peterson, *Demonic Males* (Boston, MA: Houghton Mifflin, 1996).
3. Ernst Mayr, "Can SETI Succeed? Not Likely," *Bioastronomy News*, Vol. 7 (1995).
4. George Gaylord Simpson, "The Non-prevalence of humanoids," in *This View of Life* (Chicago: University of Chicago Press, 1964).
5. Frederick Crews, "The New Attack on Evolution," *The New York Review of Books,* Vol. XLVIII (15), October 4, 2001 (parts I and II).

6. William Paley, *Natural Theology; or Evidences of the Existence and Attributes of the Deity. Collected from the Appearances of Nature* (London, F. Faulder, 1802; Oxford: J. Vincent, 1828 [2nd edition]).
7. Richard Dawkins, *The Blind Watchmaker* (New York: W. W. Norton & Co., 1986).
8. Stephen J. Gould, *Wonderful Life: The Burgess Shale and the Nature of History,* (New York: W. W. Norton, 1989).
9. Irven DeVore, ed., *Primate Behavior: Field Studies of Monkeys and Apes*, (New York: Holt, Rinehart & Winston, 1965).
10. Richard B. Lee and Irven DeVore, *Man the Hunter*, (Chicago, IL: Aldine, 1968).

The Outlook for Cosmic Company

SETH SHOSTAK

The SETI Institute, Mountain View, California 94043, USA

ABSTRACT: The last 100 million years or so has seen a continued increase in encephalization for several terrestrial species. Intelligence has survival value. Developments in astrobiology suggest that what was once considered enormously improbable, namely life, is now suspected of being ubiquitous. It may be that the evolution of intelligence is unlikely, but in a finite, breathtakingly large universe (10^{22} stars) small probability likely does not matter. Even if nature is indifferent to producing intelligence, SETI might still succeed. Biological intelligence may be rare, but it has the potential for creating engineered synthetic intelligence, capable of rapid and directed self-evolution. The galaxy could be rife with such long-lived, communicating devices, even if intelligent protoplasm is both rare and fleeting. SETI is looking for narrow-band, microwave signals that are not produced naturally. Ultimately, SETI is more exploration than experimentation.

KEYWORDS: Drake equation; probabilities; Fermi paradox; artificial intelligence; living machines

The title of this chapter is somewhat ambiguous. One wag of my acquaintance assumed that I would be speculating on ET's future financial prospects. In fact I intend only to suggest something far more modest: namely, that my day job—SETI—is not wasting its time. I will opine that there is good reason to believe that the implications of Professor DeVore's views—that we will never hear an alien transmitter because intelligence is rare—are wrong.

I am sanguine about the outlook for cosmic sentience, and I note that the public shares my view. Indeed, the public is more sanguine than I am. Surveys taken since the 1960s have repeatedly demonstrated that a large fraction of the American populace not only believes that the aliens are out there, but that they're here as well, buzzing the countryside or occasionally abducting unsuspecting folk for salacious experiments. There are many reports that bolster this belief, including 50 years of UFO sightings, abductions, implants, and the bizarre phenomenon known as crop circles. The last are particularly

Address for correspondence: Dr. Seth Shostak, Project Phoenix, The SETI Institute, 2035 Landings Drive, Mountain View, CA 94043. Voice: 650-960-4530; fax: 650-961-7099.
seth@seti.org

intriguing, not only because some people continue to believe in their extraterrestrial nature even after a pair of British gentlemen admitted to having constructed the first circles with boards and ropes, but also in the light of the extraordinary motivation required of aliens to indulge in such agrarian graffiti projects.

But the point is that, if the public is to be believed, I could have settled this debate by bringing a few alien bodies to the conference. The general populace might regard that as both obvious and easy (although it would probably incur the wrath of the feds). But let's admit that the present debate will not be settled by popular vote. The public may believe that proof of alien intelligence has been freeze-dried and stacked up by nefarious government agents. Few scientists would agree, and in any case, I have no cosmic cadavers at hand.

Professor DeVore has argued that the evolution of intelligence on Earth has been a highly contingent and undirected enterprise. He claims that it is extraordinarily unlikely that, even in a universe awash in life, intelligence will arise. Therefore, the outlook for a SETI detection is depressingly bleak (although Professor DeVore nonetheless maintains that the SETI search should continue). Can I say anything in direct refutation of his argument? Rather little. I'm not a biologist, and I therefore hesitate to contest the rarity of intelligence as argued from a biological viewpoint.

Still, I will essay a few biological points. First, I note that intelligence is a specialization, a highly adaptive specialization. Professor DeVore stresses the uniqueness of humans. But the question, of course, is not how rare is our hardware—our brain—what Philip Morrison has called "a slow speed computer working in salt water." Rather, the question is: how rare is the functionality of our brain?

It may be hard to argue its inevitability on the basis of, say, convergent evolution. Nature clearly has an interest in streamlining large, underwater animals or in producing eyes. But intelligence may be a less compelling feature, and as DeVore points out, it hasn't arisen often on this planet. On the other hand, the last 100 million years or so have seen a continued increase in encephalization for several species. This suggests at least some interest by nature in brain-power.

A second remark is that intelligence has survival value. Humans occupy virtually all the biological niches of the planet, and indeed are so successful that some regard us as too successful. The point is that survival value translates into staying power. Once started, intelligence will have an enhanced probability of enduring the catastrophes that have obliterated most species.

Third, and as a kind of cautionary footnote to biological arguments, I note that estimates of what is commonplace and what is rare can change quickly. Twenty-five years ago, DNA or any similar basis for life was argued to be so complicated that it was extraordinarily unlikely to have arisen elsewhere. One estimate was that DNA would randomly cook up in a seething primordial soup only once every 10^{42} years. That is industrial-strength pessimism. If this argument were correct, then every star in the visible universe could host a billion habitable planets, and the evolution of DNA would nowhere be duplicated. The clear implication was that biology was a fluke. Now, as the millennium ends, we are debating possible fossils from Mars and considering how we could search for water-borne creatures under Europa's icy carapace. We seem to be opening the door to a "universal biology," in the same way that we have established universal physics and chemistry. What was once considered enormously improbable is now suspected of being ubiquitous.

These and similar arguments have been extensively discussed in the SETI literature. However, rather than dwell on evolution's likely products, allow me to argue the case for detecting intelligence by pointing out that the universe offers many possibilities for providing us with a signal—more possibilities than at first may be apparent.

First, I would like to note that the usual view of aliens is highly conservative. When we consider the cut of ET's jib, we generally incline to the models offered to us by Hollywood. Cinema aliens are highly anthropomorphic (as director Frank Capra once observed, "people are most interested in people") and not very diverse. Movie aliens are like movie cowboys: their hats come in only two colors. There are good aliens, such as those in "Close Encounters of the Third Kind" and "E.T." These latter are more than anthropomorphic. They look like children, and are clearly harmless. Black-hatted aliens, on the other hand, such as those in "Independence Day," are modeled after our natural enemies: carnivores, snakes, insects, etc. These fictional aliens, both good and bad, populate our collective unconscious. They are what we expect to be keying the transmitter at the other end of a SETI signal. Their general properties are as follows:

They are soft and squishy, protoplasmic, pulsing, oozing, and mortal.

They are an animal species, and are members of a society composed of a large number of individuals.

They have evolved via Darwinian evolution.

They are living on planets; indeed, somewhat Earth-like planets (and this is why SETI researchers become so excited at the discovery of extrasolar planets).

That may be what we expect. But what is it that our search technology can actually find? What we are looking for are narrow-band, microwave signals —a type of emission that (as far as we know) is not produced naturally. So the requirements for SETI to succeed are (*a*) the existence of technology, that is, a transmitter at the other end, and (*b*) most probably a deliberately broadcast signal designed to get our attention. This latter requirement derives from the fact that advanced civilizations may be relatively "radio quiet," relying on low power (think of cellular phones) or highly directed (optical) communications. Consequently, they will not be leaking high-powered emissions into space. Furthermore, in order to inexpensively deliver an easily detected signal at interstellar distances, they will need to use reasonably large-sized transmitting antennas. Such antennas are highly directional (typical gains are 10^5 to 10^7 for terrestrial equipment).

The picture we generally have of aliens, and one whose probability Professor DeVore has addressed, is hardly either described or proscribed by these simple technical requirements. There are other possibilities. So I hope you will let me (at least momentarily and in a spirit of general amity) to grant DeVore's premise. Let us assume that high biological IQ is rare. I will continue to maintain that SETI could still find the universe rife with intelligence—with sentient entities.

The argument requires letting your imagination run free. Think in terms of deep space, but also deep time. It is worth keeping in mind that the number of stars in the visible universe is $\sim 10^{22}$, or comparable to the number of grains of sand on all the beaches of Earth. Approximately 10 percent of these resemble our sun. In addition, evidence for planets around sun-like stars is growing. At least 3–5 percent of the stars examined for extrasolar planets are found to have them. This implies the existence of billions of planetary systems in the Milky Way alone. (Within the visible universe, the number is approximately 100 billion times larger.) Finally, note that the universe is two to three times the age of Earth. We are the new kids on the block.

It may be that the evolution of intelligence is unlikely. In an infinite universe, that wouldn't matter. In a finite, but breathtakingly large universe—an old universe of 10^{22} stars—it *probably* doesn't matter.

It is conventional practice in the SETI community to parse this imposing number of possible sites for life by application of the Drake Equation. First proposed by Frank Drake in 1961, this simple formula sets out the factors that determine the number of contemporary broadcasting societies in our own galaxy (this latter limitation is mere parochialism. One could easily consider the rest of the universe, but signals from far-away galaxies are clearly more difficult to detect.) The Drake Equation estimates the number of suitable stars, multiplies this by the fraction with habitable planets, then by the fraction of those planets that have developed technologically sophisticated life. The resulting product is then multiplied by the fraction of the galaxy's lifetime during which such a technologically civilization is active (this incorporates the famous "L" term, which is the lifetime of a technological civilization).

While a reasonable approach, the Drake Equation makes the assumption that "they" are like "us": having evolved upon, and living on (or near) a biologically suitable planet. This is also the assumption made by both Hollywood and by my esteemed opponent.

But let us consider our own case. What is the long-term future of humans? To begin with, we will certainly spread beyond the bounds of our world. A single, round planet can't host us indefinitely. Our initial forays will be to obvious targets: the moon, Mars, and the space in between (populated by the rotating aluminum cans promulgated by Gerry O'Neill and Thomas Heppenheimer in the 1970s). Some time early in the next millennium we will colonize the asteroids, as envisioned by Freeman Dyson.

In other words, within a century or two, at most, humankind will be dispersed. This is a potent hedge against self-inflicted catastrophe. It will be difficult to eradicate the human species once colonies are spread throughout the solar system. While I might (with difficulty) get rid of the ants in my kitchen, I cannot eliminate the world's entire ant complement. Humans are going through a bottleneck of only a few centuries' time—a bottleneck during which we might be able to exterminate ourselves. But given the relatively short period of time involved, we may reasonably hope to survive this peril. Professor DeVore has pointed out the large number of species that have gone extinct, sometimes catastrophically, on this planet. This is referred to as "clearing the deck." Humans (and by extension, other intelligent beings) may not be very susceptible to such sudden elimination, simply because they are quickly off the deck.

The spread into the solar system will give humanity time. Certainly, it will allow sufficient time to push forward the type of research that, someday soon, may initiate the evolution of our own successors. At first, these will consist of artificial augmentations of our traditional biology. The immediate prospects are for engineered replacements for diseased or destroyed human tissue. Perhaps we will make some such constructs a permanent part of our anatomy, becoming like the Borg: half organic, half manufactured.

But this may only be a stopgap measure on the road to developing completely synthetic sentience. Machine intelligence—fully equal (and quickly superior) to human intelligence in its ability to reason, to create, to amuse—is a frequent staple of science fiction as well as being a popular prognostication of technology futurists. There seems little reason to believe that it will not be realized. Silicon life, in the form of machine intelligence, would be free not only from the frailties and brief lifetimes of biology, but, of greater import, would not be constrained to the undirected and slow change of Darwinian evolution. A machine intent upon, say, increasing its memory, could do so by simply building and inserting more circuitry. It could improve itself by design, evolving in a Lamarckian, rather than a Darwinian manner. This would, of course, quickly result in vastly improved machinery.

I do not consider this scenario unlikely. Many technologically advanced cultures will disperse, thereby ensuring their survival. Survival will then give these societies the time to engineer synthetic intelligence, capable of rapid and directed self-evolution. Biological intelligence may indeed be rare. It may be "unlikely." But even if this scenario has played out only occasionally in the ten billion years of galactic history, it would have widespread consequences. Those consequences arise because such advanced intelligences could spread out.

The classic Drake Equation, as we have noted, assumes that the extraterrestrials will remain in the solar system of their origin. This is not unreasonable, assuming that the aliens are mortal beings with lifetimes short compared to the time needed to bridge the distances between the stars. Interstellar travel for biological beings is problematic (despite the suggestions of Hollywood) given the high speed necessary to achieve a tolerably short trip. But such restraints will not apply to machine intelligence. Even at non-relativistic speeds (~1 percent the speed of light), such sentience could spread throughout the galaxy within a few tens of millions of years. This is, of course, a thousand times less than the age of the galaxy, so there has been plenty of time for it to have occurred. Intelligence could have rapidly dispersed.

There is a well-known argument against this scenario, codified as the "Fermi paradox." Despite the fact that there has been more than enough time for a thorough colonization of the galaxy, we do not see obvious evidence of alien presence nearby. This might suggest that the starfields of the Milky Way have never been pioneered by intelligent entities, either machine-like or metabolic, and humans may be the smartest things in the galaxy. In other words, the extraterrestrials are not here, and therefore not there. But many suggestions for reconciling the Fermi paradox have been offered, including several that would allow for widespread dispersal of intelligence. For example, the "colonizers" might be cryptic. This idea may be particularly applicable to machines, entities that could consist of individual, massive synthetic intelligences, and not the hordes of little gray guys that are so often portrayed as our galactic brethren. Another possibility is that our part of the galaxy is relatively poor in both material and energy resources, and consequently only infrequently traversed. In any case, absence of evidence is not evidence of absence.

The point is this: even though the evolution of intelligence may be rare, its presence in the galaxy may not be. Intelligent biology, once it reaches a level of technological sophistication comparable to our own, will be able to disperse throughout nearby space. This will protect it from extinction, either self-imposed or external, and give it enough time to engineer intelligent progeny. These sentient machines could both spread out and greatly outlive their biological creators. The galaxy could be rife with long-lived, communicating devices, even if intelligent protoplasm is both rare and fleeting.

In this way, even if Professor DeVore is correct about nature's indifference to producing intelligence, SETI might still succeed.

Consequently, it seems reasonable and proper to adopt the "nuts and bolts" approach to SETI: let's do the experiment. We should do this with no illusions. SETI is not falsifiable—a negative result does not disprove the premise of cosmic companionship, and consequently such searches are obviously more exploration than experimentation. That means that a failure to find the aliens will not settle this debate. Indeed, the only way the argument over intelligence can be settled is if we do find something. In other words, this debate is asymmetric: my position alone can be proven right. I may not win, but it's certain my opponent cannot win.

I will end by guessing (and it is only a guess) that a positive SETI detection may be no more than a decade or two away. The motivation for this suggestion (other than self-serving optimism) lies in the rapid improvement in SETI search power made possible by improved microelectronics. Every decade, these experiments become faster by a factor of a hundred. So there is at least some reason to expect that success might come soon. If so, we will get a signal, and we will know that intelligence is not restricted to Earth. But what we will not know—at least not at first, and maybe never—is who or what is running the transmitter. Personally, I do not think they will be squishy.

Theology after Contact

Religion and Extraterrestrial Intelligent Life

JOHN F. HAUGHT

Department of Theology, Georgetown University, Washington, D.C. 20057, USA

ABSTRACT: The prospect of encountering extraterrestrial intelligent life raises important questions for religion and theology. Even if an actual encounter with extraterrestrials never actually takes place, or proves impractical, terrestrial religious thought already has resources that can render intelligible and allow us theologically to appreciate such an eventuality.

KEYWORDS: radical monotheism; exo-theology; logic of achievement; existential anxiety

Religious thinkers have long entertained the idea of the existence of extraterrestrial intelligent "worlds," and not always in "heaven" (the angelic hosts), but also in "the heavens" as well.[1] However, the actual discovery of an extraterrestrial world of living and intelligent beings elsewhere in our universe would, to say the least, be a most interesting new stimulus to theology. Obviously, given the distances that separate our planet from any other possible intelligent civilizations, it is doubtful that much of an encounter is going to take place for a long time. And if and when it does, communication along the electromagnetic spectrum will be maddeningly slow. Even in the neighborhood of our own galaxy whole lifetimes would go by while initial greetings are being exchanged.

Nonetheless, even the mere entertainment of the prospect of eventual contact—whether it ever actually occurs or not—is a wholesomely expansive exercise for theology. And it seems appropriate even now to ponder some of the questions that an encounter with other worlds of intelligent beings would raise for religious thought.

What would happen to the notion of God? Would our own sense of significance in the universe be diminished? What would be the implications for those Earthly faith traditions that identify themselves as specially chosen, as

Address for correspondence: Department of Theology, Georgetown University, Washington, D.C. 20057. Voice: 202-687-6119; fax: 202-87-8000.
haughtj@georgetown.edu

people set apart (the question of religious particularity)? Would our own religions and theologies make any sense to intelligent beings from other planets? What implications would the discovery of other intelligent beings have for the large question of cosmic purpose? And does the world of religious thought even now perhaps provide us with any conceptual frameworks that would be hospitable to, and perhaps even enthusiastic about, the prospect of extraterrestrial intelligence (ETI). I shall say only a few words about the first three of these questions, and devote a bit more attention to the latter three.

WHAT WOULD HAPPEN TO THE IDEA OF GOD?

Certainly, at the very least, an encounter with alternative intelligent worlds would be one more in a series of great occasions modern cosmology has provided for theology to enlarge its sense of God and divine creativity. But contact with extraterrestrial beings (ETs) would also provide an opportunity for theology, on its part, to display the unitive power of radical monotheism. Any other intelligent cosmic provinces in this universe would obviously be grounded in the same creative principle that our earthly monotheisms posit as the source of all things "visible and invisible." Radical monotheism—with its belief that all things, all forms of life, all peoples and all worlds have a common origin and destiny (in a God who creates and encompasses all beings impartially)—is still the surest ground we have for embracing that which at first seems alien.[2] To learn to love what God loves is the vocation and the constant struggle to which our greatest religious prophets have already called us. Of course, tribalism and ethnic hatred, as well as disregard for nonhuman forms of life, still tragically persist here on Earth, but an argument could be made that this is so only because radical monotheism, which emphasizes the ontological unity underlying all diversity, still has too tenuous a hold on human awareness, including that of religious people themselves. Many people do not yet *really* believe in the ultimate unity of all beings even here in our own world. And so, the discovery of other intelligent worlds would be a wholesome new challenge to radicalize monotheism.

Viewed theistically, all galaxies and all universes are rooted in an ultimate unity of being; so our travels could never bring us into an encounter with anything completely alien to us. *Nihil alienum*. Theology's relevance to SETI lies most fundamentally in its conviction that all possible worlds have a common origin in the one God. And by virtue of the omnipresence of the one God we too would have an extended home in all possible worlds to which we might eventually travel.[3]

At the same time, the fundamental unity of being implied in the notion of divine creativity would tend, by its very nature, to unfold in an unlimited *diversity* of ways, and possibly a multitude of different "worlds." In the *Summa*

Theologiae St. Thomas Aquinas poses the childlike question as to why God created so many different kinds of beings. His answer: so that what is lacking in one thing as far as expressing the infinity of God is concerned can be supplied by something else.[4] Diversity in creation, in other words, is appropriate precisely because of the nature of an infinitely resourceful creativity. The basic theistic belief that the reality of God has already become partially manifested in the extravagant multiplicity of non-living and living beings on our own planet should already have prepared the religious mind for a disclosure of even richer diversity elsewhere—and in ways completely unfamiliar to us now. Perhaps there is no better way for religious people to prepare themselves for "exo-theology" than by developing here and now an "eco-theology" deeply appreciative of the revelatory richness of the variety of life-forms on our planet.[5]

THE QUESTION OF HUMAN IMPORTANCE

Would knowledge of the existence of more intelligent, and perhaps more ethically developed, beings elsewhere perhaps undermine our self-esteem, thus making our religions seem woefully provincial and unduly anthropocentric in convincing their devotees that they are somehow special? What would be the religious implications of an extended "Copernican principle," one whereby the Earth's intelligent occupants would be shown to be just one more "average" population in a universe of countless intelligent worlds?

In the first place, we should emphasize that it is biologically inconceivable that there would be other *humans* anywhere else in the universe; so our uniqueness as a species is virtually guaranteed in any case. "Of men elsewhere, and beyond, there will be none forever," writes Loren Eiseley. Natural selection has brought us about along roads that will "never be retraced" biologically.[6]

Second, and more to the point, however, according to the great teachers in Islam, Hinduism, Judaism, Christianity, Buddhism and other religious traditions, we express our own unique human dignity and value not by looking for signs of our mental or ethical superiority over other forms of life, but by following a path of service and even self-sacrifice with respect to the whole of life, wherever it may be present. Authentic existence, as Buddhism especially makes clear, consists of our capacity for compassion rather than the urge toward competition. The meaning of our lives according to many religious traditions consists of the opportunity to give ourselves to something larger, more important and more enduring than ourselves. Thus, it is inconceivable that the eventual encounter with beings that may in some ways be our superiors would ever render such instruction obsolete.

THE QUESTION OF RELIGIOUS PARTICULARITY

Perhaps, though, contact with ETI would be the occasion of heightened anguish to those faiths that believe they have received special election and revelation from God. Wouldn't an encounter with other forms of personal, free, and responsible beings put considerable strain on traditions that claim the status of being "a people set apart"?

The claim of special election might possibly undergo some stress after "contact." One response, of course, would be to treat ETs as potential subjects of conversion, in which case contact would simply provide new fields for missionary activity. Mary Russell conjures up such an approach—together with its potential hazards—in her interesting science-fiction novel, *The Sparrow.*[7]

However, in the context of contemporary Christian theology, at least, the idea of special election is even now being divested of the connotations of rank and privilege that it might once have suggested. Election, the sense of being specially called or set apart by God, may be understood essentially as a vocation to serve the cause of life and justice rather than being interpreted as lifting us out of our fundamental relatedness to the entire cosmic community of beings. It is worth recalling here also that in Christian belief Jesus' own sense of being called by God did not prevent him from taking on the status of a slave and of being subjected to the most humiliating destiny available during his time, that of crucifixion. In the same spirit, solidarity with Christ would continue for the Christian to mean belonging to one whose own life was itself a vulnerable openness to the estranged and alien, to what does not yet belong. After contact, "belonging to Christ" could then readily be thought of as requiring a more radical inclusiveness than before, one open to and supportive of the adventures of many intelligent worlds. Such an eventuality, once again, would not require an abandonment but instead a fuller appropriation of the central teaching and practice of the faith.

What seems to be universally applicable in Christianity (and indeed other religious traditions) is the ideal of embracing rather than eliminating diversity, an ideal that beckons and challenges, no matter how much it has been ignored in practice. The history of religion is ambiguous at best in meeting this challenge, but historically the encounter of various faiths with what they initially perceived to be alien cultures and practices has often led to the enrichment rather than the dissolution of their traditions. One may surmise that in the far distant future, if interstellar travel ever occurs, our terrestrial religions' contact with even more alien "cultures" will provide fresh challenges and opportunities for growth.

ARE EXTRATERRESTRIALS RELIGIOUS?

This brings us, however, to a fourth and perhaps more interesting question for religious thought as it hypothetically prepares for contact. Would the "others" (I prefer this designation to that of "aliens") be able to make any sense at all of our own religious life and thought? And should we expect that other intelligent beings would practice anything like what we call religion, and which might in this respect make them similar to us? Let us put aside once again the sobering probability that, because of the enormous distances they would have to traverse, any messages flowing back and forth at the speed of light would not add up to many exchanges in the course of a single human lifetime, nor would they extend very far beyond our own cosmic neighborhood. Instead let us suppose that we shall eventually be given the opportunity of prolonged conversation with other beings that impress us as being both alive and intelligent. What must their own kind of life and intelligence be like in order to allow us to share with them in a meaningful way our own deepest hopes, including ideas about "God" or "salvation"? What are some of the marks that any other conceivable instances of intelligent life in this universe would have to possess in order for us to be able to converse with them about our own religious beliefs, and that might also open us up to an understanding of theirs, if they have any?

In contemplating such questions we are reminded of just how much, in the way of both content and expression, our earthly religions borrow from the unique features of this planet, and therefore how any religions on other worlds would be idiosyncratically shaped by theirs as well. Our own persistent religious metaphors are inseparable from the experience of *Earth's* own characteristics: rotation from day to night, of the exposure to sun and moon; its deserts, oceans, rivers and streams, clouds, rain, storms and whirlwinds, grass and trees, blood and breath, soil and sexuality, maternity, paternity, sisterhood and brotherhood. Think of how prominently our experience of trees, for example, shapes religious imagery: the tree of life, the tree of "knowledge of good and evil," the Bodhi tree, the tree of the cross, the cedars of Lebanon, etc. Likewise, we should note that the very earthy occurrence of fertility, say, of inert seeds miraculously sprouting to life out of Earth's topsoil, has given us the highly significant religious metaphor of "resurrection." And the notion of "spirit," now ironically employed to refer to what is unearthly, comes from the Latin *spiritus* (in Hebrew *ruach,* and in Greek *pneuma*), a notion that originally meant the "breath of life" and which, as we now realize, requires the existence of Earth's enlivening atmosphere as its physical basis. Imagine what our religions would be like, Thomas Berry asks, if we lived on something like a lunar landscape.[8] Would not extraterrestrial ecologies breed other extraordinary blendings of land, life, and religious longing? And wouldn't we have a very difficult time connecting with them?

Difficult, perhaps, though not impossible. But in order to conceive of how we might be able to engage in anything like theological conversation with cosmic Others we need first to clarify our terms. What exactly do we mean by *life*, by *intelligence,* and by *religion*?

First, *life*. What allows us to identify living beings as "alive" at all, and thus to distinguish them from nonliving things or processes, is that they share with us humans the trait of *striving* to achieve some goal, and therefore the possibility of failing or succeeding.[9] If an entity were not recognizable as a kind of striving, or of struggling against limits of some kind, and therefore as capable of succeeding or failing in the effort, we would not properly call it living. The great Jewish philosopher Hans Jonas remarks that even the most primitive instances of metabolism are in some rudimentary way constantly "striving" against the threat of being dissolved into their inanimate surroundings.[10]

Michael Polanyi argues that we recognize the distinctive features of life primarily through a *personal* knowledge, one shaped by what he calls "the logic of achievement."[11] Living beings are capable of "achieving" in a way that does not apply to purely chemical reactions. I would suggest, then, that human persons are interested in the possibility of life elsewhere in the universe in great measure because we sense that we share something special with all other striving, struggling beings. We feel a kind of connatural relatedness to all other striving beings, a connection that we do not have with inanimate things. For we spontaneously realize that all modes of life, ours included, can in many ways either "succeed" or "fail" in a way that merely physical and chemical processes cannot.

And so, if we ever encountered life on other worlds we would call it alive (regardless of its chemical make-up) only if we recognized—through what Polanyi calls a "personal" rather than objectifying knowledge—that it participates with us in a kind of striving that risks the possibility of failure. Of course, in our search for life elsewhere we would also be on the lookout for such qualities as the transgenerational sharing of information that we find in the genetic flow of life here on Earth. We would look for open, self-organizing systems that pump energy out of their environment and so maintain themselves at a high level of complexity. But we would also look for instances of exquisite organismic fragility, beings that need to "exert" themselves in some degree even to maintain their organic identity against the constant threat of being dissolved into their inanimate surroundings. Life elsewhere as well as here, in other words, could be identified as such only if it conforms in some way to the logic of achievement. How this understanding of life bears upon the question of whether ETs are religious will become clear shortly.

Next, though, what do we mean by *intelligent* life, the special set of features for which SETI (The Search for Extraterrestrial Intelligence) professes to be looking, and which we confidently think we could identify if we ever stumbled across it. First of all, if we find intelligent *life*, then it must be manifested in some sort of *striving*; and, second, if it is *intelligent* life, it must be the kind of striving that we associate in ourselves with *a desire to know.* If the desire to know is absent then there may be life—sentient and even conscious life—but not intelligent life. Any being that is not somehow striving to achieve some goal, even if this goal is simply that of surviving, is not alive; and any being whose striving does not include the search for insight and knowledge is not intelligent, at least in the sense that we humans minimally understand the term. SETI already tacitly assumes such a notion of "intelligence" when it searches the heavens for electromagnetic signals which only a technologically sophisticated, and similarly insight-seeking and truth-desiring source is sending out.

Finally, what do we mean by *religion*? Let us understand by "religion" a specific kind of striving also. Before religion is anything else it is a manifestation of *life*, a specific kind of human life, striving toward a goal. Underneath all of its extravagant symbolic, ritualistic, doctrinal, ethical, and institutional foliage religion is an expression of life, of intelligent life—striving, exploring, hoping life. Religion, I would suggest, is intelligent life at perhaps its most intense level of striving.

The whole terrestrial religious endeavor may be thought of as a kind of "route-finding," a quest for pathways that promise to carry us through the most intractable limits on life.[12] Even from our perch here on Earth, therefore, can we not identify at least some of the most severe limits that *all* other forms of intelligent life would inevitably have to face along with us? And in identifying these limits would we not be placing ourselves and the Others within a common (hermeneutical) circle, one that would allow conversation with them in spite of broad ecological differences?

I think that *if* they possess anything like what we call intelligent life we can reasonably expect to discover that extraterrestrials at least have the *capacity* for a religious mode of venturing. Since any possible Others we shall ever encounter will be inhabitants of the same Big Bang universe that we belong to, the general features of this cosmos as made known to us by our terrestrial science will presumably also apply to them. We must expect to find, then, that any living, sentient and intelligent beings will be subject to the transience and perishability characteristic of all things positioned on the slopes of entropy. They would be subject to the physical forces that break orderly or complex arrangements down into disordered and simple ones. They too would be subject to transience and eventual perishing. They, like us, would be subject to the threat of failure, and eventually nonbeing, that every living finite being has to confront.

We may conclude, then, that since all living and intelligent beings would experience the same basic physical limits on life that we do, a meaningful exchange about religious route-finding through these limits could conceivably occur. For these Others, if they are truly striving centers, would also be in search of ways to transcend the limits on their particular forms of life. And if they are truly intelligent, they would have an awareness of their possible nonbeing. They might even have, in other words, what Paul Tillich calls "existential anxiety." Anxiety, the awareness of finitude, drives intelligent life to find a courage that can conquer the threat of nonbeing. In our human experience it is the quest for courage in the face of nonbeing that leads many of us to seek the foundational support of religious faith, and in some cases to an understanding of "God" as the source of courage to continue life's striving in the face of fate, death, guilt and meaninglessness.[13] If any Others "out there" are alive and intelligent, it would not be surprising that they too need courage. If so, they would be no less potentially religious than we are.

DOES SETI HAVE IMPLICATIONS FOR THE QUESTION OF COSMIC PURPOSE?

Whether the universe has any "point" or "purpose" to it is a question that religions must always be concerned about, perhaps above all else. Religions can put up with all kinds of scientific ideas as long as they do not contradict the sense that the whole of things is meaningful. They can survive the news that the Earth is not the center of the universe, that human beings are descended from simian ancestors, and that the universe is 15 billion years old. What they cannot abide, however, is the suspicion that the whole of things is pointless.[14]

It is worth asking, therefore, how the search for ETI might bear upon the question of cosmic purpose and, by implication, on the meaning and mission of our own lives. Any serious religious reflection on cosmology takes the question of purpose to be both unavoidable and central, and so it is especially for this reason that theological reflections on SETI seem appropriate in the context of the present book.

Generally speaking, "purpose" means orientation toward the realization of a value. So, to say that the universe has a purpose would be to imply that it is oriented toward the realization of something intrinsically good or valuable. Cosmic purpose does not have to imply a particular *finis* or end. Purpose is not identical with a predetermined plan or design, both of which tend to close off the future in a suffocating way. All we need in order to affirm cosmic purpose is an awareness that something of undeniable importance is going on in the universe, and that it is doing so in a way that is tied essentially and not just accidentally to the whole of the cosmos.

Of course, in an unfinished universe there will by definition always be ambiguity. And so here and now we will look intensely for whatever indicators we can find to support our own suspicions, whether these be pessimistic or hopeful. Accordingly, it would seem relevant to our understanding of what this universe is all about, that we try to find out whether intelligent life is abundantly distributed throughout the cosmos, or, for that matter, whether it exists only here on Earth. Certainly the existence of ETI would force us to reexamine the claim by evolutionists such as Ernst Mayr, Jacques Monod, Stephen Jay Gould, Richard Dawkins, and many others that life and intelligence are the results of utterly improbable, purely random statistical aberrations in an overwhelmingly lifeless and mindless universe. In this respect SETI would seem to have theological importance.

However, it is not good form, theologically speaking, to make the credibility of a religious sense of cosmic purpose contingent upon the vicissitudes of scientific exploration. And so the discovery of ETI cannot be looked to as a deciding factor on a question of such vital religious importance as that of cosmic purpose. Anyway, one cannot help suspecting that scientific thinkers already inclined to think of the universe as "pointless" would persist in looking for ways to understand and explain even an abundant distribution of intelligent life in the cosmos as no less the consequence of blind chance and impersonal physical laws than life and intelligence on Earth now seem to them to be. If the physics of the early universe has come upon coincidences, constants, and initial conditions predisposed toward the emergence of carbon-based life and intelligence, then scientific thinkers already conditioned to explanation in terms of "chance" and "necessity" alone will have no trouble speculatively conjuring up an infinity of mindless universes within whose amplitude our own mind-birthing cosmos can present itself as an unintelligible and impersonal accident. Some scientific thinkers have in this way already adjusted their cosmic pessimism to the Big Bang universe after the intellectually more appealing eternal universe of traditional materialism was challenged by the cosmology of Einstein, LeMaitre, and Hubble. And so there is little doubt that the discovery of ETI would scarcely change the minds of those already comfortable with the notion of an essentially mindless universe devoid of meaning.

For this reason, then, our reflections on SETI throw us back once again on the question of what our own intelligence, even if it turns out to be the sole instance of it in the cosmos, might imply as far as the character, and possible purposiveness, of the universe is concerned. As I noted earlier, any process that moves incrementally toward the establishment or intensification of intrinsic value could be called purposeful. If so, then might we not plausibly claim that a universe that proceeds over the course of its history—however long and meandering this journey through time may be—toward the establishment of intelligent life, is a purposeful one? Even if intelligent life manifests itself only on one planet, could it not still be considered a property of the cosmos as a whole, especially in the light of recent astrophysics?[15] In this case the existence of our own intelligent life would be sufficient of itself to render the universe meaningful, and the discovery of ETI would not add anything qualitatively new to this judgment.

After all, intelligence itself is the most indubitable instance we have of intrinsic value. If you find yourself doubting or denying what I have just said, it is only because you are now at this moment spontaneously acknowledging the value of your own intelligence. It is impossible for you consistently to deny the intrinsic importance of your intelligence. By issuing judgments about the truth-status of the assertions I have just made, you have already demonstrated how deeply you treasure your own mind and its capacity to understand, criticize, and know.

Now if what I have just said is correct—and you really can't doubt it without proving my point—then the existence of even one instance, or one planetary outpost, of intelligence in this vast universe might be enough to make the whole story that leads up to its existence a purposeful one, especially if that story is continuous with and ingredient in the emergence of intelligent life. Now that with the help of physics and astrophysics we understand how intricately our own intelligence is connected to the fifteen-billion-year cosmic story, and to the physical features of the universe from the very earliest microseconds of cosmic time, to assert that the universe is inherently purposeless seems arbitrary at best. In view of the spontaneous (and undeniable) valuation of your own intelligence on the one hand, and our new scientific understanding of the cosmic process constitutive of your intelligence on the other, you cannot but wonder about the coherence of any claim that the universe is inherently pointless. To argue in complete seriousness that the cosmos is ultimately unintelligible, or even to entertain doubts about the purposiveness of this patently mind-bearing universe, would at this point in our scientific understanding of the cosmos seem to sabotage the very mind that is making such an assertion.

The point to be made here with respect to SETI and cosmic purpose is that the existence of intelligent life on Earth, whether it exists elsewhere or not, may already tell us something about the essential nature of the whole universe. Perhaps we do not need to have any other instances of intelligent life to convince us that this is an essentially mind-bearing universe.

However, even aside from the point I have just made, SETI may eventually have some implications for the question of cosmic purpose. Let us recall that the modern loss of a sense of cosmic purpose is ultimately rooted in the modern expulsion of mind from nature—by Cartesian dualism, classical mechanism, and modern scientism. It is not out of science itself, but out of the assumed mindlessness of nature that the historically recent and culturally provincial idea arose that the cosmos is pointless and that the appearance of our own intelligence, therefore, is a purely accidental one. An essentially mindless universe would seem to be a purposeless one, but a universe in which intelligent life is an essential rather than accidental property could hardly be called purposeless. And so, any future discovery that instances of intelligence occur abundantly in the universe could not help but place the burden of proof upon those who see no intrinsic connection between mind and the rest of nature.

AVAILABLE FRAMEWORKS FOR A "THEOLOGY AFTER CONTACT"

Theology is typically more responsive than predictive. Of course, a few prophetic voices can read the signs of the times and issue appropriate warnings about what is to come. But by and large religious thought, undertaken as it is by finite and short-sighted humans, seldom accurately anticipates, much less prepares us for, the crises that occur in connection with unprecedented events in human history or new discoveries in the realm of science. Indeed, most of the theological content of the dominant traditions comes from religion's reaction to crises rather than anticipation of them. Undoubtedly, then, the actual shape theology would take on if we ever do encounter ETI cannot be accurately predicted here and now, but must await the event itself.

Still, I would suggest, all too briefly here, that the cosmic vision of Teilhard de Chardin as well as process theology (based on concepts of the philosopher Alfred North Whitehead) are both already inherently open to being developed into a "theology after contact." Not the least of the reasons for their adaptability is that they have already enthusiastically embraced the Darwinian portrait of life as well as the notion that the entire universe is still in the process of being created. Though Teilhard reflected only occasionally on the possibility of ETI, keeping most of his speculation firmly anchored to our planet, the general thrust of his visionary writings is cosmic in scope. As such, the urge toward increasing complexity and consciousness so evident to Teilhard in his surveys of the history of life on Earth could be a trend that is occurring throughout the cosmos. For this famous Jesuit paleontologist (1881–1955) the "point" or purpose of the universe has something to do with the emergence and intensification of "complexity-consciousness." As physical complexity increases in the universe, Teilhard claims, so does consciousness. But, he acknowledges, the cosmic evolution of consciousness is still far from being finished. Here on Earth the "noosphere," the cerebralization now taking place on a planetary scale, is still in process. And it is not inconceivable that parallel worlds of consciousness are evolving elsewhere.

Finally, contemporary "process theology" with its vision of cosmic purpose is also expansive enough to accommodate the discovery of ETI. For the "process philosopher" Alfred North Whitehead and his theological followers, the purpose of the cosmos consists of its aim toward the intensification of beauty.[17] Since—at least for Whitehead—beauty is an intrinsic value, any process that leads toward its establishment could be called "teleological," at least in a loose sense. "Beauty," in Whitehead's thought, means the "harmony of contrasts" or the "ordering of novelty," many diverse instances of which have appeared in the evolution of the cosmos and in the emergence of life, mind, and culture in our terrestrial setting.

Intelligent life, however, is only one instance of cosmic beauty. We really have no idea of the many forms the cosmic aim toward bringing about beauty might assume within the totality of the universe. Perhaps, then, SETI has set its goals too narrowly for theology. What we call intelligent life might turn out to be too trivial a notion to capture what is already "out there," or the incalculable cosmic outcomes that may yet occur in the future of this unfinished universe. The notion of "beauty," however, is encompassing enough to anticipate a wide variety of cosmic evolutionary phenomena. As we explore the universe we should ask not only about the meaning of intelligence, but also about what the existence of beauty implies as far as the essential character of the whole universe is concerned. It is clear that the universe has always been dissatisfied with the monotony of the status quo, and so has produced innumerable instances of ordered novelty. Perhaps the aim toward beauty, then, is enough to endow the universe with purpose—though it is not necessary for us to add that we would not be able to arrive at such a conclusion unless there were also intelligent subjects capable of enjoying it.

NOTES AND REFERENCES

1. See Michael J. Crowe, *The Extraterrestrial Life Debate 1750–1900* (Cambridge: Cambridge University Press, 1986); Stephen J. Dick, *Plurality of Worlds: The Origins of the Extraterrestrial Life Debate from Democritus to Kant* (Cambridge: Cambridge University Press, 1982); Ted Peters, "Exo-Theology: Speculations on Extraterrestrial Life," in James R. Lewis, ed., *The Gods Have Landed: New Religions from Other Worlds* (Albany: State University of NewYork Press, 1995), pp. 187–206.
2. See H. Richard Niebuhr, *Radical Monotheism and Western Culture* (London: Faber and Faber, 1943).
3. Roch Kereszty, as quoted by Thomas F. O'Meara, "Extraterrestrial Intelligent Life," *Theological Studies, Vol.* 60, p. 29.
4. *Summa Theologiae* I, 48, *ad* 2.
5. The term "exo-theology" (a take-off on "exo-biology," which studies the prospects of life outside of our planet) is used by Peters, p. 188.
6. Quoted by Stephen J. Dick, *Life on Other Worlds: The 20th-Century Extraterrestrial Life Debate* (Cambridge: Cambridge University Press, 1998), p. 194.
7. Mary Russell, *The Sparrow* (New York: Fawcett Columbine, 1996).
8. This question has been often raised by Thomas Berry. See his book *Dream of the Earth* (San Francisco: Sierra Club Books), p. 11.

9. See Michael Polanyi, *Personal Knowledge: Towards a Post-Critical Philosophy* (New York and Evanston: Harper & Row, 1958), pp. 327, 344.
10. Hans Jonas, *Mortality and Morality* (Evanston, IL: Northwestern University Press), p. 60.
11. Polanyi, p. 327.
12. John Bowker, *Is Anybody Out There?* (Westminster, MD.: Christian Classics, Inc., 1988), pp. 9–18; 112–143.
13. Paul Tillich, *The Courage to Be* (New Haven: Yale University Press, 1952), pp. 40–45.
14. See W. T. Stace, "Man Against Darkness," *The Atlantic Monthly,* Vol. CLXXXII (Sept. 1948), p. 54.
15. Other essays in this volume give the details of the so-called "anthropic" character of the universe.
16. See Pierre Teilhard de Chardin, *Activation of Energy*, trans. René Hague (New York: Harcourt Brace Jovanovich, 1970), pp. 99–127. In 1944 Teilhard wrote that the hypothesis of other planets inhabited by intelligent beings has a "positive likelihood," in which case "the phenomenon of life and more particularly the phenomenon of man lose something of their disturbing loneness." (p. 127) There may be many "noospheres" or "thinking planets." "It is almost more than our minds can dare to face," he continues, but the evolutionary tendency toward complexification and centration might well have a "cosmic" scope. Yet "there can still be only a single Omega," that is, a single transcendent Reality whose being enfolds the entire universe. (p. 127)
17. See Alfred North Whitehead, *Adventures of Ideas* (New York: The Free Press, 1967), esp. p 265.

Cosmic Questions and the Relationship between Science and Religion

JAMES B. MILLER

*Program of Dialogue on Science, Ethics, and Religion,
American Association for the Advancement of Science,
Washington, D.C. 20005, USA*

The three cosmic questions considered in this volume are of significantly different kinds. One is a rather straight forward empirical questions that may be able to be answered in the next century. One is a more theoretical question for which it may be very difficult to generate sufficient empirical evidence to discern which of the theoretical alternatives, if any, is most adequate. One of the questions is not strictly a scientific question at all but one on the broader philosophical or religious interpretation of the findings of science.

Are we alone? We may well have a definitive answer to that question before the end of the 21st century. Within a decade or two we are likely to have a better idea of the degree to which life in the universe is ubiquitous or rare. The more life, the more likely that some of it, given a rich ecological context, will have evolved functional capacities comparable to human intelligence. It is also worth noting in passing that even on Earth it is likely that non-human intelligence will appear in the next century. From a religious perspective, an encounter with an extraterrestrial intelligence (or a homegrown artificial intelligence) will call for an expansion of the theological horizon, but is unlikely to undermine religion, as such, except for those traditions that are constituatively committed to very small, homocentric universe.

Did the universe have a beginning? This is a theoretical question that sounds as though it were an empirical one. At present, Big Bang cosmology seems the best explanation of the development of the universe from an early hot, dense state. Evidence is mounting that some inflationary version of Big Bang cosmology will also be supported by the growing body of observational data. However, it is unlikely that any observation or set of observations will be able to definitively determine whether ours is the only universe there is, whether it is finite but has no beginning point, or whether it is but one of an

Address for correspondence: Dr. James B. Miller, Senior Program Associate, Program of Dialogue on Science, Ethics, and Religion, AAAS, 1200 New York Avenue, NW, Washington, D.C. 20005. Voice: 202-326-7044; fax: 202-289-4950.
jmiller@aaas.org

ensemble of universes evolving in some meta-spacetime. This is not to say that observational evidence will not make some of the theoretical proposals seem less plausible than others. It is to say, however, that the question of cosmic beginnings will remain a matter of theoretical judgment rather than evidentiary conclusion.

The great religions of the world have different notions of cosmic history and the situation of the universe in time. The western Abrahamic faiths have a more linear view of cosmic history, with a beginning and an end toward which universal history is moving. The eastern traditions of Hinduism and Buddhism have a more existential view of history, situating the universe temporally in a timeless present. In this context there is a complex dance between cosmological theory and theological or religious understanding. Because all religious perspectives have some assumption about the nature of the cosmos, particular theoretical alternatives about cosmic origins will be more or less congenial with particular religious traditions. Because theories tend to be underdetermined by the evidence, the judgments of cosmologists between the various theoretical alternatives may be shaded by extrascientific commitments including religious ones.

Is the universe designed? Is the cup half full or half empty? The question of cosmic design is not strictly a scientific one. It is of the nature of science to seek natural (as contrasted with transcendent or ultimate) explanations for natural phenomena. It is in the effort to discover the foundational order of the cosmos, its most basic laws, that a transcendent domain is approached (but not necessarily and transcendent orderer).

The answer to the question of cosmic design, yes or no, is a religious interpretation of what can be known or reasonably believed about the structure and history of the cosmos rather than a direct conclusion required by that structure and history. It is a religious interpretation because the religious stance, in its most generic sense, addresses the question of personal meaning in the midst of all the dimensions of our experience of life. To be sure, the credibility of particular religious claims must be judged in relation to what we know reliably about the structure and history of the cosmos. But it seems to be a distinguishing feature of human nature that we as a species do make such claims.

In the end the papers in this volume do not answer the cosmic questions they consider. In the final account, perhaps we are not all that different from our hominid ancestors who first looked at the heavens and wondered. If anything, we today have access to a richer, more varied and complex vision than they did. Yet like them we are drawn to the question: What does it all mean? Even if we say, "Nothing," we have expressed a religious stance in the midst of the cosmos.

Annotated Bibliography

[Suggestions for further reading by conference speakers]

Anandita Balslev

Sarvepalli Radhakrishnan and Charles A. Moore, eds., *A Source Book in Indian Philosophy* (Princeton, NJ: Princeton University Press, 1957). This book includes creation hymns in the Hindu tradition.

A. L Basham. *The Wonder That Was India* (London, 1954). This book provides a discussion of Puranic ideas of cosmology.

Anindita N. Balslev, *A Study of Time in Indian Philosophy* (Wiesbaden: Otto Harrassowitz, 1983). This book discusses various theories of creation, time and causality.

Anindita N. Balslev and Jitendranath Mohanty, eds., *Religion and Time* (the Netherlands: E.J. Brill, 1995). This book discusses various conceptions of time in different world religions.

John Barrow

John Barrow and Frank Tipler, *The Anthropic Cosmological Principle*, (Oxford University Press, 1988). This book offers a wide ranging account of design arguments and modern scientific developments that led to the Anthropic Principles in cosmology.

John Barrow, *The Origin of the Universe* (HarperCollins, 1997). This book is a short popular account of modern ideas about the beginning of the universe.

John Barrow, *The Artful Universe* (OUP and Little Brown, 1996). This book discusses the ways in which the fabric of the physical universe shapes some of our aesthetic inclinations.

John Barrow, *Impossibility: The Limits of Science and the Science of Limits* (Oxford University Press, 1999). This book explores the significance of various limits to knowledge and discusses fundamental limits on our ability to determine the structure of the Universe.

John Barrow, *Between Inner Space and Outer Space* (Oxford University Press, 1999). This book is a collection of essays, some written for magazines and newspapers, about important issues and developments in science. A number are concerned with cosmology and the astronomical significance of life.

David Ray Griffin

David Ray Griffin, *Evil Revisited: Responses and Reconsiderations* (Albany, NY: State University of New York Press, 1991). This book summarizes the theodicy presented in my 1976 book on this topic, *God, Power, and Evil: A Process Theodicy,* then provides responses to critiques it received plus a few second thoughts. The key notion is that, whereas critics have been right to consider traditional theism, with its doctrine of divine omnipotence, to be falsified by the world's evil, the process theism of Alfred North Whitehead and Charles Hartshorne is not thus falsified, because it regards divine power as persuasive, not all-determining.

David Ray Griffin "Professing Theology in the State University," in David Ray Griffin and Joseph C. Hough, Jr., eds., *Theology and the University* (Albany, NY: State University of New York Press, 1991), pp. 3–34. This article argues that there is no good reason, either in the nature of the university or of the U.S. Constitution, why theological views of the universe cannot be advocated in state-supported universities, if these views are advocated in terms of reason and evidence.

David Ray Griffin, "Whitehead's Deeply Ecological Worldview," in Mary Evelyn Tucker and John Grim, eds., *Worldviews and Ecology: Religion , Philosophy and the Environment* (New York: Orbis Books, 1994), pp. 190–206. This article argues that the philosophy of Alfred North Whitehead provides a position that reconciles "egalitarianism of inherent value," which is insisted on by some advocates of "deep ecology," with the commonsense view that mammals have more intrinsic value than bacteria and worms.

David Ray Griffin, "Christian Faith and Scientific Naturalism: An Appreciative Critique of Phillip Johnson's Proposal," *Christian Scholar's Review,* Vol. 28/2 (Winter 1998), pp. 308–328. Johnson, having equated "scientific naturalism" with its materialistic, atheistic version, rejects naturalism altogether in favor of a position that allows for supernatural interruptions of the evolutionary process to account for new species. I argue that by understanding "naturalism" to mean simply the denial of supernaturalism, without necessarily involving materialism and atheism, we can have a version of scientific naturalism that does justice to the various facts Johnson thinks to require supernaturalistic explanations.

David Ray Griffin, "A Richer or a Poorer Naturalism? A Critique of Willem Drees's *Religion, Science and Naturalism,*" *Zygon,* Vol. 32/4 (December 1997), pp. 593–614. Drees argues that a materialistic version of scientific naturalism can be shown to be adequate for religion, so that there is no need to turn to the "richer" naturalism provided by Whitehead. I argue that, besides not being adequate for religious and ethical purposes, the materialistic version of naturalism is not even adequate for scientific purposes.

David Ray Griffin, *Religion and Scientific Naturalism: Overcoming the Conflicts* (Albany, NY: State University of New York Press, 2000). Part I develops the points made in the critiques of Johnson and Drees at greater length, arguing that Whitehead provides a version of naturalism that is equally usable for scientific and religious purposes. Part II illustrates this thesis in terms of physics and time, the mind-body problem, parapsychology, and evolutionary theory.

Rocky Kolb

Steven Weinberg, *The First Three Minutes: A Modern View of the Origin of the Universe* (New York: Basic Books, 1993). The first modern book describing the Big Bang theory.

John D. Barrow and Joseph Silk, *The Left Hand of Creation: The Origin and Evolution of the Expanding Universe* (Oxford University Press, 1994). The story of the origin and evolution of the universe from the Big Bang to the present day.

Stephen Hawking, *A Brief History of Time* (New York: Bantam Doubleday Dell, 1998). The most popular book about modern physics. A classic!

Alan H. Guth, *The Inflationary Universe: The Quest for a New Theory of Cosmic Origins* (Helix Books, 1998). The story of the discovery of the inflationary universe, as told by the very person who made the discovery.

Rocky Kolb, *Blind Watchers of the Sky: The People and Ideas That Shaped Our View of the Universe* (Helix Books, 1997). This book describes the development of great ideas in cosmology, starting with Brahe, Kepler, Galileo, and Newton, up to the modern discoveries of the 20th century.

Lawrence Kushner

Martin Buber, *Hasidism and Modern Man,* trans. and ed. Maurice Friedman (Humanities Press International, 1988). A classic. Hasidism, an 18th century, Eastern European, revivalist movement was the last great flowering of Jewish spirituality and mysticism. Probably the best single volume on Jewish spirituality written by one of the greatest teachers of this century.

Arthur Green and Barry Holtz, *Your Word is Fire: The Hasidic Masters on Contemplative Prayer* (Woodstock, VT: Jewish Lights Publishing, 1993). An anthology of meditations on contemplative (as opposed to "petitionary," or "intercessory") prayer. Each of these extraordinary teachings are written as contemporary poetry. A guide book for meditation.

Lawrence Kushner, *The River of Light: Spirituality, Judaism and the Evolution of Consciousness* (Woodstock, VT: Jewish Lights Publishing, 1990). An attempt to synthesize some of the insights of psychoanalysis, Midrash, and the new physics.

Daniel C. Matt, *God and the Big Bang: Discovering Harmony Between Science and Spirituality* (Woodstock, VT: Jewish Lights Publishing, 1996). Matt is one of our generation's foremost interpreters of Kabbalah (the Jewish mystical tradition). In this book he creates a coherent and contemporary mystical-scientific theology.

Gershom G. Scholem, *On The Kabbalah and Its Symbolism,* trans. Ralph Manheim. (New York: Schocken, 1965.) A collections of essays on Kabbalah by its master historian. Scholarship of extraordinary depth and light.

Kenneth Nealson

This is a series of articles in a topical space magazine (*Ad Astra—Magazine of the National Space Society,* Vol. 11, No. 1, January/February 1999) written by credible authors, many of whom are involved with various aspects of space research, missions, or policy: M. K. Hobish and K. Cowing, "Astrobiology 101," *Ad Astra,* Vol. 11, pp. 20–24; M. Meyer, "Ex astra: Life from the stars," *Ad Astra,* Vol. 11, pp. 28–31; P. Boston, "The search for extremophiles on Earth and beyond," *Ad Astra,* Vol. 11, pp. 34–36; M. Race and J. D. Rummel, "Bring'em back alive—or at least carefully," *Ad Astra,* Vol. 11, pp. 37–40.

An op-ed piece discussing the general goals of astrobiology and the attempts of NASA to set up this new program as a virtual institute: "Hello, out there! The new science of astrobiology," *Newsweek*, September 21, 1998, p. 12.

Jaroslav Pelikan

Marie-Dominique Chenu. *Nature, Man, and Society in the Twelfth Century,* ed. and trans. Jerome Taylor and Lester A. Little (Toronto: University of Toronto Press, 1998). The twelfth and thirteen centuries were the heyday of medieval science and of medieval biblical exegesis (as well as of renewed contact with Jewish and Arabic scholars in both of these fields). As a result, the traditional reading both of Athens and of Jerusalem came in for critical and creative reconsideration.

Francis MacDonald Cornford, *Plato's Cosmology: The "Timaeus" or Plato Translated with a Running Commentary* (reprint edition, 1957). As Plato's most important treatise on cosmology—and as the only dialogue of Plato that was known in the West during the Middle Ages—*Timaeus* determined the vocabulary of science, philosophy, and theology, indeed, their way of framing questions. That has also made it the easiest of all the dialogues to misread by unconsciously shading its meaning toward later cosmological or theological theories. Cornford's explication of the text corrects these misreadings and opens up the vast world of this poetic-scientific classic.

Werner Jaeger, *Early Christianity and Greek Paideia* (1961). Author of the magisterial three-volume *Paideia: The Ideals of Greek Culture,* Werner Jaeger was also the editor of the works of St. Gregory of Nyssa (d. ca. 395), who exploited the "counterpoint" between Athens and Jerusalem with special brilliance. Having devoted his Gifford Lectures of 1936 to the science and theology of the pre-Socratics, Jaeger was uniquely positioned, in what turned out to be his final book, to draw together the several lines of the historical development.

Gerhard May, *Creatio ex nihilo: The Doctrine of "Creation out of Nothing" in Early Christian Thought,* trans. A. S. Worrall(T&T Clark Ltd., 1994). Neither the Hebrew verb *bara* nor the Greek verb *ktizo*, which are the principal vocables in the Bible for "create," explicity means creation "out of nothing." It is in Hellenistic Judaism that the beginnings of this idea are to be found, and then in Christian controversies, including also the controversies over whether Christ was a creature, that the concept of *creatio ex nihilo* established itself.

Jaroslav Pelikan, *Christianity and Classical Culture: The Metamorphosis of Natural Theology in the Christian Encounter with Hellenism.* Gifford Lectures at Aberdeen, 1992-1993 (New Haven, CT: Yale University Press, 1995). Opening sentence: "It remains one of the most momentous linguistic convergences in the entire history of the human mind and spirit that the New Testament happens to have been written in Greek—not in the Hebrew of Moses and the prophets, nor in the Aramaic of Jesus and his disciples, nor yet in the Latin of the *imperium Romanum*, but in the Greek of Socrates and Plato, or at any rate in a reasonably accurate facsimile thereof, disguised and even disfigured though this was in the *Koine* by the intervening centuries of Hellenistic usage."

John Polkinghorne

J. Barrow and F. Tipler, *The Anthropic Cosmological Principle* (Oxford University Press, 1986). An encyclopedic work detailing scientific insights and philosophical relating to anthropic coincidences. A mine of information, but a book that few will read cover to cover.

J. Leslie, *Universes* (Routledge, 1989). A book of modest length that accurately presents the science of anthropic issues and carefully discusses possible metaphysical consequences. A feature of Leslie's style is that he frames his philosophical arguments in terms of stories, making them readily accessible to the general reader.

J. C. Polkinghorne, *Reason and Reality* (Trinity Press International, 1991). Presents a defense of religious belief as arising from the search for motivated understanding, and so bearing a cousinly relationship to scientific belief. Chapter 6 deals particularly with anthropic issues.

Seth Shostak

Paul Davies, *Are We Alone? Philosophical Implications of the Discovery of Extraterrestrial Life* (New York: Basic Books, 1995). The popular physicist-author Paul Davies has produced a fascinating book, based on his lectures, on the meaning of a SETI success, including the sociological and religious implications. This book is lways interesting and frequently provocative.

Frank Drake and Dava Sobel, *Is Anyone Out There? The Scientific Search for Extraterrestrial Intelligence* (New York: Delacorte Press, 1992). The astronomer who pioneered the radio search for extraterrestrial civilizations provides a personal view of the SETI saga. More than a biography, this book is filled with personal experiences dating back to the beginning of the modern hunt for intelligence elsewhere.

Donald Goldsmith and Tobias Owen, *The Search for Life in the Universe* (Reading, MA: Addison-Wesley, 1992). A wide-ranging and clear review of the field, suitable for college students and interested non-specialists. Covers everything from basic astronomy to the matter of getting in touch with the extraterrestrials.

Albert Harrison, *After Contact: The Human Response to Extraterrestrial Life* (New York: Plenum Press, 1997). A psychologist looks at the likely reaction to the detection of cosmic company. Unusual in presenting a look at this subject from the point of view of a social scientist. Comparison of the effect of finding the aliens with similar events in human history.

Seth Shostak, *Sharing the Universe: Perspectives on Extraterrestrial Life* (Berkeley Hills Books, 1998). A lively and up-to-date account of the search for aliens, and their likely construction and behavior. This is the insider's view of the efforts to find the extraterrestrials, from UFOs to radio telescopes. It includes much discussion of what the aliens might be like, both physically and psychologically, and what it would mean for us to find them.

Jill Tartar

Frank Drake and Dava Sobel, *Is Anyone Out There? The Scientific Search for Extraterrestrial Intelligence* (Delta [reprint], 1994). The definitive story of the history of SETI, told by the man who lived it.

Seth Shostak, *Sharing the Universe: Perspectives on Extraterrestrial Life* (Berkeley Hills Books, 1998). This book considers what present-day science can say about the nature of cosmic company, and answers the age-old question: With what kind of neighbors do we share the universe?

Project Cyclops: A Design Study of a System for Detecting Extraterrestrial Intelligent Life (1971). Prepared under Stanford/NASA/Ames Research Center 1971 Summer Faculty Fellowship Program in Engineering Systems Design, CR I 14445. Provides technical detail on both the astronomical and signal processing sides of SETI. Reverently (and accurately) described as the "bible of SETI."

All three of the books just listed are available for purchase through the SETI Institute's website gift shop (http: www.seti.org/shop/Welcome.html).

J.C. Tarter, "Results from Project Phoenix: Looking Up from Down Under," in *Astronomical and Biochemical Origins and the Search for Life in the Universe,* C.B. Cosmovici, S. Bowyer, and D. Werthimer, eds. (Bologna: Editrice Compositori, 1997), pp. 633–643. We used the Parkes and Mopra antennas of the Australia Telescope National Facility (ATNF) and observed 209 solar-type stars in the Southern hemisphere over much of the frequency range 1200–3000 MHz. The sensitivity of these observations was sufficient to rule out any narrow-band transmitters stronger than 5 × 1012 Watts (the power of strong terrestrial radars) broadcasting during our observations. The use of two widely separated antennas, linked together as a pseudo-interferometer, proved to be an extremely effective discriminant against RFI. This mode of observing will be continued by Project Phoenix. Observations begin again in October 1996, with the 140-foot antenna at NRAO Green Bank linked to an old AT&T Earth Station at Woodbury, Georgia that has been resuscitated by students and faculty at Georgia Tech Research Corporation. Later observations will make use of the Arecibo Observatory and the Nanqay Observatory, each linked with a second site.

D.K. Cullers and R.P. Stauduhar, "Follow-Up Detection in Project Phoenix," in *Astronomical and Biochemical Origins and the Search for Life in the Universe,* C.B. Cosmovici, S. Bowyer, and D. Wertheimer, eds., Bologna: Editrice Compositori, 1997), pp. 645–651. Effective detection of extraterrestrial radio signals requires elimination of terrestrial interference. However, the sensitivity of standard, two-element interferometry depends on the geometric mean of the antenna areas. This is very disadvantageous if the antennas are not the same size. Project Phoenix performs pseudointerferometry by transmitting detailed signal descriptions to its second site, where they are used for confirmation. The primary site uses the larger antenna and square-law detectors. Use of matched filtering and a lower threshold makes possible detection with the second, much smaller, antenna. The pseudointerferometer can handle up to 12 signals having bandwidths of order I Hz in a 10 MHz band. In Australia, it allowed observing to complete in less than half the time that would have been required using a single site's stem.

Trinh Xuan Thuan

To my knowledge, there is no book that discusses in parallel cosmological concepts from both the scientific and Buddhist points of view. Explorations of parallels between science and Buddhism have focused mainly on the field of quantum mechanics and that of the mind. Here are some representative books that treat those subjects:

Fritjof Capra, *The Tao of Physics: An Exploration of the Parallels between Modern Physics and Eastern Mysticism* (Boston, MA: Shambhala, 1981). An early attempt to transcend the split between science and religion by drawing parallels between discoveries in modern physics (mainly quantum mechanics) and insights in Eastern religions and philosophies. Some of the physics described in the book are dated.

Jeremy W. Hayward, *Gentle Bridges: Conversations with the Dalai Lama on the Sciences of Mind* (Boston, MA: Shambala, 1992). Western scientists and the Dalai Lama discuss the interface between the cognitive sciences and Buddhist psychology. The themes include the mind and the brain, the self, perception, evolution, and artificial intelligence.

Jean-François Revel and Matthieu Ricard, *The Monk and the Philosopher: A Father and Son Discuss Life's Eternal Questions* (New York: Schocken, 1999). These dialogues between noted French philosopher Revel and his son, Ricard, a Tibetan monk, contrast Buddhist and European philosophy, science, psychology, ethics, political theory, and spirituality. In an intriguing twist of the age-old debate between reason and faith, Western norms of thinking confront Eastern concepts of spiritual experience.

Francesco Varela, *Sleeping, Dreaming and Dying: An Exploration of Consciousness with the Dalai Lama* (Wisdom, 1997). Dialogues between Western scientists and the Dalai Lama revolving around three key moments of consciousness: sleep, dreams, and death.

B. Allan Wallace, *Choosing Reality: A Buddhist View of Physics and the Mind* (Snow Lion Publications, 1996). An attempt to apply the mode of philosophical inquiry in the Buddhist tradition to the foundations of physics. All phenomena are seen as lending themselves to multiple interpretations, providing us with the freedom and responsibility to choose our reality within the context of valid experience.

Anna Case-Winters

"Teleological Argument for the Existence of God." This article, found in the *Encyclopedia of Philosophy,* gives a brief historical overview of the argument from design. It also takes note of its pitfalls and its critics. This is a good place to begin.

John Marks Templeton, ed., *Evidence of Purpose* (Templeton Foundation Press, 1994). This book is a helpful collection of articles by scientists from a variety of fields exploring the evidence for purpose as they see it in their work. The authors also consider the relationship between their findings and theological convictions concerning the universe as created by God.

Keith Ward, *God, Chance and Necessily* (Oneworld Publications Ltd., 1996). In this book Ward sets up an instructive argument between what he terms "scientific materialism" and his own view. His target is a whole set of perspectives that he believes set faith and science in opposition. His own view is that science in fact reveals evidence of design in nature that is best explained by the existence and activity of God. Ward takes on some key thinkers in cosmology, biology, and sociobiology and seeks to correct the weaknesses and fallacies he sees in their interpretation of the evidence. This work is more useful if one also reads the work of the thinkers with whom he is arguing.

Paul Davies, *The Mind of God* (Touchstone Books, 1993). In this book Davies makes a case for the universe as purposeful and for consciousness as a fundamental facet of reality. While he does not go so far as to draw theological conclusions that God provides the purpose or the mind behind it all, his own scientific exploration is fertile ground for theological reflection.

Index of Contributors